METHODS IN MOLECULAR BIOLOGY

For further volumes:
http://www.springer.com/series/7651

For over 35 years, biological scientists have come to rely on the research protocols and methodologies in the critically acclaimed *Methods in Molecular Biology* series. The series was the first to introduce the step-by-step protocols approach that has become the standard in all biomedical protocol publishing. Each protocol is provided in readily-reproducible step-by-step fashion, opening with an introductory overview, a list of the materials and reagents needed to complete . the experiment, and followed by a detailed procedure that is supported with a helpful notes section offering tips and tricks of the trade as well as troubleshooting advice. These hallmark features were introduced by series editor Dr. John Walker and constitute the key ingredient in ,each and every volume of the *Methods in Molecular Biology* series. Tested and trusted, comprehensive and reliable, all protocols from the series are indexed in PubMed.

Microproteins

Methods and Protocols

Edited by

Stephan Wenkel

Department of Plant Physiology
Umeå Plant Science Centre, Umeå University, Umeå, Sweden

Humana Press

Editor
Stephan Wenkel
Department of Plant Physiology
Umeå Plant Science Centre
Umeå University
Umeå, Sweden

ISSN 1064-3745 ISSN 1940-6029 (electronic)
Methods in Molecular Biology
ISBN 978-1-0716-5012-7 ISBN 978-1-0716-5013-4 (eBook)
https://doi.org/10.1007/978-1-0716-5013-4

This Humana imprint is published by the registered company Springer Science+Business Media, LLC, part of Springer Nature.
The registered company address is: 1 New York Plaza, New York, NY 10004, U.S.A.

Preface

Not long ago, small open reading frames (sORFs) and their encoded microproteins were largely dismissed as biological noise—too small to matter, too elusive to detect. Today, that perspective has shifted dramatically. Microproteins, typically less than 150 amino acids in length, have emerged as critical regulators of cellular processes, acting as modulators of transcription, signaling intermediates, and scaffolding elements across a wide range of systems.

This volume captures the dynamic and multidisciplinary nature of the microprotein field. It brings together cutting-edge methods from proteomics, transcriptomics, bioinformatics, and functional genomics, offering researchers a comprehensive toolkit for discovering and characterizing microproteins. The book is structured into four sections: detection and identification; transcriptome-based discovery; computational, structural, and evolutionary insights; and functional and biomedical relevance. Each chapter is authored by experts advancing the frontier of microprotein research, reflecting the collaborative spirit and innovation that drive this field.

I hope this compendium serves not only as a technical guide but also as an inspiration for further exploration into these once-hidden elements of the proteome. As the methods and tools presented here continue to evolve, so too will our understanding of microproteins and their roles in biology and disease.

Umeå, Sweden *Stephan Wenkel*

Contents

Contributors

IKER VALLE ARAMBURU • *Department of Molecular Biology, Umeå University, Umeå, Sweden; Laboratory for Molecular Infection Medicine Sweden (MIMS) and Umeå Center for Microbial Research (UCMR), Umeå University, Umeå, Sweden*

KATIA AVIÑA-PADILLA • *CINVESTAV-Irapuato, Guanajuato, Mexico*

ARIEL A. BAZZINI • *Stowers Institute for Medical Research, Kansas City, MO, USA; Department of Molecular and Integrative Physiology, University of Kansas Medical Center, Kansas City, KS, USA*

XIONGWEN CAO • *Shanghai Key Laboratory of Regulatory Biology, Institute of Biomedical Sciences, School of Life Sciences, East China Normal University, Shanghai, China*

DOMINIEK CATTEEUW • *iRIP Unit, Laboratory of Microbiology, Department of Biochemistry and Microbiology, Ghent University, Ghent, Belgium*

YANRAN CHEN • *Shanghai Key Laboratory of Regulatory Biology, Institute of Biomedical Sciences, School of Life Sciences, East China Normal University, Shanghai, China*

SONIA CHOTHANI • *Program in Cardiovascular and Metabolic Disorders, Duke-National University of Singapore, Singapore, Singapore; Genome Institute of Singapore (GIS), Agency for Science, Technology and Research (A*STAR), Singapore, Singapore*

NA DING • *College of Bioinformatics Science and Technology, Harbin Medical University, Harbin, China*

LARS A. EICHOLT • *Institute for Evolution and Biodiversity, University of Muenster, Muenster, Germany*

BERTRAND FABRE • *Laboratoire de Recherche en Sciences Végétales (LRSV), CNRS/UT/INPT, Auzeville-Tolosane, France*

ZI-HONG FAN • *MOE Key Laboratory of Tumor Molecular Biology, College of Life Science and Technology, Jinan University, Guangzhou, China*

RITA FERREIRA • *LAQV-REQUIMTE, Department of Chemistry, University of Aveiro, Aveiro, Portugal*

IGOR FESENKO • *Laboratory of Systemic Analysis of Proteins and Peptides, Shemyakin-Ovchinnikov Institute of Bioorganic Chemistry, Moscow, Russia*

CONRAD FIHN • *Department of Chemistry and Biochemistry, University of California, Santa Cruz, CA, USA*

DARIA GANAEVA • *Laboratory of Systemic Analysis of Proteins and Peptides, Shemyakin-Ovchinnikov Institute of Bioorganic Chemistry, Moscow, Russia*

YUAN GAO • *Department of Pathology and Molecular Medicine, Queen's University, Kingston, ON, Canada*

ANNI GE • *Department of Pathology and Molecular Medicine, Queen's University, Kingston, ON, Canada*

PATRICIA GIMENO LE PAIH • *Department of Molecular Biology, Umeå University, Umeå, Sweden; Laboratory for Molecular Infection Medicine Sweden (MIMS) and Umeå Center for Microbial Research (UCMR), Umeå University, Umeå, Sweden*

CLÉMENCE GUILLON • *Laboratoire de Recherche en Sciences Végétales (LRSV), CNRS/UT/INPT, Auzeville-Tolosane, France*

CUITONG HE • *State Key Laboratory of Medical Proteomics, National Center for Protein Sciences (Beijing), Research Unit of Proteomics & Research and Development of New Drug of Chinese Academy of Medical Sciences, Beijing, China*
QING-YU HE • *MOE Key Laboratory of Tumor Molecular Biology, College of Life Science and Technology, Jinan University, Guangzhou, China*
MARIBEL HERNÁNDEZ-ROSALES • *CINVESTAV-Irapuato, Guanajuato, Mexico*
LENA HO • *Cardiovascular and Metabolic Disorders Program, Duke-NUS Medical School, Singapore, Singapore*
POLLY YINGSHAN HSU • *Department of Biochemistry & Molecular Biology, Michigan State University, East Lansing, MI, USA*
VERONIQUE JONCKHEERE • *iRIP Unit, Laboratory of Microbiology, Department of Biochemistry and Microbiology, Ghent University, Ghent, Belgium*
ISAIAH D. KAUFMAN • *Department of Biochemistry & Molecular Biology, Michigan State University, East Lansing, MI, USA*
GORDON T. LUU • *Bruker Scientific, LLC, Billerica, MA, USA*
IRINA LYAPINA • *Laboratory of Systemic Analysis of Proteins and Peptides, Shemyakin-Ovchinnikov Institute of Bioorganic Chemistry, Moscow, Russia*
ANNA MAMAEVA • *Laboratory of Systemic Analysis of Proteins and Peptides, Shemyakin-Ovchinnikov Institute of Bioorganic Chemistry, Moscow, Russia*
DEVIKA MENON • *Program in Cardiovascular and Metabolic Disorders, Duke-National University of Singapore, Singapore, Singapore; Genome Institute of Singapore (GIS), Agency for Science, Technology and Research (A*STAR), Singapore, Singapore*
JIEMIN NAH • *Cardiovascular and Metabolic Disorders Program, Duke-NUS Medical School, Singapore, Singapore*
ZHAO PENG • *School of Life Sciences, and Hubei Key Laboratory of Genetic Regulation and Integrative Biology, Central China Normal University, Wuhan, Hubei, China*
SERGE PLAZA • *Laboratoire de Recherche en Sciences Végétales (LRSV), CNRS/UT/INPT, Auzeville-Tolosane, France*
OWEN RACKHAM • *Program in Cardiovascular and Metabolic Disorders, Duke-National University of Singapore, Singapore, Singapore; School of Biological Sciences, University of Southampton, Southampton, UK*
LAURA M. SANCHEZ • *Department of Chemistry and Biochemistry, University of California, Santa Cruz, CA, USA*
NAVEEN SHANKAR • *Department of Plant Physiology, Umeå Plant Science Centre, Umeå University, Umeå, Sweden*
ASHISH SHARMA • *CSIR-Central Institute of Medicinal and Aromatic Plants (CSIR-CIMAP), Lucknow, Uttar Pradesh, India; Academy of Scientific and Innovative Research (AcSIR), Ghaziabad, Uttar Pradesh, India*
LAURE SIMOENS • *iRIP Unit, Laboratory of Microbiology, Department of Biochemistry and Microbiology, Ghent University, Ghent, Belgium*
POOJA SRIDHARAN • *Cardiovascular and Metabolic Disorders Program, Duke-NUS Medical School, Singapore, Singapore*
PATRICE THULEAU • *Laboratoire de Recherche en Sciences Végétales (LRSV), CNRS/UT/INPT, Auzeville-Tolosane, France*
VALERIE A. TORNINI • *Department of Integrative Biology and Physiology; Institute for Society and Genetics, University of California, Los Angeles (UCLA), Los Angeles, CA, USA*
ANTHONY J. TREICHEL • *Stowers Institute for Medical Research, Kansas City, MO, USA*

PRABODH KUMAR TRIVEDI • *CSIR-Central Institute of Medicinal and Aromatic Plants (CSIR-CIMAP), Lucknow, Uttar Pradesh, India; Academy of Scientific and Innovative Research (AcSIR), Ghaziabad, Uttar Pradesh, India*

PETRA VAN DAMME • *iRIP Unit, Laboratory of Microbiology, Department of Biochemistry and Microbiology, Ghent University, Ghent, Belgium*

RUI VITORINO • *Department of Medical Sciences, Institute of Biomedicine—IBIMED, University of Aveiro, Campus Universitário de Santiago, Aveiro, Portugal; Cardiovascular R&D Centre—UnIC@RISE, Department of Surgery and Physiology, Faculty of Medicine, University of Porto, Porto, Portugal*

CUIHONG WAN • *School of Life Sciences, and Hubei Key Laboratory of Genetic Regulation and Integrative Biology, Central China Normal University, Wuhan, Hubei, China*

YANG WANG • *MOE Key Laboratory of Tumor Molecular Biology, College of Life Science and Technology, Jinan University, Guangzhou, China*

STEPHAN WENKEL • *Department of Plant Physiology, Umeå Plant Science Centre, Umeå University, Umeå, Sweden*

HSIN-YEN LARRY WU • *Department of Biochemistry & Molecular Biology, Michigan State University, East Lansing, MI, USA*

QIANRU XIAO • *School of Life Sciences, and Hubei Key Laboratory of Genetic Regulation and Integrative Biology, Central China Normal University, Wuhan, Hubei, China*

JUAN XU • *College of Bioinformatics Science and Technology, Harbin Medical University, Harbin, China*

PING XU • *State Key Laboratory of Medical Proteomics, National Center for Protein Sciences (Beijing), Research Unit of Proteomics & Research and Development of New Drug of Chinese Academy of Medical Sciences, Beijing, China; School of Basic Medicine, Anhui Medical University, Hefei, China; Research Unit of Environmental Toxicology, School of Public Health, China Medical University, Shenyang, China*

XIAOLONG YANG • *Department of Pathology and Molecular Medicine, Queen's University, Kingston, ON, Canada*

JIAXIN YU • *College of Bioinformatics Science and Technology, Harbin Medical University, Harbin, China*

JING ZHANG • *MOE Key Laboratory of Tumor Molecular Biology, College of Life Science and Technology, Jinan University, Guangzhou, China*

YAO ZHANG • *State Key Laboratory of Medical Proteomics, National Center for Protein Sciences (Beijing), Research Unit of Proteomics & Research and Development of New Drug of Chinese Academy of Medical Sciences, Beijing, China*

Part I

Detection and Identification of Microproteins

Chapter 1

Microproteins: Uncovering Hidden Layers of the Proteome

Naveen Shankar and Stephan Wenkel

Abstract

Once dismissed as nonfunctional transcriptional noise, small open reading frames (sORFs) and their encoded microproteins have rapidly emerged as key players in diverse biological processes. Ranging from just a few to about 150 amino acids in length, microproteins are now recognized for their ability to modulate cellular functions, often acting as dominant-negative regulators, scaffolds, or signaling intermediates. Initially overlooked due to technical and conceptual limitations, they are increasingly being detected thanks to advances in high-resolution proteomics, ribosome profiling, and integrative bioinformatics. In this chapter, we provide a concise overview of the discovery, origins, functions, and biological significance of microproteins. We also introduce the structure of this methods book, which captures the latest experimental and computational tools used to identify, characterize, and functionally dissect microproteins across a wide range of organisms and research disciplines. This emerging field exemplifies the power of cross-disciplinary collaboration, and we hope this volume will support and inspire further research in this exciting and rapidly expanding area.

Key words Microproteins, sORFs, lncRNAs, Evolution, Protein structure

1 What Are Microproteins?

Nucleic acids serve as the blueprint of life, driving evolution through their inheritance and recombination, and as templates for translation, they play a central role in regulating protein synthesis. In contrast, proteins constitute the functional machinery of the cell, carrying out essential biological processes-from DNA replication and gene expression regulation to the synthesis of complex biomolecules. When assigning protein-coding potential to specific genomic regions, it was traditionally assumed that proteins shorter than 100 amino acids were likely artefacts [1]. Given that the largest known proteins can exceed 30,000 amino acids in length [2], and that the average size of eukaryotic proteins is approximately 500 amino acids, this threshold aimed to ensure reliability in gene annotation. Including smaller open reading frames (ORFs), particularly those encoding peptides shorter than 100 amino acids and

Stephan Wenkel (ed.), *Microproteins: Methods and Protocols*, Methods in Molecular Biology, vol. 2992,
https://doi.org/10.1007/978-1-0716-5013-4_1,

especially those under 20 amino acids, would substantially increase the predicted number of genes [1]. Consequently, a conservative threshold of 100 codons has been widely adopted to distinguish bona fide protein-coding genes from potential artefacts.

The initial adoption of a 100-codon minimum threshold for identifying open reading frames proved effective in enriching for bona fide protein-coding genes. However, it is now well established that proteins smaller than 100 amino acids, often referred to as microproteins, are indeed synthesized, with several playing crucial biological roles. Eliminating this threshold altogether, especially when combined with the inclusion of noncanonical translation initiation sites, would generate an overwhelming number of putative novel genes, many of which are likely to be artifactual. How can this challenge be addressed? First, we must clearly define what constitutes a microprotein. Second, we need to understand their origins. And finally, with advances in analytical technologies and computational methodologies, we must consider how these small proteins can be reliably detected. This book aims to explore and provide insights into these questions, both in this introductory chapter and in the chapters that follow.

Definitive lines are hard to draw in biology, where exceptions frequently challenge the norms. Rather than strict boundaries, we frequently encounter a continuum, shades of gray that shift as new exceptions to established rules emerge. In the broadest sense, microproteins are defined as small proteins, typically ranging from just a few amino acids up to approximately 150 amino acids in length. One of the smallest known functional microproteins is tarsal-less (TAL), also known as polished rice (PRI) in drosophila, which comprises only 12 amino acids [3]. At first glance, this may seem unremarkable, peptide hormones such as oxytocin and vasopressin fall within a similar size range and are well known for their critical biological roles [4]. However, a key distinguishing feature lies in their origin. While peptide hormones are typically derived from larger pre–pro-peptide precursors through proteolytic processing, microproteins like PRI/TAL are directly translated from small open reading frames (sORFs), representing a fundamentally different mode of biogenesis.

2 Origins of Microproteins

The origins of microproteins are diverse. In the simplest case, a microprotein may be encoded by a small gene, commonly referred to as a small open reading frame (sORF) [5]. These sORFs are sometimes located near or within canonical coding regions and, depending on their position, are classified as upstream ORFs (uORFs), internal ORFs (intORFs), or downstream ORFs (dORFs). In several cases, such sORFs have been shown to function in *cis*, modulating the translation efficiency of the primary ORF and thus exerting a regulatory role. However, trans-acting functions cannot be excluded. Beyond their association with conventional ORFs, microproteins have also been identified in noncoding RNAs, such as long noncoding RNAs (lncRNAs) [6] and micro-RNAs [7]. Microproteins encoded by microRNAs, known as miPEPs, have been shown to enhance the activity of their corresponding microRNAs in *trans* [8].

Another important source of microproteins is transcript isoforms. Mechanisms such as alternative splicing, alternative transcription initiation, and alternative polyadenylation can give rise to truncated proteoforms of longer canonical proteins [9]. In addition to arising from alternative transcripts, microproteins can also be produced through alternative translation. In this case, translation is initiated from an alternative open reading frame (altORF), which begins at a start codon located outside the canonical reading frame [10]. As a result, a single mRNA molecule can give rise to multiple proteoforms. Translation in an alternative reading frame often produces protein sequences that differ significantly from the canonical protein, potentially endowing them with entirely distinct biological functions.

How can we explore and uncover this hidden microproteome? One of the most direct approaches is the detection of microproteins using mass spectrometry. Chapters 2, 3, 4, 5 and 6 of this book outline mass spectrometry-based methods for identifying peptides derived from small open reading frames. However, in contrast to sequencing-based approaches, which offer significantly higher sensitivity, mass spectrometry remains technically challenging for microprotein detection. Nevertheless, the direct detection of peptides by this method provides stronger evidence of their physical existence. Sequencing based techniques have also greatly advanced the field. Chapter 7 describes ribosome profiling, a powerful method that reveals which regions of the transcriptome are actively being translated. This technique offers insights into how translation is regulated in response to specific environmental cues. Once a microprotein has been identified, its cellular localization and interaction partners can provide important clues about its function. Chapter 15 presents strategies for investigating microproteins that localize to mitochondria. Together, these complementary approaches form a robust toolkit for uncovering and characterizing the functional landscape of the microproteome.

3 Functional Roles of Microproteins

Microproteins perform a wide range of functions. As previously described, some microproteins can act in *cis* by influencing the expression or function of a nearby larger open reading frame

(ORF). However, the majority of microproteins act in *trans*, participating in larger protein complexes to regulate diverse biological processes. A specific subset of microproteins, referred to as microProteins, has been shown to interact with transcriptional complexes and exert dominant-negative effects [11]. These microProteins are small proteins that share sequence homology with larger proteins, often transcription factors, particularly within their protein–protein interaction domains. Through these domains, microproteins can bind to transcriptional regulators and prevent transcription factors from forming functional complexes. Over the past few years, numerous microproteins have been identified and functionally characterized. Their roles span a broad spectrum of biological processes, from basic metabolism (e.g. sugar metabolism and mitochondrial function) to the regulation of organ development, proteolysis, and disease pathways. In plants, microproteins have also been discovered that regulate developmental processes and mediate adaptive growth responses. Here, some microproteins are involved in multiple processes.

4 Dissecting Microprotein Translation and Coding Potential

The discovery of microproteins is often facilitated by techniques that focus on identifying small open reading frames (sORFs). Among the technologies used to identify sORFs, ribosome profiling (also known as Ribo-seq) has emerged as a particularly powerful and widely adopted approach (see Chapter 7). Ribo-seq enables researchers to capture a snapshot of translating ribosomes by sequencing ribosome-protected fragments (RPFs), thereby allowing the precise mapping of ribosome positions across the transcriptome [12, 13]. This technique provides single-nucleotide resolution, which is crucial for distinguishing actively translated regions from nonfunctional or noise-associated signals [12, 14]. It has been instrumental in identifying novel translated regions, including those in noncanonical locations such as upstream open reading frames (uORFs), 5′ and 3′ untranslated regions (UTRs), and within long noncoding RNAs (lncRNAs) [12, 15, 16].

How can we distinguish between true translation and mere ribosome binding on an mRNA? Despite its precision, the detection of ribosomes on an mRNA alone is not sufficient to conclude that a region is being actively translated. Ribosomes can transiently associate with mRNAs in a nonproductive manner, leading to false positives in translation inference. Therefore, a more reliable indicator of translation is the presence of a codon-aligned signal known as three-nucleotide (3-nt) periodicity in ribosome footprints. This periodicity arises because ribosomes move along mRNA in steps of three nucleotides (corresponding to codons) and P-sites, where the peptidyl-tRNA is located, tend to align with the first nucleotide of each codon [12]. When plotted at nucleotide resolution, this produces a striking pattern of periodic ribosome density that is characteristic of active translation. In contrast, footprints arising from random or noncanonical binding lack this codon-level alignment, resulting in a more uniform or dispersed distribution of reads. In the ideal case, an actively translated sORF would exhibit uninterrupted 3-nt periodicity from its start codon to its stop codon. Such a pattern would provide strong evidence for consistent ribosome movement and translation along the entire length of the sORF [13, 14]. In addition, as stated before, a single mRNA molecule can produce both the canonical protein and alternative proteins from different reading frames, which can break the expected 3-nt periodicity and further complicate the interpretation of ribosome profiling data. Moreover, in practical applications, typical Ribo-seq datasets suffer from limited sequencing depth and uneven coverage. This means that many codons, particularly in lowly expressed sORFs, lack sufficient reads to assess periodicity, making reliable detection of translation across the full ORF challenging.

How can we reliably detect translation when the data is noisy or incomplete? To address this challenge, a number of statistical and computational approaches have been developed to extract meaningful signals from sparse or noisy data (*see* Chapter 8 for more details). For example, translation metrics such as the ribosome release score (RRS) have been introduced to evaluate whether ribosome occupancy significantly drops after the stop codon, a hallmark of canonical translation termination. Other methods rely on comparing the characteristics of ribosome footprints within candidate sORFs to those in known protein-coding sequences. These analyses often include read length distributions, in-frame enrichment, and frame-specific read counts. Importantly, statistical models are also employed to distinguish coding from noncoding regions by testing for periodicity using chi-square tests, t-tests, or signal decomposition techniques [12]. Some models generalize periodicity across codons or combine signals from multiple samples to enhance robustness, especially in low-coverage datasets [17]. While these methods substantially improve the ability to detect translated sORFs, they still offer indirect evidence and may fail to capture continuous translation along the full length of the sORF.

Because of these limitations, experimental validation remains essential for confirming sORF translation and assessing the coding potential of microproteins. Validation efforts aim not only to confirm that translation occurs but also to assess the biological relevance and stability of the resulting protein product. One direct method involves the detection of endogenous microproteins in cells or tissues using antibodies specific to the microprotein of interest. However, generating effective antibodies can be technically challenging, especially for microproteins, which often lack immunogenic epitopes due to their small size and low complexity [18]. As a more practical alternative, many researchers use epitope tagging strategies, where the sORF sequence is cloned with a short, detectable tag and expressed as a transgene in cultured cells [19]. This allows straightforward detection via standard techniques such as western blotting or immunofluorescence. Nonetheless, this method may not accurately replicate native expression levels, regulation, or subcellular localization. More recently, genome editing tools like CRISPR-Cas9 have increasingly been used to insert epitope or fluorescent tags directly into the endogenous loci of sORFs [18]. This allows the tracking of translation and localization in the native cellular context. While highly informative, these approaches are resource and time-intensive, and the insertion of tags can potentially interfere with protein function or expression. No single method provides a complete picture-computational predictions offer speed and scale, while experimental approaches are slow but deliver critical validation. To rigorously uncover the coding potential of sORFs and the function of the microproteins they encode, a combined strategy is often essential.

5 Structural and Evolutionary Insights

Are small proteins merely linear chains, or do they possess distinct structural features that influence their roles? Structural determination of microproteins is inherently challenging due to their short sequences, lack of known homologs, and often noncanonical origins. Conventional homology-based modeling approaches are therefore frequently insufficient for these sequences. However, recent advances in artificial intelligence-based prediction tools, such as AlphaFold2 and RoseTTAFold, have revolutionized structure prediction, enabling accurate modeling of even novel, nonhomologous peptides [20]. These tools have revealed that many microproteins adopt stable secondary structures, including alpha-helices and beta-sheets, and in some cases, form compact tertiary folds. Moreover, predicted structures have provided functional context, such as in the case of ASDURF, a uORF-encoded microprotein, which serves as a prefoldin-like subunit of the PAQosome, with implications in cellular response to stress via biogenesis of several human protein complexes [21].

A significant subset of microproteins is intrinsically disordered, especially in isolation, which confers functional advantages in regulatory roles [20]. Disordered microproteins often exhibit flexibility that allows them to engage in context-dependent interactions with multiple partners. This plasticity is crucial in processes such as transcriptional regulation, signal transduction, and protein complex assembly. Disorder also enables microproteins to undergo induced folding upon binding, a characteristic commonly observed in intrinsically disordered regions (IDRs) of the proteome (see Chapter 10 for details). Studies have shown that such disordered segments often harbor functional elements like post-translational modification sites or short linear motifs (SLiMs) [20].

Despite their limited size, microproteins frequently contain SLiMs of three to ten amino acid-long conserved regions that mediate specific protein–protein interactions. These motifs are often embedded within intrinsically disordered regions, enhancing their accessibility and interaction potential [20]. Recent experimental work employing techniques such as protein interaction screen on matrix (PRISMA) has revealed that many microproteins, particularly those of recent evolutionary origin, utilize SLiMs to engage with components involved in RNA splicing, translation regulation, and endocytosis. Examples include proline-rich motifs that interact with SH3 domains, or clathrin box motifs driving vesicular trafficking functions. These findings emphasize the ability of microproteins to integrate into existing cellular networks through compact, modular interfaces [22].

Contrary to canonical proteins, most microproteins are evolutionarily young. An analysis of over 7000 human-translated sORFs revealed that nearly 90% originated within the primate lineage and lacked detectable homologs in non-primate mammals [22]. Nevertheless, evolutionary conservation remains an important indicator of functional significance. Approximately 10% of microproteins exhibit sequence conservation across mammals, while many others maintain conserved reading frames, translation initiation sites, or critical functional motifs despite sequence divergence [22]. Conservation may also be inferred from preserved expression patterns, interactome profiles, or structural features. These patterns suggest selective pressures to retain specific biological roles such as in mitochondrial localization, splicing regulation, or signaling across species, even in the absence of primary sequence homology [20].

How are new microproteins created in the genome? Microproteins frequently arise through noncanonical gene birth mechanisms. The most common is de novo gene evolution, whereby previously noncoding genomic regions acquire translational potential through point mutations, frameshifts, or start codon gains [23]. Many microproteins also originate from uORFs, lncRNAs, or alternative reading frames within annotated genes. Among 6506 evolutionarily young human sORFs analyzed, over 4900 displayed conserved structure across primates, and about 1100 emerged de novo from noncoding sequences [22]. These findings highlight the remarkable plasticity of the genome to give rise to novel functional elements. De novo gene birth enables rapid evolutionary innovation, providing organisms with new molecular tools for adapting regulatory networks or acquiring species-specific traits. This process challenges traditional views of genome stability and supports a more dynamic model of proteome evolution.

Deciphering the structural and evolutionary features of microproteins is pivotal for understanding their physiological roles and contributions to disease. Structurally resolved microproteins can shed light on their mechanistic functions, while motif and disorder-based analyses reveal their capacity to act as dynamic regulators. The involvement of microproteins in diverse processes ranging from mitochondrial metabolism to immune responses and cancer progression underscores their clinical relevance. Furthermore, their compact size and modular interfaces make them attractive candidates for therapeutic development. Microproteins and their functional motifs could serve as novel drug targets or be engineered into peptide-based therapeutics [18, 24]. With the maturation of technologies for sORF discovery and microprotein characterization, these small yet influential regulators emerge as a promising and transformative frontier in biomedical science, offering profound implications for future therapeutic innovation.

6 Microprotein Function, Localization, and Interactomes

Historically, sORFs were overlooked in genome annotations, dismissed as too short and unlikely to produce functional proteins [25]. However, recent advances in high-throughput sequencing technologies, particularly Ribo-seq, have challenged this view by revealing that hundreds of sORFs are indeed translated across various model organisms [25–27]. While the sequencing efforts have significantly expanded the catalog of potential microproteins encoded by sORFs, functional characterization of these microproteins remains a major hurdle in elucidating their biological roles.

In recent years, CRISPR/Cas9-based knockout screens have emerged as a powerful and systematic strategy to investigate the functional significance of microproteins in human cells [19, 28]. These genome-wide or targeted pooled screens are particularly useful for identifying sORFs involved in critical cellular phenotypes, such as proliferation, survival, and drug sensitivity. In a typical pooled screening setup, a guide RNA (gRNA) library is designed to target a curated list of sORFs. This library is then introduced into Cas9-expressing cells using lentiviral vectors. After transduction, the cells are subjected to selective pressures such as nutrient deprivation, drug exposure, or other environmental stresses to uncover sORFs that are functionally important under those conditions [29]. During the screen, next-generation sequencing is used to monitor the relative abundance of each gRNA over time. A reduction in gRNA abundance indicates that knockout of the corresponding sORF impairs cell fitness, suggesting an essential function. Conversely, increased abundance of certain gRNAs may reflect a growth advantage, hinting at a negative regulatory role for the targeted sORF [29]. Thus, changes in gRNA frequency serve as a proxy for the functional relevance of sORFs and the microproteins they may encode. After the identification of candidate sORFs from knockout screens, their coding potential is experimentally validated. This is typically achieved by expressing epitope-tagged versions of the predicted microproteins using tags such as split-GFP, or FLAG [19]. Alternatively, endogenous tagging using a GFP knock-in strategy can confirm whether the sORF is indeed translated into a protein product [18]. To further test the functional relevance of these microproteins, researchers often generate constructs with mutations in the start codon and assess whether they can rescue the phenotype of the knockout [30]. Failure to rescue the phenotype with a non-translatable version confirms the role of the microprotein in mediating the observed biological effect.

Subcellular localization is another critical parameter in characterizing microprotein function (*see* also Chapter 15). Localization studies are commonly performed using fluorescence microscopy in cells expressing either epitope-tagged microproteins or GFP knock-in constructs [18, 19]. Co-staining with organelle-specific markers allows researchers to map microproteins to cellular compartments such as mitochondria, endoplasmic reticulum, or the nucleus, offering clues about their possible roles. However, one limitation of using large tags like GFP is the potential for altered microprotein stability or mislocalization. To address this, smaller peptide tags like HiBiT (1.3 kDa) have been developed for more accurate detection [31]. HiBiT tagging, which relies on its high-affinity interaction with LgBiT to form a functional NanoLuc luciferase, allows highly sensitive luminescent detection of microproteins at endogenous levels without significantly perturbing their native properties (*see* Chapter 16 for more details).

Understanding how microproteins interact with other cellular proteins is a key aspect of functional characterization (*see* Chapter 15). Proximity ligation methods based on biotin ligase enzymes, such as MicroID (TurboID) or APEX, have been used to identify interactiomes of microproteins [32, 33]. In this approach, which is fused to microproteins of interest or bait proteins and covalently biotinylate proteins in close proximity (1-10 nm) upon treatment with biotin (MicroID) or biotin phenol with hydrogen peroxide (APEX). Biotinylated proteins are subsequently purified using biotin-streptavidin-based systems and identified by MS analysis. Another popular protein–protein interaction mapping technique-protein interaction screen on a matrix (PRISMA) has recently been applied to microprotein research [22]. In PRISMA, the microprotein is broken down into overlapping short peptides that are synthesized onto cellulose membranes. These peptide arrays are incubated with cell lysates, and interacting proteins are identified via mass spectrometry. This method is particularly well-suited for detecting interactions mediated by short linear motifs in intrinsically disordered regions, features commonly found in microproteins [22]. Importantly, PRISMA does not require epitope tagging or antibody-based pulldown, making it a scalable and unbiased platform for mapping microprotein interactomes. Collectively, these approaches are reshaping our understanding of microprotein functions.

7 Relevance in Health and Disease

Could small proteins be key drivers of human disease? Emerging evidence suggests they might be-microproteins are increasingly recognized as critical players in cellular pathways commonly disrupted in disease. This is particularly evident in cancer, where numerous microproteins have been implicated as regulators of cell proliferation and survival [18]. Microproteins are important for diverse cellular mechanisms that are often dysregulated in cancer, such as mitochondrial respiration, extracellular signaling, DNA damage response, angiogenesis, metastasis, and invasion [26]. Recently, large-scale screening studies based on high-throughput CRISPR, ribosome profiling and MS technologies have uncovered microproteins implicated in various types of cancer, such as breast, colorectal, leukemia, and childhood medulloblastoma [29, 30, 34] (*see* Chapter 17). Studies have reported approximately 11% of microproteins encoded by sORFs are essential for cancer cell viability, a rate comparable to that of canonical proteins [26]. In some contexts, such as specific CRISPR knockouts in cancer versus non-cancer cell lines, up to 17% of sORFs impacted cell proliferation [26]. Additionally, many lncRNAs, previously classified as noncoding, have been identified through RNA interference screens to encode microproteins with oncogenic functions; examples include *LINC00263*, *DANCR*, and *PVT1* [26, 35]. Interestingly, genomic alterations in cancer can further drive tumorigenesis by creating novel sORFs or enhancing the translation of existing ones. A striking example is gene fusions involving the lncRNA *BCAR4*, which produce an oncogenic peptide from a previously unrecognized ORF within *BCAR4*, contributing to tumorigeneses in different cancers [36]. Together, these findings highlight that microproteins encoded by sORFs are not merely byproducts of transcriptional noise but are active contributors to tumor growth and maintenance.

Microproteins encoded by noncanonical open reading frames in lncRNAs and circRNAs are emerging as promising targets for cancer immunotherapy, especially in tumors with low mutational burden [24, 26]. These microproteins can efficiently generate MHC-I-bound peptides, making them suitable candidates for adoptive T cell therapies like TCR therapy [26]. For instance, a microprotein encoded by the lncRNA *PVT1*, frequently fused with *MYC* in colorectal cancer, produces peptides that are presented on MHC-I and recognized by CD8+ T cells, eliciting targeted immune responses [37]. Similarly, cryptic peptides translated from circular RNAs have been shown to contribute to the cancer immunopeptidome and induce antitumor immunity, with vaccination against such peptides reducing tumor growth in murine models [26]. Personalized vaccines based on microprotein expression profiles, especially when combined with checkpoint inhibitors, hold significant potential to enhance the efficacy of cancer immunotherapy.

Microproteins also hold promise as direct anticancer therapeutic targets. For example, microprotein-protein interactions might be targeted using small-molecule inhibitors or peptide-based therapeutics, as demonstrated recently for the MAVI1 microprotein [38]. An exciting frontier in cancer therapy involves harnessing molecular glues and proteolysis-targeting chimeras (PROTACs) to precisely target microproteins [26]. These innovative molecules can latch onto the unique structural features of microproteins, enabling highly specific modulation or degradation of these elusive small proteins [26]. By exploiting their distinct shapes and interactions, researchers aim to craft powerful small-molecule inhibitors or degraders that zero in on cancer-associated microproteins with remarkable precision, offering the promise of potent treatments that minimize harmful off-target effects and unlock new avenues for precision oncology. Microproteins themselves hold promise as therapeutic agents in cancer treatment, similar to miniproteins, which are small protein scaffolds being increasingly explored in drug development [26]. A notable example is OMO-103 (Omomyc), a 91-amino-acid miniprotein that effectively inhibits the oncogene MYC, once thought undruggable [39]. Inspired by such successes, microproteins could be engineered for enhanced stability, specificity, and potency to disrupt cancer pathways. Advances in AI-driven protein design and directed evolution have already optimized miniproteins and peptides. Applying these techniques to microproteins may accelerate the development of novel cancer therapeutics, including hyperstable synthetic miniprotein scaffolds for targeted drug delivery and modulation of protein interactions.

Another avenue in microprotein research is their use as biomarkers for cancer detection and progression (*see* Chapter 19). Microproteins often exhibit cancer-specific expression patterns, enabling their use in distinguishing tumor from normal tissue, predicting disease progression, and informing clinical outcomes [24]. In colorectal cancer (CRC), the HOXB-AS3 peptide, encoded by a lncRNA, acts as a tumor suppressor, with low expression levels correlating with poorer patient prognosis [40]. Conversely, SRSP, another lncORF-encoded peptide, is upregulated in CRC and associated with aggressive disease features [41]. Similar diagnostic relevance is seen in breast cancer, where reduced expression of peptides such as ASRPS and CIP2A-BP is linked to worse outcomes in triple-negative breast cancer [41]. In glioblastoma (GBM), circORF-derived peptides like SHPRH-146aa and PINT87aa show decreased expression in tumors and are associated with longer survival when expressed at higher levels [42]. Likewise, the peptide SMIM30 is associated with poor prognosis in hepatocellular carcinoma (HCC), while DDUP, from the lncRNA CTBP1-DT, promotes chemoresistance in ovarian cancer [43].

8 Emerging Frontiers, Knowledge Gaps, and Building a Microprotein-Centered Research Community

Microprotein research is rapidly accelerating, fueled by a growing appreciation of their surprising and widespread roles across diverse model organisms. But, how many small proteins with big roles are still waiting to be discovered? Advances in technologies such as deep sequencing, functional genomics, and mass spectrometry have opened new avenues for exploring microproteins, fostering opportunities for interdisciplinary research and cross-training. The role of microproteins is increasingly well characterized in areas like cancer biology. However, broader application of this research, particularly in fields like plant biology, still faces notable technical hurdles.

A major hurdle in uncovering the biological roles of microproteins is the absence of a unified, high-confidence annotation of noncoding ORFs (ncORFs). Existing studies vary drastically in the number of unannotated proteins they report, highlighting the inconsistencies in discovery approaches. Further complicating this landscape is the variability in data quality, particularly in peptide-spectrum matches from mass spectrometry, which reveals substantial differences in reliability between datasets. These issues highlight the pressing need for consistent workflows and rigorous standards in the discovery and validation of microproteins. In response, scientists in human biology have initiated collaborative efforts to consolidate evidence and create consensus resources. One such initiative compiled a unified catalog of 7264 Ribo-seq-supported ncORFs from seven previous studies, providing a foundational dataset for further exploration [44]. Additionally, experts across proteomics, immunopeptidomics, Ribo-seq analysis, and genome annotation have joined forces to establish a shared framework for protein-level validation of ncORFs [45]. These collective efforts aim to produce a robust reference landscape that supports systematic functional studies and drives biomedical discovery. These coordinated efforts aim to establish a standardized reference that not only supports functional studies in human systems but also offers a scalable model for investigating ncORFs in diverse organisms, including animals and plants.

Functional characterization of identified microproteins is essential for understanding their endogenous roles in physiology and disease. Recent advancements in CRISPR-Cas-based genome editing technologies have provided researchers with powerful tools to precisely target and modify small open reading frames (sORFs) in a variety of model organisms (*see* Chapters 13 and 14). These technologies allow for disruption or alteration of the coding potential of microproteins, offering critical insights into their biological significance within native contexts. In addition, innovations in gene tagging technologies enable the incorporation of small, minimally disruptive tags into sORFs [31, 46]. These tags are instrumental in reporting the endogenous dynamics of microproteins, including their spatial and temporal localization, expression levels, and regulatory mechanisms. Large-scale protein interaction screens further complement these approaches by identifying potential binding partners and interaction networks, shedding light on microprotein-mediated pathways and functions [22]. Together, these methodologies are transforming our understanding of microproteins, placing them at the forefront of biological discovery. The integration of gene editing, tagging, and interaction profiling opens the door to exploring previously unrecognized layers of cellular regulation, highlighting microproteins as key players in physiology and disease.

As microprotein research accelerates, an interdisciplinary framework is essential to drive progress. Coordinated efforts among academia, industry, and funders should focus on standardized ncORF annotation, integrated multi-omic databases, and high-throughput functional assays. Centralized repositories of validated reagents such as antibodies, tagged constructs, and CRISPR libraries will enhance reproducibility and accessibility. Cross-disciplinary training programs will empower emerging researchers to navigate this complex field, while targeted funding opportunities can fuel innovative breakthroughs. Open-access platforms will foster a culture of collaboration and rapid data sharing, breaking down barriers between labs and accelerating discovery. By forging strategic partnerships with translational scientists and pharmaceutical developers, we can unlock the therapeutic potential of microproteins and bring new treatments to patients. With the right infrastructure and a spirit of collaboration, microproteins have the power to transform our understanding of gene function, evolution, and disease.

References

1. Jaunbocus N, Ebenki V, Su H, Slavoff SA (2025) Eukaryotic microproteins. Annu Rev Biochem 94:1
2. Small JV, Fürst DO, Thornell L-E (1992) The cytoskeletal lattice of muscle cells. Eur J Biochem 208(3):559–572
3. Kondo T, Hashimoto Y, Kato K, Inagaki S, Hayashi S, Kageyama Y (2007) Small peptide regulators of actin-based cell morphogenesis encoded by a polycistronic mRNA. Nat Cell Biol 9(6):660–665
4. Douglas AJ, Ludwig M (2000) Quo vadis neurohypophysial hormone research? Exp Physiol 85:267S
5. Hanada K, Akiyama K, Sakurai T, Toyoda T, Shinozaki K, Shiu SH (2009) sORF finder: a program package to identify small open reading frames with high coding potential. Bioinformatics 26(3):399–400
6. Anderson DM, Anderson KM, Chang CL, Makarewich CA, Nelson BR, McAnally JR, Kasaragod P, Shelton JM, Liou J, Bassel-Duby R, Olson EN (2015) A micropeptide encoded by a putative long noncoding RNA regulates muscle performance. Cell 160(4): 595–606
7. Lauressergues D, Couzigou JM, San Clemente H, Martinez Y, Dunand C, Bécard G, Combier JP (2015) Primary transcripts of microRNAs encode regulatory peptides. Nature 520(7545):90–93
8. Thuleau P, Ormancey M, Plaza S, Combier J-P (2024) miPEPs and cPEPs as tools to monitor plant gene expression and to develop alternative strategies in agriculture. J Exp Bot: erae501. https://doi.org/10.1093/jxb/erae501
9. Edwards A, Chiurazzi MJ, Blaakmeer A, Vittozzi Y, Sharma A, Matton S, Kruusvee V, Straub D, Sessa G, Carabelli M, Morelli G, Wenkel S (2024) A shade-responsive microProtein in the Arabidopsis ATHB2 gene regulates elongation growth and root development. elife 13:RP96725
10. Cao X, Khitun A, Luo Y, Na Z, Phoodokmai T, Sappakhaw K, Olatunji E, Uttamapinant C, Slavoff SA (2021) Alt-RPL36 downregulates the PI3K-AKT-mTOR signaling pathway by interacting with TMEM24. Nat Commun 12(1):508
11. Staudt AC, Wenkel S (2011) Regulation of protein function by "microProteins". EMBO Rep 12(1):35–42
12. Chothani S, Ho L, Schafer S, Rackham O (2023) Discovering microproteins: making the most of ribosome profiling data. RNA Biol 20(1):943–954
13. Ingolia NT, Hussmann JA, Weissman JS (2019) Ribosome profiling: global views of translation. Cold Spring Harb Perspect Biol 11(5):a032698
14. Ingolia NT (2016) Ribosome footprint profiling of translation throughout the genome. Cell 165(1):22–33
15. Camarena ME, Theunissen P, Ruiz M, Ruiz-Orera J, Calvo-Serra B, Castelo R, Castro C, Sarobe P, Fortes P, Perera-Bel J, Albà MM (2024) Microproteins encoded by noncanonical ORFs are a major source of tumor-specific antigens in a liver cancer patient meta-cohort. Sci Adv 10:eadn3628
16. Rocha AL, Pai V, Perkins G, Chang T, Ma J, De Souza EV, Chu Q, Vaughan JM, Diedrich JK, Ellisman MH, Saghatelian A (2024) An inner mitochondrial membrane microprotein from the SLC35A4 upstream ORF regulates cellular metabolism. J Mol Biol 436(10):168559
17. Yang H, Li Q, Stroup EK, Wang S, Ji Z (2024) Widespread stable noncanonical peptides identified by integrated analyses of ribosome profiling and ORF features. Nat Commun 15(1):1932
18. Hassel KR, Brito-Estrada O, Makarewich CA (2023) Microproteins: overlooked regulators of physiology and disease. iScience 26(6): 106781
19. Azam S, Yang F, Wu X (2025) Finding functional microproteins. Trends Genet 41(2): 107–118
20. Mohsen JJ, Martel AA, Slavoff SA (2023) Microproteins—discovery, structure, and function. Proteomics 23(23–24):e2100211
21. Cloutier P, Poitras C, Faubert D, Bouchard A, Blanchette M, Gauthier MS, Coulombe B (2020) Upstream ORF-encoded ASDURF is a novel Prefoldin-like subunit of the PAQosome. J Proteome Res 19(1):18–27
22. Sandmann CL, Schulz JF, Ruiz-Orera J, Kirchner M, Ziehm M, Adami E, Marczenke M, Christ A, Liebe N, Greiner J, Schoenenberger A, Muecke MB, Liang N, Moritz RL, Sun Z, Deutsch EW, Gotthardt M, Mudge JM, Prensner JR, Willnow TE, Mertins P, van Heesch S, Hubner N (2023) Evolutionary origins and interactomes of human, young microproteins and small peptides translated from short open reading frames. Mol Cell 83(6):994–1011.e18
23. Carvunis AR, Rolland T, Wapinski I, Calderwood MA, Yildirim MA, Simonis N,

Charloteaux B, Hidalgo CA, Barbette J, Santhanam B, Brar GA, Weissman JS, Regev A, Thierry-Mieg N, Cusick ME, Vidal M (2012) Proto-genes and de novo gene birth. Nature 487(7407):370–374
24. Ge A, Chan C, Yang X (2024) Exploring the dark matter of human proteome: the emerging role of Non-Canonical Open Reading Frame (ncORF) in cancer diagnosis, biology, and therapy. Cancers (Basel) 16(15)
25. Petri L, Van Humbeeck A, Niu H, Ter Waarbeek C, Edwards A, Chiurazzi MJ, Vittozzi Y, Wenkel S (2024) Exploring the world of small proteins in plant biology and bioengineering. Trends Genet 41(2):170–180
26. Hofman DA, Prensner JR, van Heesch S (2024) Microproteins in cancer: identification, biological functions, and clinical implications. Trends Genet 41(2):146–161
27. Simoens L, Fijalkowski I, Van Damme P (2023) Exposing the small protein load of bacterial life. FEMS Microbiol Rev 47(6)
28. Schlesinger D, Elsässer SJ (2022) Revisiting sORFs: overcoming challenges to identify and characterize functional microproteins. FEBS J 289(1):53–74
29. Schlesinger D, Dirks C, Navarro C, Lafranchi L, Spinner A, Raja GL, Mun-Sum Tong G, Eirich J, Martinez TF, Elsässer SJ (2025) A large-scale sORF screen identifies putative microproteins involved in cancer cell fitness. iScience 28(3):111884
30. Prensner JR, Enache OM, Luria V, Krug K, Clauser KR, Dempster JM, Karger A, Wang L, Stumbraite K, Wang VM, Botta G, Lyons NJ, Goodale A, Kalani Z, Fritchman B, Brown A, Alan D, Green T, Yang X, Jaffe JD, Roth JA, Piccioni F, Kirschner MW, Ji Z, Root DE, Golub TR (2021) Noncanonical open reading frames encode functional proteins essential for cancer cell survival. Nat Biotechnol 39(6):697–704
31. Simoens L, Jonckheere V, Catteeuw D, Van Damme P (2025) Endogenous HiBiT-tagging combined with affinity complementation: a new strategy for small open reading frame-encoded polypeptide detection in bacteria. Adv Sample Prep 13:100143
32. Chu Q, Rathore A, Diedrich JK, Donaldson CJ, Yates JR, Saghatelian A (2017) Identification of microprotein-protein interactions via APEX tagging. Biochemistry 56(26): 3299–3306
33. Na Z, Dai X, Zheng SJ, Bryant CJ, Loh KH, Su H, Luo Y, Buhagiar AF, Cao X, Baserga SJ, Chen S, Slavoff SA (2022) Mapping subcellular localizations of unannotated microproteins and alternative proteins with MicroID. Mol Cell 82(15):2900–2911.e7
34. Hofman DA, Ruiz-Orera J, Yannuzzi I, Murugesan R, Brown A, Clauser KR, Condurat AL, van Dinter JT, Engels SAG, Goodale A, van der Lugt J, Abid T, Wang L, Zhou KN, Vogelzang J, Ligon KL, Phoenix TN, Roth JA, Root DE, Hubner N, Golub TR, Bandopadhayay P, van Heesch S, Prensner JR (2024) Translation of non-canonical open reading frames as a cancer cell survival mechanism in childhood medulloblastoma. Mol Cell 84(2):261–276.e18
35. Liu SJ, Horlbeck MA, Cho SW, Birk HS, Malatesta M, He D, Attenello FJ, Villalta JE, Cho MY, Chen Y, Mandegar MA, Olvera MP, Gilbert LA, Conklin BR, Chang HY, Weissman JS, Lim DA (2017) CRISPRi-based genome-scale identification of functional long noncoding RNA loci in human cells. Science (1979) 355(6320):aah7111
36. Nickless A, Zhang J, Othoum G, Webster J, Inkman MJ, Coonrod E, Fontes S, Rozycki EB, Maher CA, White NM (2022) Pan-cancer analysis reveals recurrent BCAR4 gene fusions across solid tumors. Mol Cancer Res 20(10): 1481–1488
37. Hung KL, Yost KE, Xie L, Shi Q, Helmsauer K, Luebeck J, Schöpflin R, Lange JT, Chamorro González R, Weiser NE, Chen C, Valieva ME, Wong ITL, Wu S, Dehkordi SR, Duffy CV, Kraft K, Tang J, Belk JA, Rose JC, Corces MR, Granja JM, Li R, Rajkumar U, Friedlein J, Bagchi A, Satpathy AT, Tjian R, Mundlos S, Bafna V, Henssen AG, Mischel PS, Liu Z, Chang HY (2021) ecDNA hubs drive cooperative intermolecular oncogene expression. Nature 600(7890):731–736
38. Shi T-T, Huang Y, Li Y, Dai X-L, He Y-H, Ding J-C, Ran T, Shi Y, Yuan Q, Li W-J, Liu W (2023) MAVI1, an endoplasmic reticulum-localized microprotein, suppresses antiviral innate immune response by targeting MAVS on mitochondrion. Sci Adv 9:eadg7053
39. Garralda E, Beaulieu ME, Moreno V, Casacuberta-Serra S, Martínez-Martín S, Foradada L, Alonso G, Massó Vallés D, López-Estévez S, Jauset T, Corral de la Fuente E, Doger B, Hernández T, Perez-Lopez R, Arqués O, Castillo Cano V, Morales J, Whitfield JR, Niewel M, Soucek L, Calvo E (2024) MYC targeting by OMO-103 in solid tumors: a phase 1 trial. Nat Med 30(3): 762–771
40. Xing Y, Sun X, Li F, Jiang X, Jiang A, Li X, Lv R, Shao L (2021) Long non-coding RNA (lncRNA) HOXB-AS3 promotes cell proliferation and inhibits apoptosis by regulating ADAM9 expression through targeting miR-498-5p in endometrial carcinoma. J Int

Med Res 49(6). https://doi.org/10.1177/03000605211013548

41. Wang Y, Wu S, Zhu X, Zhang L, Deng J, Li F, Guo B, Zhang S, Wu R, Zhang Z, Wang K, Lu J, Zhou Y (2020) LncRNA-encoded polypeptide ASRPS inhibits triple-negative breast cancer angiogenesis. J Exp Med 217(3). https://doi.org/10.1084/jem.20190950
42. Zhang M, Huang N, Yang X, Luo J, Yan S, Xiao F, Chen W, Gao X, Zhao K, Zhou H, Li Z, Ming L, Xie B, Zhang N (2018) A novel protein encoded by the circular form of the SHPRH gene suppresses glioma tumorigenesis. Oncogene 37(13):1805–1814
43. Ren L, Qing X, Wei J, Mo H, Liu Y, Zhi Y, Lu W, Zheng M, Zhang W, Chen Y, Zhang Y, Pan T, Zhong Q, Li R, Zhang X, Ruan X, Yu R, Li J (2023) The DDUP protein encoded by the DNA damage-induced CTBP1-DT lncRNA confers cisplatin resistance in ovarian cancer. Cell Death Dis 14(8):568
44. Mudge JM, Ruiz-Orera J, Prensner JR, Brunet MA, Calvet F, Jungreis I, Gonzalez JM, Magrane M, Martinez TF, Schulz JF, Yang YT, Albà MM, Aspden JL, Baranov PV, Bazzini AA, Bruford E, Martin MJ, Calviello L, Carvunis AR, Chen J, Couso JP, Deutsch EW, Flicek P, Frankish A, Gerstein M, Hubner N, Ingolia NT, Kellis M, Menschaert G, Moritz RL, Ohler U, Roucou X, Saghatelian A, Weissman JS, van Heesch S (2022) Standardized annotation of translated open reading frames. Nat Biotechnol 40(7):994–999
45. Deutsch EW, Kok LW, Mudge JM, Ruiz-Orera J, Fierro-Monti I, Sun Z, Abelin JG, Alba MM, Aspden JL, Bazzini AA, Bruford EA, Brunet MA, Calviello L, Carr SA, Carvunis A-R, Chothani S, Clauwaert J, Dean K, Faridi P, Frankish A, Hubner N, Ingolia NT, Magrane M, Martin MJ, Martinez TF, Menschaert G, Ohler U, Orchard S, Rackham O, Roucou X, Slavoff SA, Valen E, Wacholder A, Weissman JS, Wu W, Xie Z, Choudhary J, Bassani-Sternberg M, Vizcaíno JA, Ternette N, Moritz RL, Prensner JR, van Heesch S (2024) High-quality peptide evidence for annotating non-canonical open reading frames as human proteins. bioRxiv. https://doi.org/10.1101/2024.09.09.612016
46. Lafranchi L, Schlesinger D, Kimler KJ, Elsässer SJ (2020) Universal single-residue terminal labels for fluorescent live cell imaging of microproteins. J Am Chem Soc 142(47):20080–20087

Chapter 2

Proteomic Detection of Small and Alternative Open Reading Frame-Encoded Microproteins and Alternative Proteins

Yanran Chen and Xiongwen Cao

Abstract

RIBO-seq and proteomics have revealed that mammalian genomes harbor thousands of unannotated small and alternative open reading frames (sm/alt-ORFs <150 codons) that are translated into microproteins and alternative proteins (micro/alt-proteins). Several dozens of micro/alt-proteins have been characterized at the molecular level and demonstrated to play important biological roles. However, the overwhelming majority of micro/alt-proteins remain undefined, possibly because they are excluded from protein databases and cannot be detected by a classical proteomic workflow. Here, we introduce a proteomic pipeline to detect unannotated micro/alt-proteins in cultured cells or tissues from total protein extraction, gel-based size selection, digestion, and liquid chromatography-mass spectrometry proteomics (LC-MS/MS), to data analysis and manual validation. This approach is able to identify unannotated micro/alt-proteins from cultured cells and (patho)physiological tissues.

Key words Proteomics, Unannotated protein, Small open reading frame (smORF), Alternative open reading frame (alt-ORF), Microprotein, Alternative protein (alt-protein)

1 Introduction

Recent proteogenomics have revealed thousands of previously unannotated small and alternative open reading frames (smORFs and alt-ORFs, fewer than 150 codons) in both human and model organisms [1–3]. These smORFs and alt-ORFs are found in long noncoding RNAs, 5′ and 3′ UTR of mRNAs (upstream and downstream ORFs, uORFs and dORFs, respectively), and in frame-shifted ORFs overlapping annotated protein-coding sequences [4–6]. A recent CRISPR-based knockout screen revealed that about 10% (57 out of 553) sm/alt-ORFs are required for cell survival, compared to 17% for annotated ORFs [7], suggesting that sm/alt-ORFs may also play important biological roles. Several dozens of smORF-encoded polypeptides (SEPs, also known as microproteins) and alt-ORF-encoded proteins (alt-proteins, or small proteins) have been defined at the molecular level [8],

Stephan Wenkel (ed.), *Microproteins: Methods and Protocols*, Methods in Molecular Biology, vol. 2992,
https://doi.org/10.1007/978-1-0716-5013-4_2,

including MPEP, which promotes glioblastoma growth by binding and activating TRKB [9]; and EMBOW, which functions as an important regulator of WDR5 [10]. These studies demonstrate that identification and functional studies of micro/alt-proteins represent a major opportunity to gain insights into biology.

Currently, two approaches are popularly used to identify micro/alt-proteins: ribosome foot printing followed by deep sequencing (RIBO-seq) and liquid chromatography–mass spectrometry proteomics (LC-MS/MS), though additional experimental validations are required to assign a micro/alt-protein due to false positive and false negative detections for both methods. RIBO-seq is powerful to identify putative sm/alt-ORFs by sequencing ribosome-bound RNA [11, 12]. However, ribosome occupancy is not sufficient to classify transcripts as coding or noncoding, because ribosome bound may affect the stability or localization of an RNA instead of translation [13]. And in some cases, the RIBO-seq signal may be generated from RNA-protein interactions [14]. In contrast, micro/alt-proteins generate few tryptic peptides due to their small size; thus sm/alt-ORFs are likely under-detected by LC-MS/MS [1], especially in protein-rich samples, such as clinical tissues. In summary, RIBO-seq detects many more putative sm/alt-ORFs than LC-MS/MS, while proteomics provides the most direct evidence of micro/alt-proteins.

Here, we describe a proteomic micro/alt-protein discovery pipeline applied to cells in culture or tissues (Fig. 1). First, total proteins are extracted from cells or tissues. Then, small proteins are enriched via gel-based size-selection, followed by in-gel digestion and analyzed with LC-MS/MS. Resulting peptide spectra are searched against a three-frame translation database. Finally, annotated peptide sequences are excluded computationally, and the remaining sequences are validated manually. This approach has proved powerful for the detection of unannotated micro/alt-proteins from cell lines and clinical tissues [5, 6].

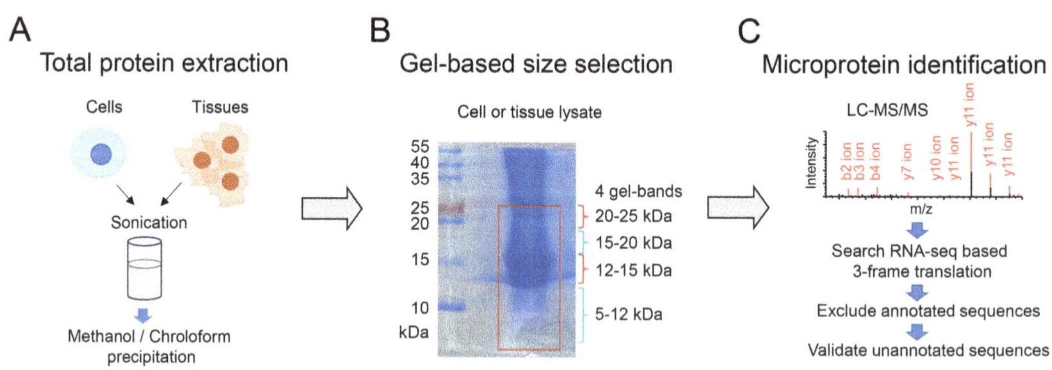

Fig. 1 Schematic overview of proteomic detection of small and alternative open reading frame encoded micro/alt-proteins (**a**) Total protein extraction (**b**) Gel-based size-selection (**c**) Microprotein identification

2 Materials

2.1 Protein Extraction and Size-Selection

1. Sonicator, and equipment for standard immunoblotting procedures.
2. RIPA buffer: 10 mmol/L Tris–HCl (pH = 7.4), 1% Triton X-100 (v/v), 0.1% sodium deoxycholate (w/v), 0.1% SDS (w/v), 140 mmol/L NaCl. Store at 4 °C.
3. 4 × SDS loading buffer: 240 mmol/L Tris–HCl (pH = 6.8), 40% glycerol (v/v), 8% SDS (w/v), 5% β-mercaptoethanol (v/v), 0.04% bromophenol blue (w/v) (Optional). Leave one aliquot at room temperature (RT) for current use and store the remaining aliquots at −20 °C.
4. Tricine-gel buffer: 3 mol/L Tris–HCl (pH = 8.45), 0.3% SDS (w/v). Store at RT.
5. Casting 16% Tricine gel:

	For 2 separating gel	For 2 stacking gels
40% acrylamide (29:1)	6 mL	1.05 mL
Tricine-gel buffer	5 mL	2.5 mL
70% glycerol	2 mL	0 mL
ddH2O	2 mL	6.7 mL
10% APS	66 μL	80 μL
TEMED	6.6 μL	8 μL

6. 10 × anode Tricine gel running buffer (the lower electrode buffer): 1 mol/L Tris-HCl (pH = 8.9). Store at RT.
7. 10 × cathode Tricine gel running buffer (the upper electrode buffer): 1 mol/L Tris-HCl (pH = 8.25), 1 mol/L Tricine, 1% SDS (w/v). Store at RT.
8. Coomassie stain buffer: Weigh 2 g Coomassie blue R250 and transfer it into a glass bottle, add 500 mL methanol, 100 mL acetic acid, and 400 mL ddH$_2$O. Store at RT.
9. Coomassie destain buffer: Mix 500 mL methanol with 100 mL acetic acid and 400 mL ddH$_2$O. Store at RT.

2.2 Reduction, Alkylation, and Trypsin Digestion

1. Tris(2-carboxyethyl)phosphine hydrochloride (TCEP) reduction buffer: 5 mmol/L TCEP, 100 mmol/L NH_4HCO_3 in ddH2O. Store at RT.
2. 2-Iodoacetamide (IAA) alkylation buffer: 100 mmol/L IAA, 100 mmol/L NH_4HCO_3 in ddH2O. Store at RT.

3. 50 mmol/L NH_4HCO_3/acetonitrile solution: Mix 50 mL 50 mmol/L NH4HCO3 in ddH2O with 50 mL acetonitrile. Store at RT.
4. 10 mmol/L NH_4HCO_3/acetonitrile solution: Mix 50 mL 10 mmol/L NH4HCO3 in ddH2O with 50 mL acetonitrile. Store at RT.
5. Digestion buffer: Mix 9 mL of the 50 mmol/L NH_4HCO_3 in ddH2O with 1 mL acetonitrile. Store at RT.
6. Peptide extraction buffer: Mix 66.7 mL acetonitrile with 31.4 mL ddH_2O and 1.89 mL formic acid. Store at RT.
7. SpeedVac vacuum concentrator.

2.3 Tryptic Peptides Clean-Up, and LC-MS/MS

1. Water-saturated ethyl acetate solution: Mix 5 mL ddH_2O with 45 mL ethyl acetate. Store at RT.
2. 5% acetonitrile +0.5% Trifluoroacetic acid wash buffer: Mix 5 mL acetonitrile with 94.5 mL ddH_2O and 0.5 mL Trifluoroacetic acid. Store at RT.
3. 70% acetonitrile elution buffer: Mix 30 mL ddH_2O with 70 mL acetonitrile. Store at RT.
4. Solvent A (2% acetonitrile with 0.1% formic acid): Mix 49 mL ddH_2O with 1 mL acetonitrile and 50 μL formic acid. Store at RT.
5. Solvent B (98% acetonitrile with 0.1% formic acid): Mix 1 mL ddH_2O with 49 mL acetonitrile and 50 μL formic acid. Store at RT.
6. Q exactive hydrid quadrupole orbitrap.

3 Methods

3.1 Total Protein Extraction

1. Harvest 10^7–10^8 cells of interest, or collect 400–600 mg tissues of interest.
2. Add 1 mL RIPA buffer with protease inhibitor cocktails. For the cell pellet sample, pipette up and down to suspend the cells, and sonicate the cells (power 60 w, 1 s ON, 2 s OFF, 40 cycles). For the tissue sample, cut the tissues into small pieces with a clean scissor, and sonicate the tissues (power 80 w, 1 s ON, 2 s OFF, 40 cycles).
3. Centrifuge the lysates at 18,407 × g at 4 °C for 10 min, and transfer the supernatant to a new 1.5-mL tube.
4. Transfer 100 μL of the lysate to a new 1.5-mL tube, add 400 μL of methanol, and vortex for 5 s.
5. Add 100 μL of chloroform and vortex for 5 s.

6. Add 300 μL of ddH_2O and vortex for 5 s, and centrifugate at 18,407 × g at RT for 2 min.
7. Remove and discard the top layer.
8. Add 400 μL of methanol to the remaining bottom layer, vortex for 5 s, followed by centrifugation at 18,407 × g at RT for 2 min.
9. Remove the supernatant without disturbing the protein pellet, and dry it at RT for 5 min.
10. Add 50 μL of 1× SDS loading buffer without dye, and boil at 100 °C for 10 min (*see* **Note 1**).
11. Measure the protein concentration with Nanodrop A280.

3.2 Tricine Gel-Based Size-Selection

1. Set 16% Tricine gel within the running chamber, and fill the chamber with 1 × Tricine SDS running buffer, add 1 × cathode buffer as the upper electrode buffer and 1 × anode buffer as the lower electrode buffer (*see* **Note 2**).
2. Combine 400–600 μg total proteins in lysate with 5 μL of 4 × SDS loading buffer with dye, vortex for 5 s, and load the sample into one well of the Tricine gel under the 1 × cathode buffer.
3. Run the gel for 1 h at 60 V, followed by 2 h at 90 V, or until the dye front reaches the bottom edge of the gel.
4. Disassemble the gel plates, remove the gel and transfer it into a clean plastic 15-cm petri dish (*see* **Note 3**).
5. Add 30 mL Coomassie stain buffer to the petri dish, and rock for 30 min at RT (*see* **Note 4**).
6. Remove and discard staining solution.
7. Add 30 mL Coomassie destain buffer, and rock for 2 h at RT (*see* **Note 5**).
8. Discard the Coomassie destain buffer. And repeat **step 7** one time.
9. Use clean razor blades to excise gel bands between 5–10, 10–15, 15–20, and 20–25 kDa size marks. Further cut each of the four bands into smaller pieces and place pieces into four clean 1.5-mL plastic tubes.

3.3 Reduction, Alkylation, and In-Gel Trypsin Digestion

1. Add 400 μL of freshly prepared 5 mmol/L TCEP in 100 mmol/L NH_4HCO_3 to cover the gel pieces, incubate at 60 °C for 10 min, then put the tubes on ice to cool down.
2. Discard the TCEP solution, add 400 μL of freshly prepared 100 mmol/L IAA in 100 mmol/L NH_4HCO_3 to cover the gel pieces, and incubate at 37 °C for 15 min in the dark.

3. Discard the IAA solution, add 500 μL of 50 mmol/L NH_4HCO_3/acetonitrile wash solution and rotate for 30 min.
4. Discard the first wash, and repeat **step 3** one more time.
5. Discard the second wash, add 500 μL of 10 mmol/L NH_4HCO_3/acetonitrile ash solution and rotate for 30 min.
6. Discard the third wash, add 500 μL of neat acetonitrile, vortex 5 s, and allow the gel pieces to shrink for 1 min.
7. Remove all of the acetonitrile, and speedvac the gel pieces to dryness at RT. This step may take up to 1 h.
8. Add 400 μL of freshly prepared trypsin solution (1:150 dilution of the 0.5 mg/mL trypsin with digestion buffer) to fully cover the gel pieces, and incubate at 37 °C for 14–16 h (*see* **Note 6**).
9. Spin down briefly and transfer the digestion buffer into a clean 1.5-mL tube.
10. Add 200 μL of peptide extraction buffer and rotate at RT for 15 min to extract tryptic peptides from the gel.
11. Spin down briefly, and combine the extraction buffer with the digestion buffer solution.
12. Repeat **steps 10–11** one more time, and speedvac the combined solution to dryness, which contains the digested peptides. This step may take up to 5 h.

3.4 Detergent Removal and Digested Peptides Desalting

1. Add 100 μL of 200 mmol/L NH_4HCO_3 in ddH2O to the digested peptides, and vortex for 10 s to resolve the peptides.
2. Add 1 mL of water-saturated ethyl acetate to the digested peptides solution, vortex for 10 s, and centrifuge at 18,407 × g at RT for 1 min. Remove and discard the upper layer. This step is to remove remained detergent.
3. Repeat **step 2** two more times. The remaining bottom layer contains the digested peptides.
4. Speedvac the solution to dryness at RT. This step may take up to 3 h.
5. Add 180 μL of 5% acetonitrile +0.5% trifluoroacetic acid in ddH_2O to the digested peptides, and vortex for 10 s.
6. Add 180 μL of 50% methanol in ddH_2O to a C18 spin column, followed with centrifugation at 200 × g at 10 °C for 2 min, and discard the flow-through.
7. Add 180 μL of 5% acetonitrile +0.5% trifluoroacetic acid in ddH_2O to the C18 spin column, followed with centrifugation at 200 × g at 10 °C for 2 min, and discard the flow-through. Repeat this step one more time.

8. Load the digested peptide solution onto the C18 spin column, followed by centrifugation at 200 × g at 10 °C for 2 min, and discard the flow-through.
9. Wash the C18 spin column with 180 μL of 5% acetonitrile +0.5% Trifluoroacetic acid in ddH_2O, centrifuge at 200 × g at 10 °C for 2 min, and discard the flow-through. Repeat this step one more time.
10. Elute the digested peptides with 45 μL of 70% acetonitrile in ddH_2O, centrifuge at 200 × g at 10 °C for 2 min, and retain the eluate.
11. Repeat **step 10** one more time, and combine the two eluates.
12. Speedvac the solution to dryness at RT. This step may take up to 3 h.

3.5 Mass Spectrometry Analysis of Micro/Alt-Proteins

1. Add 35 μL of solvent A (2% acetonitrile +0.1% formic acid in ddH_2O) to the digested peptides, centrifuge at 18,407 × g at 4 °C for 30 min, and discard the pellet.
2. Inject 5 μL of digested peptides into a prepacked column attached to an EASY-nLC 1000 (Thermo Scientific) in-line with a Thermo Scientific Q Exactive Hybrid QuadrupoleOrbitrap mass spectrometer (Thermo Scientific).
3. Separate the peptides with a 120-min gradient as follows (solvent A: 2% acetonitrile with 0.1% formic acid; solvent B: 98% acetonitrile with 0.1% formic acid): set the flow rate to 0.25 μL/min, start with 3% B, and gradients 3% B to 8% B over 5 min, 8% B to 28% B over 95 min, and 28% B to 95% B over 10 min. Maintain the isocratic flow at 95% B for 5 min, then gradient from 95% B to 3% B over 2 min, and maintain the isocratic flow at 3% B for 3 min.
4. Set the mass range to 150–2000 *m/z* with a resolution of 70,000, and the automatic gain control (AGC) target to 3 × 10^6. Collect the MS/MS data with a top 12 high-collisional energy dissociation (HCD) fragmentation method in data-dependent mode, and set the normalized collision energy to 27.0 eV, isolation window to 2.0 *m/z*, MS/MS resolution to 17,500, and dynamic exclusion to 15 s.

3.6 Micro/Alt-Protein Identification

1. Convert raw files to Mascot generic file (MGF) files using Mascot (version 2.5.1), and search the data against a database comprised of a three-frame translation of assembled transcripts from GENCODE or RNA-seq data from cells of interest, plus a contaminants database provided by Mascot, as previously described [15].
2. Set carbamidomethylation of cysteine as a fixed modification, and N-terminal acetylation and oxidation of methionine as

variable modifications. Set enzyme specificity to semiTrypsin, and maximum missed cleavage to 2. Set the peptide mass tolerance to 20 ppm, and MS/MS tolerance to 0.02 Da. Set peptide charge to 2+, 3+, and 4+. Click Run.

3. Once the run is finished, identify and validate the unannotated peptides as previously described [16]. Briefly, peptide-spectral matches to annotated proteins from Uniprot are removed with a custom script available via Zenodo at (10.5281/zenodo.5921116). And peptide-spectral matches to known contaminant protein sequences, which can be downloaded from the free MaxQuant software under MaxQuant/bin/conf/contaminants.fasta, are removed with the same custom script. The remaining peptide-spectral matches are subjected to NCBI protein BLAST against the protein database of related species. Remove any peptides found to be less than two amino acids different from a known protein, to eliminate false positive matches caused by post-translational modifications or point mutations (*see* **Note 7**).
4. Inspect the MS/MS spectra of the remaining unmatched peptides manually in Mascot. Peptide-spectral matches with the following five features should be discarded:
   (a) Peptides with fewer than five consecutive b- or y-ions.
   (b) Peptides with either only b- or y-ions.
   (c) Peptides with Mascot score less than 50.
   (d) Peptides with Mass error larger than 20 ppm.
   (e) Peptides with unmatched charge state.

   A high-quality MS/MS spectrum from microprotein EMBOW [10] is shown in Fig. 2.

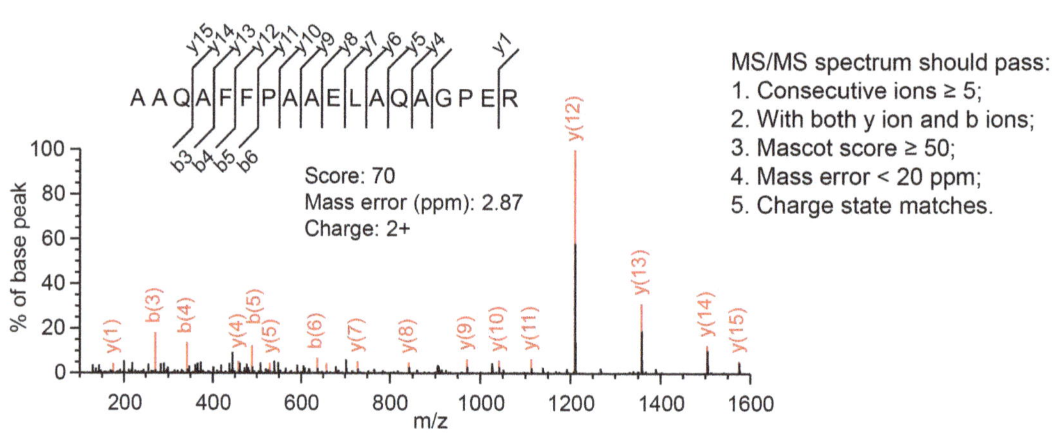

Fig. 2 High-quality MS/MS spectrum from EMBOW

5. Extract transcript sequences corresponding to the remaining peptides that pass the filters described above, which can be found in the transcript database used to generate the 3-frame search database. Copy and paste the transcript sequence to the ExPASy translate tool, and translate it into amino acid sequences in three frames. Identify the small ORF encoding each unannotated tryptic peptide by searching ORF containing the unannotated tryptic peptide that includes a start codon or a near-cognate start codon within a Kozak consensus sequence and a stop codon.

3.7 Validate the Expression of Micro/Alt-Protein

1. Construct a cDNA clone that contains the full 5'UTR and the putative sm/alt-ORF with a FLAG tag appended to the C-terminus of the micro/alt-protein into pcDNA3.1 expression vector.
2. Transfect the cDNA construct into HEK 293T cells with standard transfection procedures.
3. Harvest cells 48 h post-transfection, and determine protein expression with anti-FLAG antibody using standard immunoblotting procedures.
4. To define the full coding sequence of a sm/alt-ORF, the putative start codon can be deleted or mutated to TAA, followed by transfection and immunoblotting to determine protein expression.
5. Additional experimental validations are required to determine the endogenous expression of a micro/alt-protein, including genomic tagging and specific antibodies.

4 Notes

1. Boiling protein pellets after methanol/chloroform precipitation may take more than 10 min to fully denature the proteins; add more SDS loading buffer as needed. And to make 1 × SDS loading buffer without dye, just dilute the 4 × SDS loading buffer without dye using ddH_2O.
2. The anode and cathode Tricine gel buffers are different, and should not be mixed during running, make sure the chamber is not leaking before fill it with gel buffers.
3. Once the Tricine gel running is finished, perform all procedures with new, clean, dust-free materials while wearing clean gloves to avoid contamination from keratin and other proteins.
4. Use fresh Coomassie stain and destain buffers to avoid contamination. And use a fresh blade for each cut to avoid cross-contamination between samples.

5. All solutions used after destaining should be stored in glass bottles to avoid plasticizer contamination.
6. Normally, 400 μL digestion buffer containing trypsin is enough to cover the gel pieces. In some cases, after 1 h incubation, more digestion buffer with trypsin may be needed to fully cover the gel pieces.
7. Amino acids I and L are isobaric and are distinguishable by mass spectrometry, so peptides differing from an annotated protein with two amino acids, while one of them is I or L, should also be discarded.

Acknowledgments

The work was funded by the National Key Research and Development Program of China (2023YFC2508100/2023YFC2508104), the National Natural Science Foundation of China (32370807 to X.C.), and Shanghai Pujiang Program (22PJ1402600 to X.C.). Y.C. was supported by the National Natural Science Foundation of China (82403049 to Y.C.).

References

1. Martinez TF, Lyons-Abbott S, Bookout AL et al (2023) Profiling mouse brown and white adipocytes to identify metabolically relevant small ORFs and functional microproteins. Cell Metab 35:166–183 e111
2. Chen J, Brunner AD, Cogan JZ et al (2020) Pervasive functional translation of noncanonical human open reading frames. Science 367: 1140–1146
3. Martinez TF, Chu Q, Donaldson C et al (2019) Accurate annotation of human protein-coding small open reading frames. Nat Chem Biol 16(4):458–468
4. Slavoff SA, Mitchell AJ, Schwaid AG et al (2013) Peptidomic discovery of short open reading frame-encoded peptides in human cells. Nat Chem Biol 9:59–64
5. Li Q, Liu F, Ma X et al (2024) Proteomic profiling of unannotated microproteins in human placenta reveals XRCC6P1 as a potential negative regulator of translation. J Proteome Res 23:4005–4013
6. Cao X, Khitun A, Na Z et al (2020) Comparative proteomic profiling of unannotated microproteins and alternative proteins in human cell lines. J Proteome Res 19:3418–3426
7. Prensner JR, Enache OM, Luria V et al (2021) Noncanonical open reading frames encode functional proteins essential for cancer cell survival. Nat Biotechnol 39:697–704
8. Wright BW, Yi Z, Weissman JS, Chen J (2022) The dark proteome: translation from noncanonical open reading frames. Trends Cell Biol 32:243–258
9. Li F, Yang K, Gao X et al (2024) A peptide encoded by upstream open reading frame of MYC binds to tropomyosin receptor kinase B and promotes glioblastoma growth in mice. Sci Transl Med 16:eadk9524
10. Chen Y, Su H, Zhao J et al (2023) Unannotated microprotein EMBOW regulates the interactome and chromatin and mitotic functions of WDR5. Cell Rep 42:113145
11. Ingolia NT, Lareau LF, Weissman JS (2011) Ribosome profiling of mouse embryonic stem cells reveals the complexity and dynamics of mammalian proteomes. Cell 147:789–802
12. Lee S, Liu B, Lee S et al (2012) Global mapping of translation initiation sites in mammalian cells at single-nucleotide resolution. Proc Natl Acad Sci USA 109:E2424–E2432
13. Guttman M, Russell P, Ingolia NT, Weissman JS, Lander ES (2013) Ribosome profiling provides evidence that large noncoding RNAs do not encode proteins. Cell 154:240–251

14. Ji Z, Song R, Regev A, Struhl K (2015) Many lncRNAs, 5'UTRs, and pseudogenes are translated and some are likely to express functional proteins. elife 4:e08890
15. Cao X, Chen Y, Khitun A, Slavoff SA (2023) BONCAT-based profiling of nascent small and alternative open reading frame-encoded proteins. Bio Protoc 13:e4585
16. Khitun A, Slavoff SA (2019) Proteomic detection and validation of translated small open reading frames. Curr Protoc Chem Biol 11:e77

Chapter 3

Identification of Canonical and Alternative Proteins by Mass Spectrometry

Clémence Guillon, Patrice Thuleau, Serge Plaza, and Bertrand Fabre

Abstract

Alternative proteins (AltProts) produced from short open reading frames have been missed when genomes were annotated primarily due to their small sizes. In addition, their limited size and low abundance further challenge their identification using conventional proteomics. Several specific workflows have been developed to identify AltProts by mass spectrometry. However, the different protocols rely on the separation of AltProts from canonical proteins, thus preventing concomitant identification of both types of molecules within a single sample. Here, we present an experimental strategy for simultaneous identification of both canonical proteins and AltProts within a single mass spectrometry analysis. This approach includes a simple protein extraction method and an analysis pipeline dedicated to the identification of AltProts.

Key words Alternative proteins, Short open reading frame-encoded polypeptide, Microprotein, Peptide, Short open reading frame, Peptidomics, Proteomics, Mass spectrometry, Data-dependent acquisition

1 Introduction

Historically, following the initial characterization and annotation of eukaryotic genomes, an open reading frame (ORF) was deemed translatable based on several criteria, such as: a length of more than 100 codons, the presence of an ATG initiator codon and termination with a stop codon. Because start and stop codons appear too frequently in random DNA sequences, leading to the annotation of many sORFs that are artifacts, the criterion of 100 contiguous codons was used to select functional ORFs for genome annotation. Thus, ORFs below 100 codons were systematically excluded from annotations [1].

Continuous improvements in mass spectrometry technologies, associated with the development of ribosome profiling and advances in RNA sequencing techniques, have challenged some of these historical criteria for defining an ORF. Indeed, over the past

Stephan Wenkel (ed.), *Microproteins: Methods and Protocols*, Methods in Molecular Biology, vol. 2992,
https://doi.org/10.1007/978-1-0716-5013-4_3,

decade, numerous studies have reported the discovery of many small proteins that do not conform to the traditional criteria for genome annotation. It is now well established that thousands of unannotated ORFs in eukaryotes can be translated into functional proteins [2]. It is important to note that many of these sORFs were detected through ribosomal profiling experiments, demonstrating ribosome occupancy on the ORF, but not proving that a stable translation product is produced and functional [3]. Therefore, mass spectrometry constitutes an essential path to identify these unannotated proteins.

Although the scientific community has recently addressed the nomenclature for the various types of ORFs encoding small proteins [4], a universal term for this particular class of proteins is still lacking. These proteins are currently grouped under multiple terms, including alternative proteins (AltProts) [5], *s*hort ORF-*en*coded *p*olypeptides (SEPs) [6], or microproteins [7]. In the present article, we will use the term AltProts to refer to these small proteins. AltProts can originate from various sources, including long noncoding RNAs, untranslated regions of messenger RNAs (such as the 3′ and 5′ UTRs), or ORFs embedded within canonical ORFs but translated in a different reading frame [2]. The identification of AltProts has been significantly facilitated by the development of specialized databases, such as OpenProt [8] or sORF.org [9] (as well as others [10, 11]). These databases catalog tens of thousands of ORFs that potentially encode AltProts in different organisms. However, determining which of these ORFs are effectively translated and produce functional proteins remains a challenge. Although the alternative proteome still remains poorly understood to date, several research groups have already highlighted its potential by investigating the role of specific AltProts. For instance, hundreds of AltProts have been identified in various organisms and seem to be involved in key cellular processes and in various human pathologies [12–14].

Because AltProts are typically shorter and less abundant than canonical proteins, their identification usually requires specific small protein enrichment methods, rigorously annotated databases and dedicated bioinformatics analyses [15]. Nevertheless, despite important efforts from several research groups around the world, there is still no universal mass spectrometry-based workflow for the identification of AltProts [15].

Here, we propose a comprehensive experimental workflow to determine the composition of both canonical proteins and AltProts in a sample in a single analysis. We present a simple protein extraction method, as well as dedicated computational tools and criteria for identifying AltProts by mass spectrometry.

2 Materials

2.1 Cell Culture and Biological Material

1. Cell culture line: HeLa cells.
2. Tissue culture dishes, 150 mm diameter, surface treated, sterile (TC ClearLine 143 × 21.9 mm) (Thermo Fisher Scientific, Waltham, USA).
3. DMEM (Dulbecco's Modified Eagle Medium) Gibco™ (Thermo Fisher Scientific, Waltham, USA).
4. Fetal bovine serum (FBS) Gibco™ (Thermo Fisher Scientific, Waltham, USA).
5. Penicillin (MERCK, Burlington, USA).
6. Streptomycin (MERCK, Burlington, USA).
7. Trypsin (MERCK, Burlington, USA).
8. Phosphate buffered saline (PBS) pH 7.4 (MERCK, Burlington, USA).
9. Incubator at 37 °C with 5% CO_2.

2.2 Protein Extraction

1. Lysis buffer: ammonium bicarbonate 50 mM, sodium deoxycolate 1% (MERCK, Burlington, USA).
2. Protein quantification: BCA protein assay (Pierce™, Waltham, Massachusetts, USA).

2.3 Sample Preparation for Mass Spectrometry

1. Dithiothreitol (DDT) 1 M (MERCK, Burlington, USA).
2. 2-Chloroacetamide 600 mM (MERCK, Burlington, USA), freshly prepared and kept in the dark.
3. Ammonium bicarbonate 50 mM (MERCK, Burlington, USA).
4. Sequencing grade modified Trypsin (trypsin/protein ratio = 1/50) (Promega, Madison, USA).
5. Trifluoroacetic acid (TFA) (MERCK, Burlington, USA).
6. Desalting C18 column: (Sep-Pak C18 1 cc Vac Cartridge, 100 mg sorbent per cartridge, 55–105 μm) (Pierce™, Waltham, Massachusetts, USA).
7. High pH reversed-phase fractionation (Pierce™ high pH reversed-phase peptide fractionation kit) (Pierce™, Waltham, Massachusetts, USA).

2.4 Data Dependent Acquisition (DDA) Mass Spectrometry

1. Formic acid (FA) (MERCK, Burlington, USA).
2. Buffer A: Acetonitrile 5%, formic acid 0.2% (MERCK, Burlington, USA).
3. Buffer B: Acetonitrile 80%, formic acid 0.2%.

4. Nano-LC–MS/MS using a nanoRS UHPLC system coupled to an Orbitrap Exploris 480 mass spectrometer using FAIMS Pro Duo interface (Thermo Fisher Scientific, Bremen, Germany).
5. C18 column (PEPMAP C18 2 μm 75 μm × 500 mm, Thermo Fisher Scientific, Bremen, Germany).

2.5 Bioinformatic Analysis of DDA Data

1. *Uniprot database* (June 2023, 81,791 sequences) (database containing annotated canonical proteins).
2. *OpenProt database* (version 1.6) (database containing predicted alternative proteins).
3. Mascot search engine (version 2.8).
4. Percolator.
5. *Proline* (version 2.1).
6. *Uniprot peptide search* (https://www.uniprot.org/peptide-search).

3 Methods

3.1 Cell Culture and Biological Material

1. Grow HeLa cells in sterile conditions in DMEM supplemented with 10% FBS, 100 U/mL penicillin and 100 μg/mL streptomycin in an incubator at 37 °C and 5% CO_2 and at a maximum confluency of 80%.
2. Seed five million cells onto tissue culture dishes containing 15 ml of DMEM supplemented with 10% FBS, 100 U/mL penicillin and 100 μg/mL streptomycin at 37 °C, 5% CO_2 until 80% confluence is achieved (approx. 48 h to 72 h).
3. Remove culture medium. Rinse cells with PBS and then overlay cells with 5 ml trypsin for approximately 5 min at 37 °C, 5% CO_2, to detach cells from the petri dish. Resuspend cells and transfer them into a 15 ml Falcon. Centrifuge for 5 min at 500 g at room temperature (RT).
4. Discard supernatant and wash the cell pellet in 1 ml of PBS. Centrifuge 5 min at 500 g at RT.
5. Discard supernatant (at this stage, cell pellets can be frozen at −80 °C). Several biological replicates are required to perform any quantitative analysis later in the workflow.

3.2 Protein Extraction

1. Resuspend cells by pipetting back and forth in 200 μl of lysis buffer (*see* **Notes 1** and **2**).
2. Heat for 5 min at 95 °C and sonicate the samples.
3. Centrifuge the samples 10 min at 14,000 g at RT to eliminate cellular debris.

4. Recover the supernatant into new tubes.
5. Quantify proteins using the BCA protein assay kit (Pierce™, Waltham, Massachusetts, USA), as recommended by the supplier.

3.3 Sample Preparation for Mass Spectrometry

1. Use 100 μg of proteins of each sample for the following steps.
2. Protein reduction: Supplement protein extracts with DTT at 10 mM final concentration. Heat for 10 min at 95 °C.
3. Protein alkylation: After allowing samples to cool down to RT, supplement protein extracts with freshly prepared chloroacetamide at 30 mM final concentration (prepare a 600 mM stock solution and place immediately in the dark). Incubate 30 min at RT in the dark.
4. Protein digestion: dilute protein extracts five-fold with 50 mM ammonium bicarbonate containing trypsin to digest proteins into peptides (1 μg trypsin per 50 μg protein). Incubate overnight (O/N) at 37 °C.
5. Precipitation of sodium deoxycholate: Acidify samples with TFA at a final concentration of 0.4%. Centrifuge for 10 min at 14,000 g at 4 °C. Recover the supernatant and discard the pellet.
6. Desalting: Desalt samples on a C18 column, as recommended by the supplier and dried down with speed vac.
7. Resuspend samples with 200 μl of 0.1%TFA. Sonicate in a bath for 5 min. If necessary, centrifuge for 10 min at 14,000 g at 4 °C to remove any remaining sodium deoxycholate.
8. At this stage of the protocol, there are two options:
   (a) Carry out an exhaustive study of the microproteome/proteome: The following steps are performed on each sample.
   (b) Build a spectral library to couple your experimentation to a data-independent acquisition (DIA) quantitative approach (*see* **Note 3**) before the next step.
9. Perform peptide fractionation using high pH reverse-phase separation: high pH reversed-phase fractionation columns are used according to supplier recommendations. Here, only elution buffers with 5%, 10%, 15%, 20%, and 50% of acetonitrile are used (*see* **Note 4**).
10. Dry samples using speed vac.

3.4 Data Dependent Acquisition (DDA) Mass Spectrometry

1. Resuspend previously dried peptides in 0.2% acid formic to a final concentration of 500 ng/μl. The peptides are analyzed with a nano-LC–MS/MS approach using a UHPLC nanoRS system coupled to a Thermo Fisher Scientific Orbitrap Exploris 480 mass spectrometer with a FAIMS Pro Duo interface.

2. Inject 1 μl of sample onto an analytical C18 column heated at 45 °C. Calibrate mobile phase flow rate to 300 nl/min. Perform an 85 min gradient using buffers A and B as follows: 0–25% buffer B for 58 min, 25–40% buffer B for 20 min, 40–90% buffer B for 2 min, and 90% buffer B for 5 min.
3. The Orbitrap Exploris 480 mass spectrometer is set up as follows: the FAIMS mode is set to a gas flow of 3.9 L/min, two compensation voltages are used, −45 and −60 v, 0.8 and 0.7 s cycle times per CV. The ion selection ranges between 375 and 1200 m/z for MS scan survey with a resolution of 60,000 and a targeted AGC (automatic gain control) set at 300% over a maximum injection time of 50 ms. A minimum threshold of 5×10^3 is set for ion fragmentation. The number of ions selected is determined by the "TopSpeed" algorithm. The most abundant precursors are fragmented by high-energy collision-induced dissociation, with a normalized collision energy of 30%. The target AGC value is set at 100%, and the injection time is fixed at its maximum. MS2 resolution is set at 15,000. The precursor isolation width is set at 1.6 m/z, and the dynamic exclusion window at 45 s to prevent repetitive selection of the same peptide.

3.5 Bioinformatic Analysis of DDA Data

Perform the analysis of raw data generated by the mass spectrometer as follows:

1. *Use the Mascot search engine* to assign the MS/MS spectra to peptides and proteins (*see* **Note 5**). Configure Mascot with default parameters, specifying N-terminal acetylation and methionine oxidation as variable modifications, and carbamidomethylation of cysteines as a fixed modification.
2. Use a database resulting from the combination of two databases: the *Uniprot database* to allow the identification of canonical proteins, and the *OpenProt database*, including AltProts and isoforms (*see* **Notes 6** and **7**). Add a database containing common contaminants to the search. Rescore *peptide-spectrum matches (PSMs)* obtained through mascot using *percolator*. Then, import *all data into proline*. Filter the identifications with an *FDR of 1% for PSMs* and *5% for proteins*.
3. *PSMs for AltProts* require *manual validation* by the operator according to the following criteria (*see* **Note 8**):
   (a) The cumulative *y and b ions* must represent at least *75% of the sequence coverage of the peptide*.
   (b) Eliminate noisy spectra and spectra with many unassigned intense peaks.

(c) The *proteotypicity* of each peptide sequence must be validated using the *Uniprot peptide search engine* (against all proteomes). Importantly, the "Treat isoleucine and leucine as equivalent" module must be enabled for this analysis.

4 Notes

1. This protocol can be adapted to other biological materials beyond cultured HeLa cells. Other cell lines can be used by adjusting the culture conditions. Additionally, materials other than cultured cells can also be used. However, the extraction method might need to be adapted to the type of material used. For instance, when using animal or plant tissues, the extraction protocol should begin with a tissue grinding step in liquid nitrogen.
2. This protocol has the advantage of identifying both canonical proteins and AltProts present in a sample. However, if necessary, specific approaches for AltProt identification can be applied, particularly the use of enrichment methods for small proteins during extraction, which can facilitate the identification of AltProts, as described previously [16–18]. It should be noted that small protein enrichment may not allow the identification of many canonical proteins.
3. We describe here an approach using a data-dependent acquisition (DDA) method coupled with high-pH reverse-phase fractionation of the samples in order to provide the most comprehensive description of the proteome in a given condition. This approach can be coupled with a quantitative strategy, such as the use of Data Independent Acquisition (DIA) on the same samples. The DDA-acquired data will allow the generation of an experimental spectral library, which can then be used for the analysis of DIA data under similar experimental conditions [19]. In this situation, to generate the spectral library, we advise using a pool of each replicate used for the quantitative data analysis as starting material, as described in [19]. One half of each replicate is set aside for DIA quantitative analysis, and the second half is dedicated to the creation of a spectral library. To do the latter, pool together a similar amount of each replicate of a sample to reach a final quantity of 100 μg of peptides. This pool of peptides is then used for peptide separation by high pH fractionation.
4. In this protocol, during high pH reversed-phase fractionation, we chose to collect only five fractions out of the eight possible fractions according to the supplier's recommendations. The

number of fractions was reduced to minimize the number of samples injected. By increasing the number of fractions, you should theoretically increase the number of final identified proteins (canonical and AltProts).

5. In this protocol, mascot and proline are used as the search engine and validation/visualization software, respectively. This combination was shown to provide a good number of identifications in comparison to other software while limiting false positive identifications [20]. In addition, a viewer of PSMs is implemented in Proline, which is useful to evaluate MS/MS spectra and see potential other peptides matching a spectrum. These modules are not present in all the other software dedicated to the analysis of proteomics data. As an alternative, we would encourage to use of several search engines as it generally increases the number of identifications, as it is implemented in OpenProt, for example, through the use of Peptide Shaker. Importantly, such a combination of search engines is supported in Proline 2.1 (through the integration of a Peptide Shaker module). However, we had some problems using software such as MaxQuant [21] with large databases. Indeed, very few identifications were obtained with MaxQuant with the Human OpenProt database (around 500,000 sequences) with a PSM FDR at 1%. This was not the case using the *Drosophila melanogaster* OpenProt database (around 70,000 sequences) with MaxQuant in a previous study [16].
6. Here, we use OpenProt as the database to identify alternative proteins in human cells. OpenProt is a well-curated database predicting AltProts from RNAseq data. Currently, AltProts predictions are provided for 12 species in OpenProt. Among these 12 species, only eukaryotes are represented, and only one plant species, *Arabidopsis thaliana*. Before starting your experiment, you need to verify the availability of a suitable alternative protein database for your model species. If the organism of interest is not present in OpenProt, it is possible to predict alternative ORFs from RNAseq data if available, using dedicated workflows [22].
7. In this protocol, we use a database that is a combination of Uniprot and OpenProt. Although OpenProt includes the canonical proteome, we chose to add the latest version of Uniprot as we observed more false positives when using the RefProts part of OpenProt in previous analysis.
8. Although the manual validation of the PSMs for AltProts and checking the proteotypicity of the identified peptides are *time-consuming* at present (as these tasks are not yet automated), we believe that they are *essential* for genuine identification of AltProts and avoiding many canonical proteins' peptides mis-assigned to AltProts.

Acknowledgments

This work has been supported by the Fondation ARC pour la recherche sur le cancer and the Fondation Toulouse Cancer Santé and the French ANR project AltFUNC (ANR-23-CE44-0028).

References

1. Basrai MA, Hieter P, Boeke JD (1997) Small open reading frames: beautiful needles in the haystack. Genome Res 7:768–771
2. Brunet MA, Leblanc S, Roucou X (2020) Reconsidering proteomic diversity with functional investigation of small ORFs and alternative ORFs. Exp Cell Res 393:112057
3. Alagar Boopathy LR, Beadle E, Garcia-Bueno Rico A et al (2023) Proteostasis regulation through ribosome quality control and no-go-decay. Wiley Interdiscip Rev RNA 14:e1809
4. Mudge JM, Ruiz-Orera J, Prensner JR et al (2022) Standardized annotation of translated open reading frames. Nat Biotechnol 40:994–999
5. Leblanc S, Yala F, Provencher N et al (2024) OpenProt 2.0 builds a path to the functional characterization of alternative proteins. Nucleic Acids Res 52:D522–D528
6. Slavoff SA, Mitchell AJ, Schwaid AG et al (2013) Peptidomic discovery of short open reading frame-encoded peptides in human cells. Nat Chem Biol 9:59–64
7. Azam S, Yang F, Wu X (2025) Finding functional microproteins. Trends Genet 41:107
8. Brunet MA, Lucier JF, Levesque M et al (2021) OpenProt 2021: deeper functional annotation of the coding potential of eukaryotic genomes. Nucleic Acids Res 49:D380–D388
9. Olexiouk V, Van Criekinge W, Menschaert G (2018) An update on sORFs.org: a repository of small ORFs identified by ribosome profiling. Nucleic Acids Res 46:D497–D502
10. Choteau SA, Wagner A, Pierre P et al (2021) MetamORF: a repository of unique short open reading frames identified by both experimental and computational approaches for gene and metagene analyses. Database (Oxford) 2021:baab032
11. Hazarika RR, De Coninck B, Yamamoto LR et al (2017) ARA-PEPs: a repository of putative sORF-encoded peptides in Arabidopsis thaliana. BMC Bioinformatics 18:37
12. Andrews SJ, Rothnagel JA (2014) Emerging evidence for functional peptides encoded by short open reading frames. Nat Rev Genet 15:193–204
13. Baena-Angulo C, Platero AI, Couso JP (2024) Cis to trans: small ORF functions emerging through evolution. Trends Genet 41(2):119–131
14. Dong X, Zhang K, Xun C et al (2023) Small open reading frame-encoded micro-peptides: an emerging protein world. Int J Mol Sci 24:10562
15. Fabre B, Combier JP, Plaza S (2021) Recent advances in mass spectrometry-based peptidomics workflows to identify short-open-reading-frame-encoded peptides and explore their functions. Curr Opin Chem Biol 60:122–130
16. Fabre B, Choteau SA, Duboé C et al (2022) In depth exploration of the alternative proteome of Drosophila melanogaster. Front Cell Dev Biol 10:901351
17. Cardon T, Hervé F, Delcourt V et al (2020) Optimized sample preparation workflow for improved identification of ghost proteins. Anal Chem 92:1122–1129
18. Ma J, Diedrich JK, Jungreis I et al (2016) Improved identification and analysis of small open reading frame encoded polypeptides. Anal Chem 88:3967–3975
19. Guillon C, Pichereaux C, Lazar I et al (2024) Mass spectrometry-based workflow for the identification and quantification of alternative and canonical proteins in pancreatic cancer cells. Cells 13:1966
20. Bouyssié D, Hesse AM, Mouton-Barbosa E et al (2020) Proline: an efficient and user-friendly software suite for large-scale proteomics. Bioinformatics 36:3148–3155
21. Cox J, Hein MY, Luber CA et al (2014) Accurate proteome-wide label-free quantification by delayed normalization and maximal peptide ratio extraction, termed MaxLFQ. Mol Cell Proteomics 13:2513–2526
22. Garcia-del Rio DF, Derhourhi M, Bonnefond A et al (2024) Deciphering the ghost proteome in ovarian cancer cells by deep proteogenomic characterization. Cell Death Dis 15:712

Chapter 4

Identification and Characterization of Microproteins/ MicroORFs

Na Ding, Juan Xu, and Jiaxin Yu

Abstract

Microproteins and microOpen Reading Frames (microORFs) are small proteins encoded by short, often overlooked, coding regions in the genome. These peptides are typically under 100 amino acids in length and are believed to play critical roles in cellular processes such as stress response, metabolic regulation, and protein quality control. However, their discovery has been hindered by limitations in traditional proteomics and gene annotation pipelines. In this chapter, we present a novel method for the systematic identification of microproteins and microORFs across a range of eukaryotic species using an integrative approach that combines ribosome profiling, mass spectrometry-based proteomics, and advanced bioinformatics tools. Our method enhances the sensitivity of microprotein detection and allows for the validation of candidate microORFs in vivo. This methodology provides a comprehensive platform for microprotein discovery, offering new insights into gene regulation and cellular function.

Key words Microproteins, MicroORFs, Ribosome Profiling, Mass Spectrometry, Small Peptides, Bioinformatics, Cellular Function

1 Introduction

Microproteins and microORFs are small peptides, typically fewer than 100 amino acids in length, that are found from canonical protein-coding genes to noncoding genes to intergenic transcribed regions (Fig. 1) [1–4]. Despite their small size, these microproteins can play essential roles in various cellular processes [5, 6], such as stress response [7], gene regulation [8], and metabolic control [9, 10]. Historically, microproteins have been difficult to detect due to their low abundance, small size, and the fact that many of their encoding regions do not fit the canonical patterns recognized by traditional gene annotation pipelines. As a result, many of these small proteins were considered biologically irrelevant or simply went unnoticed. However, recent advances in ribosome profiling and mass spectrometry-based proteomics have opened up new avenues for discovering and characterizing microproteins [11–

Stephan Wenkel (ed.), *Microproteins: Methods and Protocols*, Methods in Molecular Biology, vol. 2992,
https://doi.org/10.1007/978-1-0716-5013-4_4,

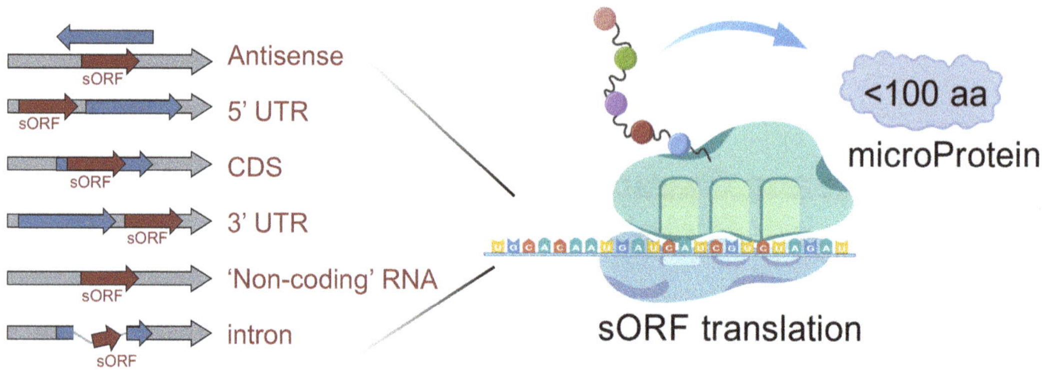

Fig. 1 The transcriptome includes a wide variety of sORFs, which can be translated into microproteins, distributed across various genomic regions

13], providing a deeper understanding of their biological significance.

Ribosome profiling, which involves deep sequencing of ribosome-protected RNA fragments (RPFs), allows for a genome-wide view of translation, offering unprecedented resolution in detecting small ORFs that are often missed by conventional transcriptomics [14, 15]. Ribosome profiling enables the identification of short, unannotated ORFs, including microORFs, by mapping ribosome occupancy across the entire transcriptome [16]. When paired with mass spectrometry (MS), this approach provides a powerful method for validating the presence of the predicted microproteins [17]. MS-based proteomics enables the identification of peptides by measuring their mass-to-charge ratio (m/z), and when coupled with liquid chromatography (LC), it allows for sensitive detection and quantification of peptides, including small microproteins [18, 19]. By integrating these two techniques, it is possible to identify and confirm the translation of previously unannotated and low-abundance microproteins (Fig. 2) [20].

The method described in this protocol combines ribosome profiling and mass spectrometry to systematically identify and validate microproteins across a variety of eukaryotic species. This integrative approach not only improves the sensitivity of detecting microproteins but also facilitates their functional annotation. The key steps involved in this process include the isolation and sequencing of ribosome footprints to identify candidate microORFs, followed by protein extraction and peptide analysis via mass spectrometry to confirm the presence of corresponding microproteins. Data integration between ribosome profiling and MS ensures that only true microproteins are validated and that their biological relevance can be assessed through bioinformatics tools. The integration of ribosome profiling and mass spectrometry offers a

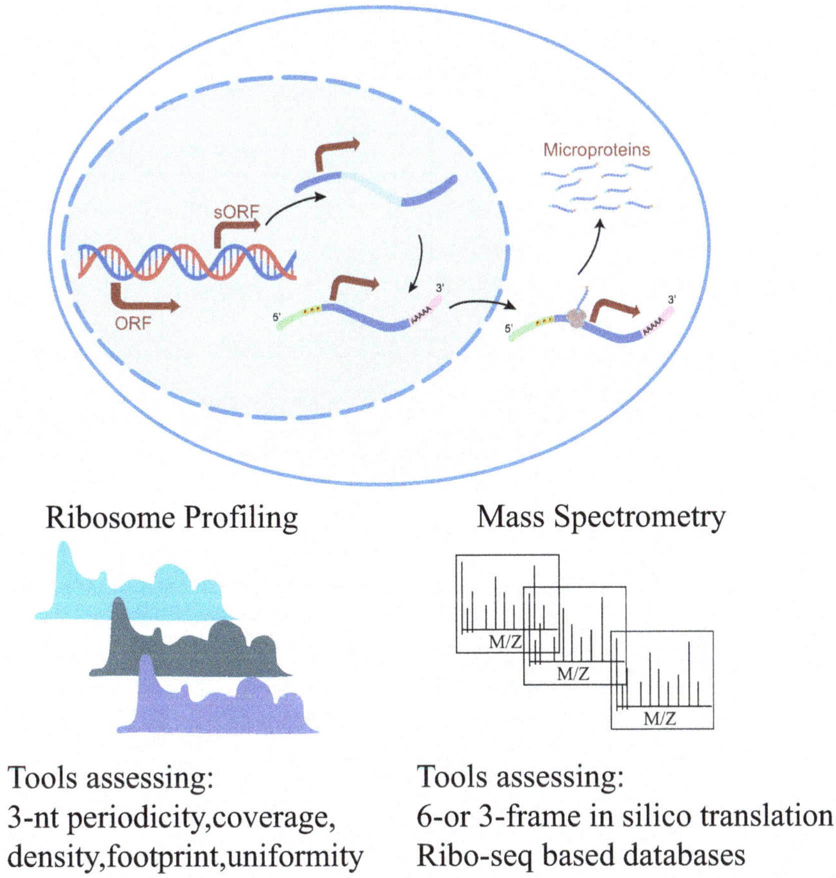

Fig. 2 The two primary approaches for quantifying and characterizing sORFs/microproteins rely on ribosome profiling and mass spectrometry

comprehensive, sensitive, and high-throughput method for uncovering the hidden layers of the proteome. This approach has broad applications (Fig. 3), including domain search and the model organisms such as *Saccharomyces cerevisiae* (yeast) [21], *Mus musculus* (mouse) [9], and human cell lines (*Homo sapiens*) [22]. By identifying novel microproteins that are involved in key biological processes such as cellular stress responses, metabolic regulation, and signal transduction, this methodology provides valuable insights into the molecular mechanisms underlying various diseases, including cancer and neurodegenerative disorders. Additionally, this approach can potentially lead to the identification of new therapeutic targets, making it a promising tool for both basic research and clinical applications.

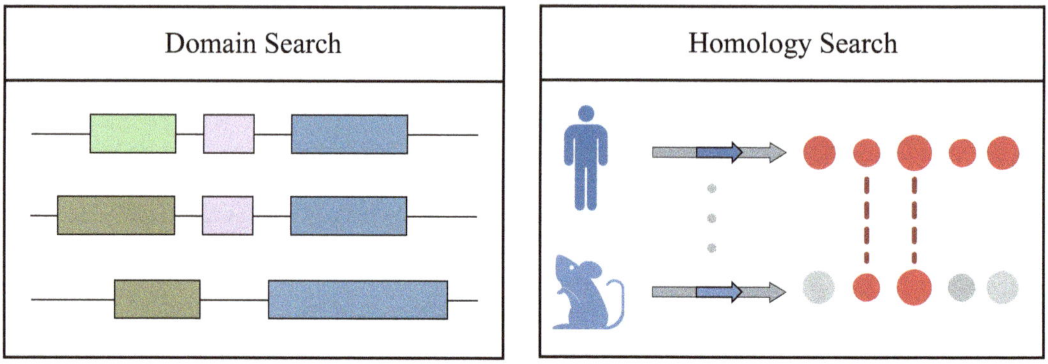

Fig. 3 Further validation and analysis of microprotein identification methods

2 Materials

This method requires several key materials, including cell lines or model organisms (e.g., *Saccharomyces cerevisiae*, *Mus musculus*, and *Homo sapiens*) for biological sample preparation. RNA extraction kits (e.g., TRIzol and Qiagen RNeasy) are used for ribosome profiling, while protein extraction kits (e.g., RIPA buffer) are employed for mass spectrometry analysis. Ribosome profiling and mass spectrometry sample preparation kits, such as those from Illumina or Thermo Fisher, ensure high-quality RNA and protein isolation. For 231metry analysis, LC–MS systems (e.g., Thermo Fisher Orbitrap) are essential for peptide separation and detection. Antibodies for protein validation (e.g., GAPDH and microprotein-specific antibodies if available) and bioinformatics tools like MaxQuant and TransDecoder for data analysis further support the discovery and functional annotation of microproteins.

2.1 Cell Lines and Organisms

1. *Saccharomyces cerevisiae* (yeast): commonly used model organism for functional genomics and proteomics.
2. *Mus musculus* (mouse): representative mammalian model for studying microprotein functions In Vivo.
3. *Homo sapiens* (human): human cell lines (e.g., HEK293T and A549) for studying microproteins in a human context (*see* **Note 1**).

2.2 Reagents and Kits

1. RNA Extraction Kits: Standard kits for total RNA isolation from cultured cells or tissues (e.g., TRIzol reagent and Qiagen RNeasy Kit).
2. Ribosome Profiling Kits: Commercial kits for ribosome footprint preparation (e.g., Illumina Ribosome Profiling Kit and Bioo Scientific Ribosome Profiling Kit).

3. Protein Extraction Kits: Kits for total protein extraction from cells or tissues (e.g., RIPA buffer and Thermo Fisher Pierce Protein Extraction Kit).
4. Protein Enrichment Kits: Kits for enriching small proteins (e.g., peptide isolation kits and MS-specific protein enrichment kits like Thermo Fisher Pierce Protein Concentrator) [23].
5. Mass Spectrometry Sample Preparation Kits: Tryptic digestion kits for protein digestion (e.g., Thermo Fisher Pierce Mass Spec Sample Prep Kit).
6. Chromatography Reagents: LC–MS-grade solvents for liquid chromatography (e.g., acetonitrile, water, and formic acid).

2.3 Antibodies (for Validation)

1. Microprotein-Specific Antibodies: If available, antibodies raised against predicted microproteins for validation via Western blot or immunofluorescence.
2. General Protein Markers: Antibodies to common cellular proteins for normalization and control (e.g., GAPDH and actin).

2.4 Bio-informatics Tools

1. Ribosome Profiling Analysis Software: Dedicated tools such as riboSeqR, riboPlot, or custom scripts (e.g., in R or Python) for processing ribosome profiling data and identifying candidate microORFs.
2. Proteomics Data Analysis Software: MaxQuant, Mascot, or other software tools for protein identification and quantification from mass spectrometry data [24].
3. ORF Prediction Tools: TransDecoder, ORF Finder, or custom scripts for predicting novel ORFs from ribosome profiling data.
4. Functional Annotation Tools: InterProScan, Pfam, and other tools for domain prediction and functional annotation of identified peptides.

3 Methods

The method combines ribosome profiling and mass spectrometry to identify and validate microproteins. Ribosome profiling involves isolating ribosome-protected RNA fragments from cell samples, sequencing them to map ribosome positions, and identifying small ORFs that may correspond to microproteins. Protein extraction follows, with tryptic digestion of proteins into peptides. These peptides are analyzed by high-resolution liquid chromatography–mass spectrometry (LC–MS) to detect and quantify the microproteins. The data from both techniques are integrated using bioinformatics tools to confirm the presence of microproteins, and functional annotation is performed to predict their biological roles.

3.1 Ribosome Profiling

1. Harvest cells in log phase, wash, and lyse to isolate ribosome-protected RNA fragments (RPFs).
2. Treat the lysate with RNase I to digest non-protected RNA regions, followed by ribosome footprint isolation through sucrose gradient centrifugation.
3. Convert RPFs into cDNA using reverse transcription and sequence the cDNA libraries using high-throughput RNA sequencing (*see* **Note 2**).
4. Map ribosome footprints to the genome or transcriptome using tools like STAR or Bowtie.
5. Identify small ORFs (microORFs) with significant ribosome occupancy, particularly those falling outside the traditional coding regions.
6. Filter out noise and assign reading frames to predict potential microproteins.
7. Compare ribosome profiling data with existing transcriptome and proteome databases to identify novel microORFs not previously annotated.
8. Cross-reference with known microORFs in other species to improve prediction accuracy.

3.2 Mass Spectrometry (MS)-Based Proteomics

1. Extract total proteins from cultured cells or tissues using RIPA buffer or other protein extraction reagents.
2. Digest extracted proteins using trypsin or other enzymes to generate peptides.
3. Clean up digestion products using solid-phase extraction or peptide enrichment methods.
4. Inject peptide mixtures into an LC–MS system for separation and analysis (*see* **Note 3**).
5. Utilize a high-resolution MS system (e.g., Orbitrap or Q-TOF) to analyze peptide mass-to-charge (m/z) ratios (*see* **Note 4**).
6. Identify peptides by comparing observed m/z ratios with theoretical peptide sequences generated from genome or transcriptome data (*see* **Note 5**).
7. Use software such as MaxQuant or Mascot to match observed MS spectra to predicted peptides from the ribosome profiling data.
8. Validate novel microproteins by comparing them to known databases and checking their presence across different biological replicates (*see* **Note 6**).
9. Use protein domain prediction tools to annotate identified peptides and infer potential functions (*see* **Note 7**).

10. Validate the biological relevance of detected microproteins through network analysis and pathway enrichment tools.

3.3 Integration of Ribosome Profiling and MS Data

1. Combine ribosome profiling data (which identifies potential microORFs) with mass spectrometry results (which confirm the presence of the microproteins).
2. Cross-check the novel peptides with known functional microproteins or hypothetical sequences predicted from genomic databases.
3. Use additional computational tools to correlate the expression levels of microproteins with specific biological conditions or stress responses.

3.4 Post-Translational Modifications

1. If relevant, perform additional analysis using phosphoproteomics or other techniques (e.g., ubiquitination or acetylation studies) to explore the functional roles of identified microproteins.
2. Analyze the impact of specific post-translational modifications on the stability, localization, and function of microproteins (*see* **Note 8**).

4 Notes

1. The choice of model organism and cell line significantly impacts the identification of microproteins. Therefore, selecting the appropriate biological system based on research goals is important.
2. Ribosome profiling can be sensitive to experimental conditions, and optimizing the RNase digestion step is crucial to ensure proper fragment size distribution.
3. Microproteins are often low in abundance, so sample enrichment techniques (such as peptide concentration or fractionation) can significantly enhance detection sensitivity.
4. MS sensitivity needs to be carefully calibrated to detect low-abundance peptides effectively.
5. Integrating ribosome profiling data with MS data provides a powerful approach to confirm the existence of microproteins and to predict their functional roles.
6. However, caution should be exercised when interpreting low-confidence microproteins or peptides, as sequencing or experimental artifacts can sometimes lead to false positives.
7. The pipeline can be adapted for use across different species to identify conserved and species-specific microproteins, providing deeper.

8. Since microproteins may be subject to post-translational modifications (e.g., phosphorylation and ubiquitination), analyzing these modifications using specialized techniques (e.g., phosphoproteomics) will enhance the understanding of their functional roles.

Acknowledgments

This work was supported by the National Natural Science Foundation of China [32300541]; the Natural Science Foundation of Heilongjiang Province [LH2023C063]; the China Postdoctoral Science Foundation [2023 T160177, 2021 M693831]; and the Heilongjiang Postdoctoral Fund [LBH-Z20071].

References

1. Couso JP, Patraquim P (2017) Classification and function of small open reading frames. Nat Rev Mol Cell Biol 18:575–589
2. Magny EG, Pueyo JI, Pearl FM et al (2013) Conserved regulation of cardiac calcium uptake by peptides encoded in small open reading frames. Science 341:1116–1120
3. Lv D, Li D, Cai Y et al (2024) CancerProteome: a resource to functionally decipher the proteome landscape in cancer. Nucleic Acids Res 52:D1155–D1162
4. Liang Y, Lv D, Liu K et al (2024) MicroProteinDB: a database to provide knowledge on sequences, structures and function of ncRNA-derived microproteins. Comput Biol Med 177: 108660
5. Guo X, Chavez A, Tung A et al (2018) High-throughput creation and functional profiling of DNA sequence variant libraries using CRISPR-Cas9 in yeast. Nat Biotechnol 36:540–546
6. Zlotorynski E (2020) The functions of short ORFs and their microproteins. Nat Rev Mol Cell Biol 21:252–253
7. Fesenko I, Shabalina SA, Mamaeva A et al (2021) A vast pool of lineage-specific microproteins encoded by long non-coding RNAs in plants. Nucleic Acids Res 49:10328–10346
8. Pang Y, Liu Z, Han H et al (2020) Peptide SMIM30 promotes HCC development by inducing SRC/YES1 membrane anchoring and MAPK pathway activation. J Hepatol 73: 1155–1169
9. Martinez TF, Lyons-Abbott S, Bookout AL et al (2023) Profiling mouse brown and white adipocytes to identify metabolically relevant small ORFs and functional microproteins. Cell Metab 35:166–183 e111
10. Grubb A, Jensson O, Gudmundsson G et al (1984) Abnormal metabolism of gamma-trace alkaline microprotein. The basic defect in hereditary cerebral hemorrhage with amyloidosis. N Engl J Med 311:1547–1549
11. Wang Y, Tang Y, Xie Z et al (2025) RPFdb v3.0: an enhanced repository for ribosome profiling data and related content. Nucleic Acids Res 53:D293–D298
12. Luo X, Huang Y, Li H et al (2022) SPENCER: a comprehensive database for small peptides encoded by noncoding RNAs in cancer patients. Nucleic Acids Res 50:D1373–D1381
13. Kustatscher G, Grabowski P, Schrader TA et al (2019) Co-regulation map of the human proteome enables identification of protein functions. Nat Biotechnol 37:1361–1371
14. Erhard F, Halenius A, Zimmermann C et al (2018) Improved Ribo-seq enables identification of cryptic translation events. Nat Methods 15:363–366
15. Calviello L, Mukherjee N, Wyler E et al (2016) Detecting actively translated open reading frames in ribosome profiling data. Nat Methods 13:165–170
16. Ouspenskaia T, Law T, Clauser KR et al (2022) Unannotated proteins expand the MHC-I-restricted immunopeptidome in cancer. Nat Biotechnol 40:209–217
17. Sun L, Wang W, Han C et al (2021) The oncomicropeptide APPLE promotes hematopoietic malignancy by enhancing translation initiation. Mol Cell 81:4493–4508 e4499

18. Van Heesch S, Witte F, Schneider-Lunitz V et al (2019) The translational landscape of the human heart. Cell 178:242–260 e229
19. Ingolia NT (2016) Ribosome footprint profiling of translation throughout the genome. Cell 165:22–33
20. Jiang S, Li H, Zhang L et al (2025) Generic Diagramming Platform (GDP): a comprehensive database of high-quality biomedical graphics. Nucleic Acids Res 53:D1670–D1676
21. Papadopoulos C, Arbes H, Cornu D et al (2024) The ribosome profiling landscape of yeast reveals a high diversity in pervasive translation. Genome Biol 25:268
22. Pennisi E (2024) 'Dark proteome' survey reveals thousands of new human genes. Science 386:951–952
23. Hadjeras L, Bartel J, Maier LK et al (2023) Revealing the small proteome of Haloferax volcanii by combining ribosome profiling and small-protein optimized mass spectrometry. Microlife 4:uqad001
24. Cai Y, Lv D, Li D et al (2023) IEAtlas: an atlas of HLA-presented immune epitopes derived from non-coding regions. Nucleic Acids Res 51:D409–D417

Chapter 5

Proteogenomic Identification of Annotated and Novel smORFs Via Microprotein Enrichment

Cuitong He, Yao Zhang, and Ping Xu

Abstract

Microproteins, encoded by small open reading frames (smORFs or sORFs) of less than 100 amino acids, play critical roles in gene expression, cellular signaling and metabolism, and hold potential as therapeutic targets. Despite recent genome annotation efforts, many smORFs remain unannotated, leaving a significant gap in our understanding of microprotein structure and function. Efficient microprotein identification is essential for functional studies and improving genome annotation, but remains a technical challenge. Here, we describe a proteogenomic approach that combines microprotein enrichment with mass spectrometry to identify both annotated and novel smORFs. Using this approach, we have successfully identified and validated several novel smORFs in the fully annotated yeast genome. While yeast serves as a model system, this protocol is adaptable to samples from different species, providing a robust framework for microprotein discovery.

Key words Microprotein, smORF, Proteogenomics, Microprotein enrichment, Mass spectrometry

1 Introduction

Despite extensive research into the structure and function of thousands of proteins, smORF-encoded microproteins have been largely overlooked. When annotating the yeast genome, researchers set a minimum threshold of 100 codons for ORF annotation to increase confidence, as annotating all theoretical ORFs with 2–99 codons would generate an additional 260,000 ORFs [1, 2]. This threshold was also chosen because ORFs longer than 100 codons are more likely to encode stable proteins [2]. Consequently, this 100-codon minimum has been widely applied to the annotation of other eukaryotic genomes [3].

However, through gene knockout studies [4], transcriptome sequencing [5] and proteomic analyses [6], numerous unannotated genes, including smORFs, have been progressively discovered, validated, and re-annotated. For example, recent studies have

Stephan Wenkel (ed.), *Microproteins: Methods and Protocols*, Methods in Molecular Biology, vol. 2992,
https://doi.org/10.1007/978-1-0716-5013-4_5,

identified hundreds of unannotated smORFs in mammalian genomes that encode functional microproteins [7–11].

Microproteins play critical roles in diverse biological processes, such as regulating protein interactions and enzyme activity by occupying binding sites on other proteins [12], acting as cytokines in signal transduction and intercellular communication [13], and serving as post-transcriptional regulators of gene expression [14]. Some microproteins are now being investigated as potential therapeutic targets for diseases such as heart failure, obesity, and cancer [15].

Identification of microproteins is essential for their functional characterization. Ribosome profiling (ribo-seq) is a powerful tool for smORF discovery, as it identifies translated mRNAs and provides information on translation efficiency and start sites by sequencing ribosome-protected mRNA fragments [16]. However, ribo-seq results require experimental validation, as ribosomes can also bind to mRNA without active translation [17].

Mass spectrometry (MS)-based proteomics is another untargeted, high-throughput approach for microprotein identification [18–21]. However, according to the UniProt database, most annotated microproteins across species lack protein-level evidence and functional studies. Efficient MS-based identification of microproteins remains a challenge, primarily because tryptic digestion of microproteins often yields only one unique peptide for identification [22]. If these peptides co-elute with high-abundance peptides from larger proteins, the latter can suppress the detection of low-abundance peptides derived from microproteins [23]. Therefore, enrichment of microproteins is a critical strategy for their efficient identification. Various chemical and biophysical methods have been developed, including solid phase extraction, peptide gels, GELFrEE and organic solvent or surfactant extraction [24].

Given the diverse physicochemical properties of microproteins, such as size, hydrophobicity, and charge, comprehensive enrichment of complex samples in a single step is challenging. Combining multiple enrichment methods can improve efficiency. Here, we present four complementary microprotein enrichment strategies and a proteogenomic approach to identify annotated and novel smORFs. Using this protocol, we previously identified 31 missing proteins in yeast and discovered three unannotated smORFs (YKL104W-A, YHR052C-B, and YHR054C-B) [25]. These newly characterized smORFs have since been incorporated into the *Saccharomyces* Genome Database (SGD) and UniProt. Notably, Moretti-Horten et al. demonstrated that one of these smORF-encoded proteins, YPR010C-A, functions as an integral inner mitochondrial membrane. Their work revealed its physical interaction with mitochondrial respiratory chain complex IV and its role as a negative regulator of complex IV assembly [26]. These findings not only underscore the robustness of our methodology but also highlight the critical biological functions performed by these previously unannotated smORFs.

2 Materials

2.1 Cell Culture

1. Yeast peptone dextrose (YPD) medium: 1% (w/v) yeast extract, 2% (w/v) Bacto-peptone, and 2% (w/v) dextrose (*see* **Note 1**).
2. 0.01% (w/v) sodium azide (NaN_3).

2.2 Microprotein Enrichment

1. Lysis Buffer I: 8 M urea, 50 mM ammonium bicarbonate (NH_4HCO_3), 50 mM iodoacetamide (IAA), and 1 mM phenylmethanesulfonyl fluoride (PMSF).
2. Glass beads: 0.5 mm diameter (Sigma–Aldrich).
3. Coomassie Brilliant Blue G250.
4. Scion Image software (Version 4.0.3.2, National Institutes of Health, USA).
5. 5 mM dithiothreitol (DTT) (*see* **Note 2**).
6. 20 mM iodoacetamide (IAA) (*see* **Note 3**).
7. Membrane molecular weight cut-off (MWCO) filter: 30 kDa (Millipore).
8. Lysis Buffer II: 50 mM HCl, 0.1% (v/v) β-mercaptoethanol (β-ME), 0.05% (v/v) Triton X-100.
9. 1×SDS gel-loading buffer: 50 mM Tris–HCl (pH 6.8), 2% (w/v) SDS, 10% (w/v) glycerol, 0.1% (w/v) Bromophenol Blue.

2.3 Protein Digestion and Sample Preparation for LC-MS/MS

1. 1.5 mL Eppendorf tubes.
2. Destaining solution: 30% (v/v) acetonitrile (ACN), 500 mM NH_4HCO_3.
3. Trypsin solution: 12.5 ng/μL trypsin, 5% (v/v) ACN, 50 mM NH_4HCO_3 (pH 8.0).
4. SpeedVac concentrator (Thermo Fisher Scientific).
5. Microcon YM-3 centrifugal filter unit (Millipore).
6. Substitution buffer: 8 M urea in 0.1 M Tris–HCl (pH 8.5).
7. Eppendorf Thermomixer.
8. Self-packing C18 microcolumns.
9. Methanol.
10. Buffer A: 1% (v/v) ACN, 0.1% (v/v) trifluoroacetic acid (TFA), 0.5% glacial acetic acid.
11. Buffer B: 80% (v/v) ACN, 1% (v/v) formic acid.
12. Sample buffer: 1% (v/v) ACN, 1% (v/v) formic acid.

2.4 LC-MS/MS

All reagents in this section are MS grade:

1. Loading buffer: 1% (v/v) ACN, 1% (v/v) formic acid (FA).
2. Ultra-performance liquid chromatography (UPLC) system (nanoAcquity, Waters, Milford, MA, USA).
3. C18 analytical column: 100 Å pore size, 3 μm particle size, 15 cm length, 75 μm inner diameter (self-made).
4. MS buffer A: 2% (v/v) ACN and 0.1% (v/v) FA.
5. MS buffer B: 0.1% (v/v) FA in ACN.
6. LTQ-Orbitrap Velos mass spectrometer (Thermo Electron, San Jose, CA, USA).

2.5 Software for MS Data Analysis

1. MaxQuant (version 1.5.6.0).
2. pFind (version 3.1.2).
3. pAnno (version 1.0).

3 Methods

3.1 Cell Culture

1. Grow *Saccharomyces cerevisiae* cells in YPD medium at 30 °C and harvest at mid-exponential phase (OD_{600} = 1.5).
2. Wash the collected cells with 0.01% (w/v) NaN_3 and store at −80 °C.

3.2 Microprotein Enrichment

Microproteins are enriched using four complementary methods.

3.2.1 Urea_Tricine Method

1. Resuspend 30 OD_{600} units of cells in 250 μL Lysis Buffer I and lyse by vortexing with glass beads (30 s vortex, 30 s on ice, 10 cycles).
2. Centrifuge at 17,000 × g for 20 min at 4 °C and collect the supernatant.
3. Determine protein concentration using a gel-based method [27]. Concentrate proteins on a short 9% SDS gel (~2 mm), stain with Coomassie Brilliant Blue G250, and quantify using Scion Image software based on dye absorbance (*see* **Note 4**).
4. Reduce 80 μg of proteins with 5 mM DTT (45 °C, 30 min) and alkylate with 20 mM IAA (room temperature, 30 min in the dark).
5. Separate the reduced and alkylated proteins by 12% Tricine–SDS-PAGE. Stop electrophoresis when the samples migrate 6.5 cm into the separating gel (*see* **Note 5**).
6. Excise the gel region below 30 kDa and divide into 13 fractions (*see* **Note 6**).

3.2.2 Urea_MWCO Method

1. Take the supernatant from **step 2** of Subheading 3.2.1 (centrifuged lysate in lysis buffer I) and load it onto a 30 kDa MWCO filter.
2. Centrifuge at 14,000 × g for 20 min at room temperature and collect the flow-through.
3. Reduce and alkylate the flow-through as described in Subheading 3.2.1.

3.2.3 HCl_Tricine Method

1. Resuspend 60 OD_{600} units of cells in 350 μL Lysis Buffer II and lyse by vortexing with glass beads (30 s vortex, 30 s on ice, 10 cycles).
2. Centrifuge at 17,000 × g for 20 min at 4 °C and transfer the supernatant to a new tube.
3. Repeat the lysis and centrifugation steps twice to maximize microprotein extraction (*see* **Note** 7).
4. Pool the extracts, lyophilize, and resuspend in 1× SDS gel-loading buffer.
5. Reduce and alkylate the sample as described in Subheading 3.2.1.
6. Separate the reduced and alkylated proteins by 12% Tricine–SDS-PAGE. Stop electrophoresis when the samples migrate 3 cm into the separating gel.
7. Excise the gel region below 30 kDa and divide into five fractions (*see* **Note 8**).

3.2.4 HCl_MWCO Method

1. Take the pooled, lyophilized, and SDS buffer-resuspended extracts from **step 4** of Subheading 3.2.3 and load them onto a 30 kDa MWCO filter.
2. Centrifuge at 14,000 × g for 20 min at room temperature and collect the flow-through.
3. Reduce and alkylate the flow-through as described in Subheading 3.2.1.

3.3 Protein Digestion and Sample Preparation for LC-MS/MS

Proteins enriched by the Urea_Tricine and HCl_Tricine methods are digested in-gel, while those enriched by the Urea_MWCO and HCl_MWCO methods are digested using Filter-Aided Sample Preparation (FASP).

3.3.1 In-Gel Digestion

1. Cut gel bands into ~1 mm^3 cubes and transfer to a 1.5 mL Eppendorf tube (*see* **Note 9**).
2. Destain the gel cubes by adding 300 μL destaining solution, vortexing for 30 min, and centrifuging at 14,000 × g for 1 min. Discard the supernatant and repeat until the gel is colorless.

3. Dehydrate the gel cubes by adding 300 μL ACN, vortexing for 10 min, and centrifuging at 14,000 × g for 1 min. Repeat until the gel cubes turn white and shrink.
4. Dry the gel cubes in a SpeedVac concentrator at 600 rpm for 10 min.
5. Digest the proteins by incubating the gel cubes with trypsin solution overnight at 37 °C (*see* **Note 10**).
6. Extract peptides by adding 60 μL extraction solution, centrifuging at 14,000 × g for 1 min, and incubating for 4 min. Repeat four times and pool the supernatants.
7. Further extract peptides by adding 60 μL 100% ACN, vortexing for 20 min, and centrifuging at 14,000 × g for 1 min. Repeat twice and pool the supernatants.
8. Dry the combined supernatants in a SpeedVac concentrator at 45 °C.

3.3.2 FASP Digestion

1. Take the reduced and alkylated flow-through from **step 3** of Subheading 3.2.2 (Urea_MWCO method) or **step 3** of Subheading 3.2.4 (HCl_MWCO method) and load it onto a Microcon YM-3 filter unit. Centrifuge at 14,000 × g for 20 min.
2. Add 300 μL substitution buffer and centrifuge at 14,000 × g for 15 min. Discard the flow-through and repeat twice (*see* **Note 11**).
3. Add 300 μL 50 mM NH_4HCO_3 and centrifuge at 14,000 × g for 15 min. Discard the flow-through and repeat twice.
4. Add 50 μL 50 mM NH_4HCO_3 containing trypsin (enzyme-to-protein ratio 1:50) and mix at 600 rpm for 1 min in a thermomixer.
5. Incubate the filter unit at 37 °C for 12 h.
6. Transfer the filter unit to a new collection tube and centrifuge at 14,000 × g for 10 min.
7. Add 100 μl. 50 mM NH_4HCO_3 and centrifuge at 14,000 × g for 10 min.
8. Dry the flow-through in a SpeedVac concentrator at 45 °C.

3.3.3 Desalting with C18 Microcolumns

1. Prepare C18 microcolumns in-house using an established protocol [28].
2. Activate the column with 20 μL methanol.
3. Wash the column with 20 μL Buffer B.
4. Equilibrate the column with 20 μL Buffer A, repeating once.

5. Redissolve the dried FASP-digested peptides (from **step 8** of Subheading 3.3.2) in 20 μL Buffer A, then load the sample onto the column and reload the flow-through twice.
6. Wash the column with 20 μL Buffer A.
7. Elute peptides with 20 μL Buffer B.
8. Dry the eluted peptides in a SpeedVac concentrator.

3.4 LC-MS/MS Analysis

1. Redissolve the dried peptides (from either Subheading 3.3.1 or 3.3.3 final steps) in loading buffer and centrifuge at 12,000 × g for 15 min at 4 °C. Transfer the supernatant to a sample vial.
2. Analyze the peptides using a nanoUPLC system equipped with a self-packed C18 column (75 μm i.d. × 15 cm, 3 μm C_{18} reverse phase fused silica) at a flow rate of 300 nL/min with a 60 min gradient: 2–4% Buffer B in 6 min, 4–10% in 2 min, 10–25% in 27 min, 25–35% in 20 min, and 35–80% in 5 min.
3. Ionize the eluted peptides at 1.5 kV and analyze using an LTQ-Orbitrap Velos mass spectrometer.
4. Acquire full-scan mass spectra (MS_1) over a range of 300–1600 Da with a resolution of 30,000 at m/z 400. Set the automatic gain control (AGC) to 1×10^6 and the maximum injection time (MIT) to 150 ms.
5. Acquire tandem mass spectra (MS_2) in data-dependent mode, selecting the 20 most intense ions for fragmentation. Set the MS_2 trigger threshold to 2000 counts, isolation width to 2 m/z, normalized collision energy to 35%, AGC to 1×10^4, and MIT to 30 ms. Exclude unassigned and singly charged ions, and apply a dynamic exclusion of 40 s.

3.5 Identification of Annotated Microproteins

1. Analyze MS data using MaxQuant.
2. Search against the *S. cerevisiae* S288C FASTA database (6721 proteins) from Swiss-Prot.
3. Set search parameters: trypsin digestion with up to two missed cleavages; fixed modification of cysteine carbamidomethylation (+57.021464 Da); variable modification of methionine oxidation (+15.994915 Da); precursor and fragment mass tolerances of 20 ppm and 0.5 Da, respectively; minimum peptide length of seven amino acids; and an false discovery rate (FDR) of 1% at both peptide and protein levels, estimated using a target-decoy approach.

3.6 Identification of Novel Microproteins

1. Analyze MS data using pFind.
2. Search against a six-frame translation database of the *S. cerevisiae* S288C genome and an N-terminal peptide database containing tryptic peptides starting with methionine. The six-frame database includes ORFs of six or more amino acids, totaling 875,609 entries.

3. Set search parameters as described in Subheading 3.5.
4. Perform quality control of candidate novel peptides using pAnno's QC module.
5. Filter novel peptides by manual spectrum inspection (*see* **Note 12**).
6. Synthesize high-quality peptides and validate their spectra using pFind and pBuild [29] (*see* **Note 13**).
7. Map novel peptides to the genome using TBLASTN to identify the chromosomal location and sequence of the encoded microproteins.

4 Notes

1. For all buffers and solutions in this protocol, Milli-Q water is used as the solvent unless otherwise specified.
2. DTT is sensitive to oxygen and should be prepared immediately before use.
3. IAA is hazardous and should be handled with caution. IAA solutions must be stored in the dark to prevent degradation.
4. Protein quantification can also be performed using the BCA method. The gel-based quantification method offers the advantage of visualizing protein concentration through the grayscale intensity of protein bands on the gel.
5. Detailed protocols for Tricine–SDS-PAGE gel preparation and electrophoresis can be found in the literature [30]. Microprotein bands may disperse during electrophoresis. To mitigate this issue, pre-cool the Tricine gel at 4 °C before use or place cold water-moistened paper on the outer surface of the gel plates to enhance heat dissipation during electrophoresis.
6. Dividing gel lanes below 30 kDa into multiple fractions improves microprotein identification coverage. The number of fractions is flexible and should be adjusted based on protein concentration. High-protein regions should be divided into multiple fractions. Avoid using only gel lanes below 10 kDa, as some microproteins may migrate to higher molecular weight regions during electrophoresis.
7. Multiple extraction steps enhance the completeness of microprotein recovery.
8. The number of gel fractions can be adjusted based on the protein amount and distribution.
9. Gel cubes should not exceed ~1 mm^3 in size, as larger cubes may reduce peptide recovery efficiency.

10. Gel cubes will swell upon absorbing trypsin solution. After swelling, add an appropriate volume of trypsin solution to fully submerge the gel cubes. To minimize evaporation, add 15 μL of 50 mM NH_4HCO_3 to the surface of the gel cubes.
11. This step is specific to proteins enriched by the HCl_MWCO method. The acidic nature of these samples requires buffer exchange to 50 mM NH_4HCO_3 to ensure optimal trypsin digestion conditions.
12. To validate novel peptides, manually inspect the quality of their spectra by evaluating fragment ion intensity and the presence of matching b/y ions. Credible novel peptides should meet the following criteria: (1) Unique peptide identification; (2) Peptide length ≥ 9 amino acids; (3) High-quality spectra with minimal impurity peaks and at least three consecutive b/y ion matches; (4) Isobaric sequence filtering to account for potential ambiguities (e.g., I = L, Q[Deamidated] = E, GG = N).
13. The cosine similarity score between synthetic and endogenous spectra is calculated as follows: (1) Remove precursor ion peaks from the initial spectrum; (2) Calculate theoretical m/z values for b^+, y^+, $b–H_2O^+$, $y–H_2O^+$, $b–NH_3^+$, $y–NH_3^+$, b^{++}, and y^{++} ions, and extract matching peaks from the experimental spectrum. (3) Convert the m/z values of selected peaks to integers, and define the intensity of each integer m/z peak as the sum of intensities within its window. (4) Align the dimensions of synthetic and endogenous spectra, filling missing values with zero; (5) Calculate the cosine similarity score between the two spectra.

Acknowledgments

This research was supported by the National Basic Research Programs (2022YFA1304600 and 2024YFC3405400), CAMS Innovation Fund for Medical Sciences (2019-I2M5-017 and 2022-I2M-C&T-B-082), the National Science Foundation (32141003, 32101190, and 32371503), Beijing Science Foundation (L246037), and the Foundation of State Key Lab of Proteomics (2021-NCPSB-001, SKLP-C202002 & SKLPY201901).

References

1. Harrison PM, Kumar A, Lang N et al (2002) A question of size: the eukaryotic proteome and the problems in defining it. Nucleic Acids Res 30:1083–1090
2. Basrai MA, Hieter P, Boeke JD (1997) Small open reading frames: beautiful needles in the haystack. Genome Res 7:768–771
3. Storz G, Wolf YI, Ramamurthi KS (2014) Small proteins can no longer be ignored. Annu Rev Biochem 83:753–777
4. Giaever G, Chu AM, Ni L et al (2002) Functional profiling of the Saccharomyces cerevisiae genome. Nature 418:387–391

5. Nagalakshmi U, Wang Z, Waern K et al (2008) The transcriptional landscape of the yeast genome defined by RNA sequencing. Science 320:1344–1349
6. Oshiro G, Wodicka LM, Washburn MP et al (2002) Parallel identification of new genes in Saccharomyces cerevisiae. Genome Res 12: 1210–1220
7. Ruiz-Orera J, Messeguer X, Subirana JA et al (2014) Long non-coding RNAs as a source of new peptides. elife 3:e03523
8. Bazzini AA, Johnstone TG, Christiano R et al (2014) Identification of small ORFs in vertebrates using ribosome footprinting and evolutionary conservation. EMBO J 33:981–993
9. Mackowiak SD, Zauber H, Bielow C et al (2015) Extensive identification and analysis of conserved small ORFs in animals. Genome Biol 16:179
10. Saghatelian A, Couso JP (2015) Discovery and characterization of smORF-encoded bioactive polypeptides. Nat Chem Biol 11:909–916
11. Martinez TF, Chu Q, Donaldson C et al (2020) Accurate annotation of human protein-coding small open reading frames. Nat Chem Biol 16:458–468
12. Zanet J, Benrabah E, Li T et al (2015) Pri sORF peptides induce selective proteasome-mediated protein processing. Science 349: 1356–1358
13. Costa LM, Marshall E, Tesfaye M et al (2014) Central cell-derived peptides regulate early embryo patterning in flowering plants. Science 344:168–172
14. Cabrera-Quio LE, Herberg S, Pauli A (2016) Decoding sORF translation - from small proteins to gene regulation. RNA Biol 13:1051–1059
15. Hassel KR, Brito-Estrada O, Makarewich CA (2023) Microproteins: overlooked regulators of physiology and disease. iScience 26:106781
16. Ingolia NT, Ghaemmaghami S, Newman JR et al (2009) Genome-wide analysis in vivo of translation with nucleotide resolution using ribosome profiling. Science 324:218–223
17. Guttman M, Russell P, Ingolia NT et al (2013) Ribosome profiling provides evidence that large noncoding RNAs do not encode proteins. Cell 154:240–251
18. Cassidy L, Kaulich PT, Maaß S et al (2021) Bottom-up and top-down proteomic approaches for the identification, characterization, and quantification of the low molecular weight proteome with focus on short open reading frame-encoded peptides. Proteomics 21:e2100008
19. Fabre B, Combier JP, Plaza S (2021) Recent advances in mass spectrometry-based peptidomics workflows to identify short-open-reading-frame-encoded peptides and explore their functions. Curr Opin Chem Biol 60:122–130
20. Khitun A, Slavoff SA (2019) Proteomic detection and validation of translated small open reading frames. Curr Protoc Chem Biol 11:e77
21. Chen Y, Cao X, Loh KH et al (2023) Chemical labeling and proteomics for characterization of unannotated small and alternative open reading frame-encoded polypeptides. Biochem Soc Trans 51:1071–1082
22. Slavoff SA, Mitchell AJ, Schwaid AG et al (2013) Peptidomic discovery of short open reading frame-encoded peptides in human cells. Nat Chem Biol 9:59–64
23. Ting YS, Egertson JD, Payne SH et al (2015) Peptide-centric proteome analysis: an alternative strategy for the analysis of tandem mass spectrometry data. Mol Cell Proteomics 14: 2301–2307
24. Mohsen JJ, Martel AA, Slavoff SA (2023) Microproteins-discovery, structure, and function. Proteomics 23:e2100211
25. He C, Jia C, Zhang Y et al (2018) Enrichment-based proteogenomics identifies microproteins, missing proteins, and novel smORFs in Saccharomyces cerevisiae. J Proteome Res 17: 2335–2344
26. Moretti-Horten DN, Peselj C, Taskin AA et al (2024) Synchronized assembly of the oxidative phosphorylation system controls mitochondrial respiration in yeast. Dev Cell 59:1043–1057
27. Xu P, Duong DM, Peng J (2009) Systematical optimization of reverse-phase chromatography for shotgun proteomics. J Proteome Res 8: 3944–3950
28. Zhai L, Chang C, Li N et al (2013) Systematic research on the pretreatment of peptides for quantitative proteomics using a C18 microcolumn. Proteomics 13:2229–2237
29. Fu Y, Yang Q, Sun R et al (2004) Exploiting the kernel trick to correlate fragment ions for peptide identification via tandem mass spectrometry. Bioinformatics 20:1948–1954
30. Haider SR, Reid HJ, Sharp BL (2019) Tricine-SDS-PAGE. Methods Mol Biol 1855:151–160

Chapter 6

Identification of Peptides Encoded by Small Open Reading Frames in Microalgae by Mass Spectrometry

Zhao Peng, Qianru Xiao, and Cuihong Wan

Abstract

SEPs are important as co-operators of various biological processes with canonical proteins. In microalgae, they are reported to be involved in photosynthesis, nitrogen metabolism, stress response, and so on. Because of low abundance and small molecular weight, identifying SEPs in proteomics is usually challenging, requiring specially designed technical strategies for sample handling and data analysis. Here, we present two method examples for SEP in microalgae (one cyanobacterium and one green alga), from SEP extraction and enrichment to database search, aiming to achieve an ideal identification of SEPs while reducing the interference of canonical proteins. This chapter will provide a reference procedure for SEP exploration in microalgae.

Key words SEPs, Microalgae, Cyanobacteria, Green algae, Small protein enrichment, Mass spectrometry, *Synechocystis* sp., PCC 6803, *Chlamydomonas reinhardtii*

1 Introduction

The encoded products of small open reading frames (smORFs) are less than 100 amino acids in length, called smORF-encoded peptides (SEPs) [1, 2]. With the rapid development of omics technologies, more and more smORFs have been uncovered in various species, and their encoded SEPs are found to carry essential roles in various physiological processes [3, 4].

Microalgae are a diverse group of autotrophic microorganisms, widely distributed in different aquatic environments as primary producers through photosynthesis [5]. As a general term in taxonomy, microalgae are either prokaryotic or eukaryotic, unicellular or simple multicellular, ranging in size from 0.2 μm to 100 μm, with filamentous forms. Most microalgae possess a simple structure and rapid metabolism, enabling good adaptability to external environmental changes. Besides oxygen and carbohydrates, microalgae generate rich and miscellaneous products, attracting interest in

Stephan Wenkel (ed.), *Microproteins: Methods and Protocols*, Methods in Molecular Biology, vol. 2992,
https://doi.org/10.1007/978-1-0716-5013-4_6,

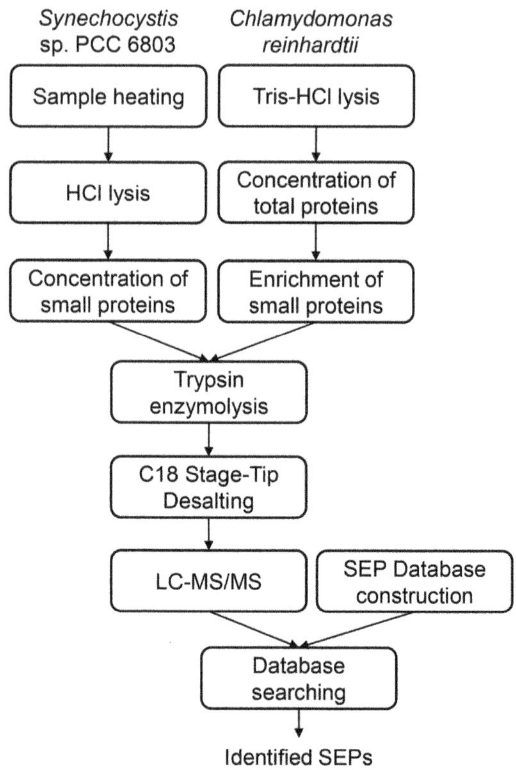

Fig. 1 Schematic overview of the workflow

pharmacology, agriculture, and ecology [6, 7]. They are also important in biological research, commonly serving as representatives of single-celled plants. They have been known as model organisms for photosynthesis and a promising and proven material applied in bioenergy [8, 9].

SEP studies introduce a new perspective to explore these ancient and unique microorganisms' life activities comprehensively. Some SEPs are involved in microalgae biological pathways, such as glucose metabolism [10–12]. Genome annotation, transcriptomic data, and prediction in silico have excavated SEP information in microalgae while lacking protein-level evidence [13, 14]. Proteomics is supposed to identify SEPs and validate their expression in vivo. However, SEP is often overshadowed by canonical proteins in mass spectrometry due to their small molecular weight and low abundance [15]. Species-specific extraction and enrichment strategies have been documented to solve the problem and enhance the identification of SEP [16–18].

Here, we established a workflow to identify SEPs in microalgae, using the cyanobacterium *Synechocystis* sp. PCC 6803 and the green alga *Chlamydomonas reinhardtii* as materials (Fig. 1). The workflow introduces experimental details, including SEP extraction, enrichment, detection, and data processing, devoted to acquiring

SEP information covered by the large abundance of canonical proteins (Fig. 1). Although the methods are designed according to the characteristics of the target algae, they are expected to be optimized and adapted to identify SEPs in other prokaryotic and eukaryotic algae.

2 Materials

2.1 Equipment

1. Orbitrap Explorer M480 high-resolution mass spectrometer (Thermo Scientific, USA).
2. Ultrasonic cell crusher.
3. Self-made C18 column: 15 cm, 75 μm, 3 μm, and 100 Å.
4. High-speed refrigerated centrifuge.
5. Multifunctional enzyme marker.
6. Electrophoresis apparatus.
7. Thermostatic oscillator.
8. Centrifugal concentration drying system.
9. Thermostatic metal bath instrument.
10. 3 M C18 SEPs diaphragm (Empore, Germany).
11. Ultraviolet spectrophotometer.

2.2 Reagents and Buffers

1. BG-11 liquid medium: In 1 L deionized H_2O, dissolve 1.5 g $NaNO_3$, 0.04 g $KaHPO_4{\cdot}3H_2O$, 0.075 g $MgSO_4{\cdot}7H_2O$, 0.036 g $CaCl_2{\cdot}H_2O$, 0.006 g Citric acid, 0.006 g Ferric ammonium citrate, 0.001 g EDTA (disodium magnesium salt), 0.02 g Na_2CO_3, and add 1 ml Trace metal mix A5 + Co. Adjust pH to 7.4 after autoclaving and cooling.
2. Trace metal mix A5 + Co: In 1 L deionized H_2O, dissolve 2.86 g H_3BO_3, 1.81 g $MnCl_2{\cdot}4H_2O$, 0.222 g $ZnSO_4{\cdot}7H_2O$, 0.390 g $Na_2MoO_4{\cdot}2H_20$, 0.079 g $CuSO_4{\cdot}5H_20$, and 0.0494 g $Co(NO_3)_2{\cdot}6H_2O$.
3. Tris-Acetate-Phosphate (TAP) liquid medium: In 1 L deionized H_2O, dissolve 2.42 g Tris powder, 0.38 g NH_4Cl, 0.10 g $MgSO_4{\cdot}7H_2O$, 57.00 mg $CaCl_2{\cdot}2H_2O$, 0.11 g K_2HPO_4, 54.00 mg KH_2PO_4, 50.00 mg $EDTANa_2$, 22.00 mg $ZnSO_4{\cdot}7H_2O$, 11.40 mg H_3BO_3, 5.06 mg $MnCl_2{\cdot}4H_2O$, 1.61 mg $CoCl_2{\cdot}6H_2O$, 1.57 mg $CuSO_4{\cdot}5H_2O$, 1.10 mg $(NH_4)_2Mo_7O_{24}{\cdot}4H_2O$, 4.99 mg $FeSO_4{\cdot}7H_2O$ and add 1 mL glacial acetic acid.
4. PBS buffer: 136.89 mM NaCl, 2.67 mM KCl, 8.10 mM Na_2HPO_4, 1.76 mM KH_2PO_4, pH 7.2–7.4.
5. HCl cell lysis buffer: Each 10 mL buffer contains HCl 43 μL, DTT 0.05 g, and DDM 0.1 g.

6. Tris–HCl cell lysis buffer: Each 10 mL buffer contains Tris–HCl (pH 8.9) 1.21 g, DTT 0.1 g and DDM 0.1 g, with one protease inhibitor cocktail tablet added.
7. Digestion buffer: 50 mM NH_4HCO_3.
8. Trypsin, sequencing grade (V5111, Promega, USA).
9. Acidified isopropyl alcohol: 1:1 mix of isopropyl alcohol and 12 mM HCl, precooled.
10. Desalting elution buffer A (6 mL acetonitrile (0.1% formic acid) and 4 mL H_2O (0.1% formic acid)).
11. Desalting elution buffer B: 8 mL acetonitrile (0.1% formic acid) and 2 mL H_2O (0.1% formic acid) mixed.
12. Mobile phase buffer A: 0.1% formic acid.
13. Mobile phase buffer B: acetonitrile 0.1% formic acid dissolved in acetonitrile.
14. Bradford quantitative kit (Beyotime Biotechnology, China).

2.3 Data Analysis Tools

1. Proteome Discoverer, version 2.4 (Thermo Fisher Scientific, USA).
2. sORF finder [19].
3. Bedtools [20].
4. Seqkit [21].
5. ProsmORF-pred [22].
6. ThermoRawFileParser [23].
7. MSGF+ [24].
8. pFind [25].
9. Peptide match—command line tool [26].

3 Methods

The procedures for *Synechocystis* sp. *PCC 6803* and *Chlamydomonas reinhardtii* differ in protein extraction but are the same in the subsequent steps (*see* **Note 1**).

3.1 Extraction of Small Proteins from Cyanobacteria

1. Inoculate and incubate the algal strain in the liquid medium till OD_{730} 0.6–0.8. Harvest and divide the algal cells into centrifuge tubes with 50 mL each. Centrifuge at 3400 × g for 5 min and discard the supernatant. Wash the sample with PBS buffer three times, then collect the precipitation (*see* **Note 2**).
2. Resuspend the precipitation with 300 μL boiling water. Heat the sample at 95 °C for 10 min. Then centrifuge at 16,100 × g at 4 °C for 10 min, and discard the supernatant.

3. Redissolve the precipitation with 500 μL of Tris–HCl cell lysis buffer, and perform ultrasonication on ice for 8 min. The parameters for ultrasonication are set as: 5 s on, 9 s off, 30% power for Biosafer 650–92 (*see* **Note 3**).
4. Incubate the sample on ice for 1–2 h. Then centrifuge at 13,700 × g for 10 min, and collect the supernatant.
5. Remove lipid from the sample as below (*see* **Note 4**): Transfer the supernatant to a new centrifuge tube. Add chloroform and H_2O in the ratio of 450:125:500 (sample/chloroform/H_2O, v/v/v), and centrifuge at 14000 × g for 10 min (*see* **Note 5**).
6. Transfer the upper water phase to a new centrifuge tube. Quantify protein using the Bradford kit and a standard curve of gradient protein concentration (*see* **Note 6**).
7. For quality control, load 20 μg of protein into a 16% Tricine gel and run PAGE at 120 V for 80 min. The remaining protein sample is dried and concentrated, and can be stored at −20 °C until use.

3.2 Extraction and Enrichment of Small Proteins from Green Algae

1. Inoculate the algal strain in the liquid medium. Cells are cultured at 22 °C and agitated at 120 rpm under continuous light of ~50 μmol photon m^{-1} s^{-2}. Collect the algae samples when the cell density reaches 4–5 × 10^6 cells/ml, centrifuge at 3000 × g for 5 min, and discard the supernatant.
2. Resuspend the precipitation with 800 μL of the 20 mM Tris–HCl cell lysis buffer on ice (*see* **Note 7**).
3. Perform ultrasonication on ice for 10 min. The parameters for ultrasonication are set as: 5 s on, 9 s off, 30% power (*see* **Note 3**).
4. Incubate the sample on ice at 4 °C for 1–2 h. Then centrifuge at 13,700 × g for 10 min.
5. Transfer the supernatant to a new centrifuge tube, and quantify protein using the Bradford kit. Divide the sample into 50 μg per tube.
6. Add pre-cooled acetone to the sample in the ratio of 4:1 (v/v) (*see* **Note 8**). Incubate the sample overnight at −20 °C (*see* **Note 9**).
7. Centrifuge at 16,100 × g at 4 °C for 15 min and discard the supernatant. Wash the sample with 300 μL of pre-cooled methanol. Discard the supernatant and dry the sample in a ventilated place at room temperature (*see* **Note 10**).
8. Redissolve the protein precipitation with 300 μL of acidified isopropyl alcohol after protein extraction (*see* **Note 11**).

9. After full dissolution of the protein precipitation, incubate the sample at 4 °C for 30 min. Centrifuge at 13,700 × g at 4 °C for 10 min.
10. Transfer the supernatant to a new centrifuge tube. For quality control, load 20 μg of protein into 16% Tricine gel and run PAGE at 120 V for 80 min. The remaining protein sample is dried and concentrated, and can be stored at −20 °C until use.

3.3 Trypsin Enzymolysis

1. Redissolve 50 μg of the protein sample in 50 μL digestion buffer. Add DTT to a final concentration of 10 mM. Incubate the sample at 37 °C for 30 min.
2. Add IAM to a final concentration of 15 mM. Incubate the sample at 37 °C for 30 min in the dark.
3. Add trypsin to the protein sample in the ratio of 1:50 (w/w). Incubate the sample overnight at 37 °C for enzymolysis (*see* **Note 12**).
4. Add formic acid to a final concentration of 1% to halt the enzymolysis. Transfer the supernatant to a new tube and centrifuge at 13,700 × g at room temperature for 5 min (*see* **Note 13**). The digested peptides in the supernatant are ready for the subsequent desalting (*see* **Note 14**).

3.4 Solid Phase Extraction for Desalting

1. To prepare the column for desalting, gently push two C18 diaphragms into the 200 μl pipette tip to produce a desalting column.
2. Add 100 μL acetonitrile (0.1% formic acid) into the column, and centrifuge at 400 × g at room temperature for 3 min. Add 100 μL 0.1% formic acid into the column, centrifuge at room temperature at 400 × g for 3 min, and repeat once.
3. Add the peptide sample after digestion into the column, centrifuge at 400 × g at room temperature for 4 min, and repeat twice.
4. Add 100 μL 0.1% formic acid into the column, centrifuge at 400 × g at room temperature for 3 min, and repeat twice (*see* **Note 15**).
5. Add 50 μL elution buffer A into the column, centrifuge at 400 × g at room temperature for 3 min, and collect the eluent sample. Then add 50 μL elution buffer A into the column, centrifuge at 400 × g at room temperature for 3 min, and collect the eluent sample.
6. Combine two eluent samples together. Dry and store the sample at −20 °C for LC–MS/MS.

3.5 LC–MS/MS Analysis

1. Redissolve the peptide sample in 15 μL 0.1% formic acid, and centrifuge at 13,700 × g at room temperature for 10 min. Load the supernatant into the LC–MS/MS system (*see* **Note 16**).
2. MS analysis is performed with a sample size of 2 μL. Peptide separation was performed on the self-made C18 column at a flow rate of 500 nl/min with mobile phase ratios ranging from 5% to 80% buffer B for 120 min.
3. Parameters are set as below in data-dependent acquisition mode (DDA): For the primary spectrum, scan range is set as 350–2000 m/z, the resolution 70,000, the AGC target value 3×10^6, and the maximum injection time 100 ms. It is followed by high-energy collision dissociation (HCD) with 32% normalized collision energy. For the secondary spectrum, the AGC target value is set as 5×10^4, the maximum injection time is 35 ms, and the dynamic exclusion time 40 s.

3.6 Database Construction

1. The search of MS raw data is against a combined database including canonical proteins and SEPs. Canonical proteins in *Synechocystis* sp. PCC 6803 are achieved from UniProt, and those in *C. reinhardtii* are achieved from Phytozome.
2. The database of SEPs is constructed through prediction from algal genomes. Genome assembly and annotation in *Synechocystis* sp. PCC 6803 is achieved from NCBI genome, and that in *C. reinhardtii* is achieved from Phytozome (*see* **Note 17**).
3. To predict SEPs from the algal genome using the software sORF finder, prepare the following FASTA files: genome sequences, CDSs, AND non-CDSs. Genome sequences and CDSs are provided from the genome annotation. Extract non-CDSs from genome sequences using the software bedtools ("intersect" function), based on the location information of non-CDSs (introns or intergenic sequences) from the genome annotation GTF/GFF file (*see* **Note 18**).
4. Use "make_model.pl", "simulate.pl", and "search_sORF.pl", referring to the manual to generate the FASTA file of smORFs (*see* **Notes 19** and **20**). Then use the software seqkit ("translate" function) to translate smORFs into SEPs.
5. For cyanobacteria, the web resource ProsmORF-pred can be an optional and convenient choice. Use the genome sequences FASTA file and the genome annotation GTF/GFF file as input to get SEPs.

3.7 Database Searching

1. Two strategies can be applied in database searching: (1) Search directly against the combination of canonical proteins and SEPs. (2) Search against canonical proteins first, filter out the matching spectra, and search against SEPs using the remaining spectra.

2. To process the first strategy (direct searching against a combined database), there are several tools. Take Proteome Discoverer as an example. Parameters are set as below: enzyme is set as trypsin, missed cleavage 2, precursor mass tolerance 10 ppm, fragment mass tolerance 0.02 Da. Set methionine oxidation and N-terminus acetylation as dynamic modifications, and carbamidomethylation as a static modification. The FDR is set to 1%. Collect SEPs with more than 2 PSMs (peptide-spectrum matches).
3. To process the second strategy, firstly search against canonical proteins using the RAW spectra file, and record the scans of the matched spectra. Transform the RAW file into the MGF file using the software ThermoRawFileParser. Filter out the matched spectra from the MGF file according to their scans as labels. At last, search against SEPs using the MGF file of the remaining spectra. As Proteome Discoverer cannot process MGF files, use MSGF+ or pFind instead.
4. After acquiring spectra of peptides matching with SEPs, use Peptide Match—Command Line Tool to search the peptides against the database of canonical proteins, and filter out those peptides existing in canonical proteins already.

4 Notes

1. Compared with cyanobacteria, the eukaryotic microalgae *Chlamydomonas reinhardtii* possesses a more complex structure, and protein extraction in *C. reinhardtii* is disturbed by numerous organelle contents and secondary metabolites. With preliminary experiments, we found Tris–HCl lysis buffer and acetone enrichment promote the extraction of small proteins in *C. reinhardtii*, while HCl lysis buffer is better for *Synechocystis* sp. PCC 6803.
2. Algal samples should be collected in the logarithmic growth phase (OD_{730} 0.6–0.8), which will affect the amount of small protein extracted and identified. If the fresh algal samples are not used for protein extraction immediately, they should be freeze-dried under liquid nitrogen conditions, stored at −80 °C, and thawed at 4 °C before use.
3. To avoid foaming during ultrasonication, ensure that the ultrasonic probe is in the middle of the tube, and be ready to tune the tube position to the abnormal sound of ultrasonication.
4. Lipid removal helps to get rid of impurities for some cyanobacteria with a high lipid content. This step can be omitted if the following procedures are not disturbed by lipid.

5. Chloroform is toxic and volatile, and should be treated in a fume hood.
6. Pre-cool the dye in advance and ensure that the R^2 of the standard curve is greater than 0.96. A low concentration of a small protein might cause inaccurate quantification. Try to reduce the volume of the solution appropriately if encountering this problem.
7. For protease cocktail tablets, use one tablet for every 10 mL of lysis buffer or refer to the manual.
8. The organic reagent is pre-cooled before use to ensure the sample is not affected by temperature.
9. The incubation step with acetone is used to remove impurity from the sample, and should not be less than 6 h to avoid protein loss in the subsequent supernatant discarding.
10. While washing the sample, be gentle and careful to avoid protein loss. Ensure that the sample is dried and free of residual methanol.
11. It may be difficult to completely dissolve the protein precipitation. Use a warm water bath or ultrasonication to promote dissolution. And consider moderately reducing the time of incubation with acetone in the previous step.
12. In addition to trypsin, other enzymes can also be used to digest the protein sample into peptides with different cutting sites and assemble overlapping sequences. Be sure to use trypsin (or other enzymes) specifically applied for mass spectrometry. The time of enzymolysis is recommended to be between 12 and 16 h, or as referred to in the manual.
13. After enzymolysis, it is necessary to centrifuge and carefully collect the supernatant in order to avoid accidental clogging of the column by insoluble proteins in the subsequent desalting.
14. It is recommended to desalt immediately after enzymolysis. Otherwise, dry and store the sample at −20 °C, and thaw it at room temperature before use.
15. If the liquid is difficult to flow through the column during centrifugation, appropriately increase the centrifuging time. The centrifuging speed should be maintained to prevent protein loss.
16. Be aware that uncarefully sample loading will possibly cause bubbles and affect MS detection.
17. In general, we collect genomic and proteomic datasets from NCBI Genome and UniProt. However, specialized resources more applicable for some species are also a good choice, such as Phytozome for *C. reinhardtii*.

18. Use introns as non-CDSs for eukaryotic algae, and intergenic sequences for cyanobacteria.
19. The parameter P(coding) for sORF finder is set as the default 0.5, or around the ratio of CDS size to genome size.
20. Be aware that there is a mistake in the original "search_sORF.pl" leading to a wrong label for the start site of smORFs. Download an updated version of "search_sORF.pl" at https://github.com/bite123/search_sORF.pl-update to avoid the issue.

References

1. Couso JP, Patraquim P (2017) Classification and function of small open reading frames. Nat Rev Mol Cell Bio 18(9):575–589
2. Plaza S, Menschaert G, Payre F (2017) In search of lost small peptides. Annu Rev Cell Dev Biol 33:391–416
3. Andrews SJ, Rothnagel JA (2014) Emerging evidence for functional peptides encoded by short open reading frames. Nat Rev Genet 15(3):193–204
4. Vitorino R et al (2021) The role of micropeptides in biology. Cell Mol Life Sci 78(7): 3285–3298
5. Dolganyuk V et al (2020) Microalgae: a promising source of valuable bioproducts. Biomolecules 10(8):1153
6. Bule MH et al (2018) Microalgae as a source of high-value bioactive compounds. Front Biosci 10(2):197–216
7. Olasehinde TA, Olaniran AO, Okoh AI (2017) Therapeutic potentials of microalgae in the treatment of Alzheimer's disease. Molecules 22(3):480
8. Förster B, Mathesius U, Pogson BJ (2006) Comparative proteomics of high light stress in the model alga. Proteomics 6(15):4309–4320
9. Lakatos GE et al (2019) Bioethanol production from microalgae polysaccharides. Folia Microbiol 64(5):627–644
10. Collier JL, Grossman AR (1994) A small polypeptide triggers complete degradation of light-harvesting Phycobiliproteins in nutrient-deprived cyanobacteria. EMBO J 13(5): 1039–1047
11. de Alvarenga LV, Hess WR, Hagemann M (2020) AcnSP - a novel small protein regulator of Aconitase activity in the cyanobacterium sp. PCC 6803. Front Microbiol 11:1445
12. Selim KA et al (2018) PII-like signaling protein SbtB links cAMP sensing with cyanobacterial inorganic carbon response. Proc Natl Acad Sci 115(21):E4861–E4869
13. Baumgartner D et al (2016) Small proteins in cyanobacteria provide a paradigm for the functional analysis of the bacterial micro-proteome. BMC Microbiol 16:285
14. Peng Z et al (2022) Identification and analysis of smORFs in Chlamydomonas reinhardtii. Genomics 114(5):110444
15. Khitun A, Slavoff SA (2019) Proteomic detection and validation of translated small open reading frames. Curr Protoc Chem Biol 11(4):e77
16. Cardon T et al (2020) Optimized sample preparation workflow for improved identification of ghost proteins. Anal Chem 92(1):1122–1129
17. Fijalkowski I, Peeters MKR, Van Damme P (2021) Small protein enrichment improves proteomics detection of sORF encoded polypeptides. Front Genet 12:713400
18. He CT, Jia CX, Zhang Y, Xu P (2018) Enrichment-based Proteogenomics identifies microproteins, missing proteins, and novel smORFs in. J Proteome Res 17(7):2335–2344
19. Hanada K et al (2010) sORF finder: a program package to identify small open reading frames with high coding potential. Bioinformatics 26(3):399–400
20. Quinlan AR, Hall IM (2010) BEDTools: a flexible suite of utilities for comparing genomic features. Bioinformatics 26(6):841–842
21. Shen W, Le S, Li Y, Hu FQ (2016) SeqKit: a cross-platform and ultrafast toolkit for FASTA/Q file manipulation. PLoS One 11(10):e0163962
22. Khanduja A, Kumar M, Mohanty D (2023) ProsmORF-pred: a machine learning-based method for the identification of small ORFs in prokaryotic genomes. Brief Bioinform 24(3):bbad101

23. Hulstaert N et al (2020) ThermoRawFileParser: modular, scalable, and cross-platform RAW file conversion. J Proteome Res 19(1): 537–542
24. Kim S, Pevzner PA (2014) MS-GF plus makes progress towards a universal database search tool for proteomics. Nat Commun 5:5277
25. Chi H et al (2018) Comprehensive identification of peptides in tandem mass spectra using an efficient open search engine. Nat Biotechnol 36(11):1059
26. Bateman A et al (2023) UniProt: the universal protein knowledgebase in 2023. Nucleic Acids Res 51(D1):D523–D531

Part II

Translation and Transcriptome-Based Discovery

Chapter 7

Identifying and Visualizing Novel Small Open Reading Frames (sORFs) Via Ribo-Seq, Transcriptome Assembly, and *ggRibo*

Hsin-Yen Larry Wu, Isaiah D. Kaufman, and Polly Yingshan Hsu

Abstract

Small ORFs (sORFs) may encode signaling peptides or microproteins that regulate diverse developmental and environmental responses. Additionally, their translation may play a crucial role in the biogenesis of noncoding RNAs. Despite their importance, sORFs are typically absent from genome annotations, and their prediction remains a significant challenge. Here, we present a bioinformatics pipeline for the identification and visualization of novel translated sORFs using Ribo-seq data in Arabidopsis. This pipeline integrates: (1) Ribo-seq/RNA-seq preprocessing and mapping, (2) (optional) transcriptome assembly to detect unannotated transcripts, (3) ORF discovery using *RiboTaper*, and (4) the gene viewer *ggRibo* for high-resolution visualization of individual sORF expression. By combining these modules, our approach provides a powerful framework for analyzing Ribo-seq data, facilitating the discovery, visualization, and future functional characterization of hidden sORFs. This pipeline can be applied to different organisms for systematic identification and visualization of novel sORFs.

Key words Ribosome profiling, Ribo-seq, Transcriptome assembly, *RiboTaper*, *ggRibo*, Small ORFs

1 Introduction

Small open reading frames (sORFs) can encode microproteins or signaling peptides that play important roles in various physiological, developmental, and stress responses [1–4]. The act of translation from sORFs may also serve regulatory functions [5–10]. Despite their importance, sORFs are often missing from genome annotations. This omission occurs because computational predictions typically exclude ORFs shorter than 100 codons to minimize false-positive identifications [11–13]. As a result, transcripts encoding sORFs are often misclassified as noncoding RNAs (ncRNAs). Moreover, only a limited number of sORFs have been identified through forward genetic screens, likely due to the random nature of traditional mutagenesis methods, which makes it less

Stephan Wenkel (ed.), *Microproteins: Methods and Protocols*, Methods in Molecular Biology, vol. 2992,
https://doi.org/10.1007/978-1-0716-5013-4_7,

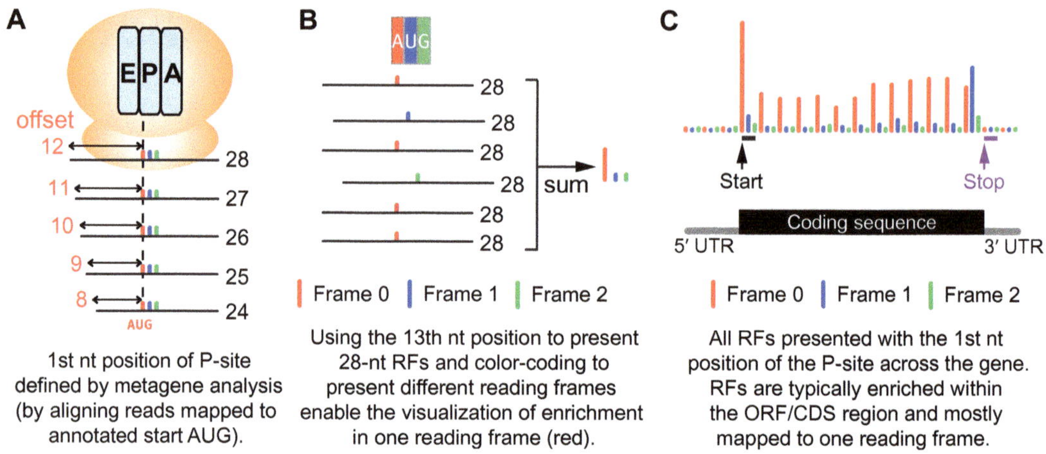

Fig. 1 Selecting one representative nucleotide position for RPFs reveals 3-nt periodicity along the coding sequence. (This figure is adapted from [27]). (**a**) *Identifying the P-site position within ribosomes* by aligning RPFs mapped to annotated start AUG codons. This analysis reveals that for 28-nt RPFs, the P-site is located between 13th and 15th nt, meaning the first nt position of the P-site is offset by 12 nt from the 5′ end of the RPF in Arabidopsis. Similarly, for 27-, 26-, 25-, 24-nt RPFs, their first nt positions of the P-site are offset by 11, 10, 9, 8 nt, respectively, from their 5′ ends. (**b**) *Color-coding RPFs based on the reading frame in which the first nt position of their P-site maps.* The primary reading frame (frame 0) is colored red, while the other two frames are colored blue and green. Summing up six RPFs reveals that four belong to the red reading frame, one to the blue frame, and one to the green frame. (**c**) *The P-site plot for an entire transcript* reveals strong 3-nt periodicity, showing enrichment of RPFs in the primary reading frame (red) within the coding sequence

likely to hit small genomic regions. Emerging evidence suggests that many sORFs are de novo genes or evolutionary young, exhibiting no sequence conservation across different lineages [14–16]. Together, the combination of their short length, few experimental examples, and low sequence conservation limits the power of homology searches and bioinformatic predictions for sORFs [17].

Ribo-seq has emerged as a high-throughput experimental approach for uncovering novel translated sORFs [18, 19]. It utilizes deep sequencing of ribosome-protected mRNA fragments (RPFs) to map and quantify ribosome occupancy on mRNAs [20]. A hallmark of high-quality Ribo-seq data is its ability to capture the 3-nucleotide (3-nt) periodicity of translating ribosomes, reflecting their codon-by-codon progression during mRNA translation [5, 20]. Figure 1 illustrates how Ribo-seq reads can be analyzed to reveal this 3-nt periodicity. This periodicity has been integrated into mainstream ORF discovery software to identify novel translated ORFs, including sORFs [21–26]. Additionally, RNA-seq followed by transcriptome assembly can be used to identify sORFs within previously unannotated transcripts [10, 15]. Considerations for Ribo-seq and RNA-seq experiments for ORF discovery are discussed *see* **Notes 1–3**.

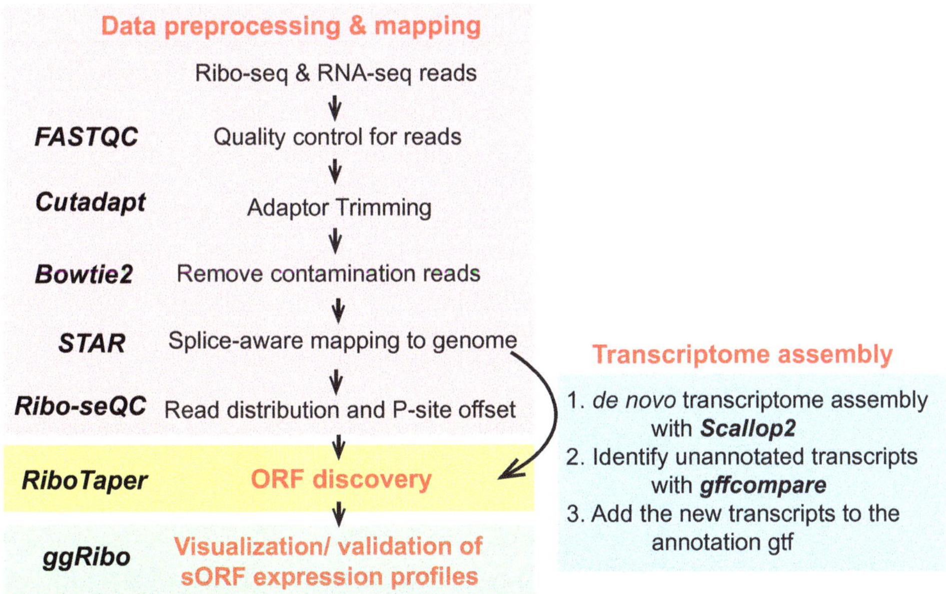

Fig. 2 The ORF discovery and visualization pipeline consists of four modules. The pipeline begins with Ribo-seq/RNA-seq data preprocessing and mapping, followed by an optional transcriptome assembly step to identify unannotated transcripts. The mapped reads are then used in the ORF discovery tool *RiboTaper*. Finally, the expression profiles of genes or sORFs of interest can be visualized/validated using the *ggRibo* R package

Here, we provide a step-by-step guide for identifying novel translated ORFs using Ribo-seq data, focusing on sORFs. The pipeline presented consists of four major modules (Fig. 2): (1) preprocessing and mapping of Ribo-seq and RNA-seq data, (2) (Optional) reference-guided de novo transcriptome assembly using RNA-seq to identify unannotated transcripts, (3) ORF discovery using *RiboTaper* [22], and (4) visualization/validation of individual sORF expression using *ggRibo* [27].

2 Materials

The example data in Arabidopsis and the required software are described below. All steps are run in bash on a Linux OS, unless otherwise noted as running in R. Since the Ribo-seq analysis requires high computational power, using a Linux server or cluster is highly recommended. Updates and the complete pipeline are available on GitHub (https://github.com/hsinyenwu/Riboseq_pipeline).

2.1 Data Preprocessing and Mapping

2.1.1 Data and Software Required

1. Ribo-seq FASTQ files from single-end 50 bp illumina sequencing. Example Ribo-seq data may be downloaded from the NCBI Short Read Archive (SRA) SRR15700008 [10] (download instructions described in Subheading 3.1, **step 1**).
2. RNA-seq FASTQ files from paired-end 100 bp (or longer) illumina sequencing. Example RNA-seq data may be downloaded from the NCBI SRA SRR15700002 [10] (download instructions described in Subheading 3.1, **step 1**).
3. Genome sequences in FASTA format and the FASTA index file. The Arabidopsis genome sequence may be downloaded from Mendeley Data (*see* Subheading 2.4).
4. Transcriptome annotation file in GTF format. Each gene should contain gene_biotype information in the ninth column. The gtf should at least contain two biotypes: "protein coding" and "noncoding." Arabidopsis transcriptome annotation may be downloaded from Mendeley Data (*see* Subheading 2.4).
5. Sequences of contaminant RNAs (rRNA, tRNA, snRNA, and snoRNA) in FASTA format. Arabidopsis contaminant RNAs may be downloaded from Mendeley Data (*see* Subheading 2.4).
6. *cutadapt* [28].
7. *FastQC* (https://www.bioinformatics.babraham.ac.uk/projects/fastqc/).
8. *bowtie2* [29].
9. *STAR* [30].
10. *Ribo-seQC* [31].

2.2 Reference-Guided Transcriptome Assembly

2.2.1 Software Required

1. *Scallop2* [32].
2. *gffcompare* [33].
3. *rtracklayer* in R [34].

2.3 ORF Discovery Using RiboTaper

2.3.1 Software Required

1. *RiboTaper* [22].

2.4 Visualization/Validation of Individual sORF Expression Using ggRibo

2.4.1 Software and Data Required

1. *ggRibo* (https://github.com/hsinyenwu/ggRibo) [27] in R.
2. An example dataset could be downloaded from Mendeley Data (https://data.mendeley.com/datasets/wm6cs4zbtw/1).

3 Methods

3.1 Data Preprocessing and Mapping

Here, we use *FastQC* to evaluate the sequencing reads between steps, *cutadapt* to trim the adapter sequence included during the library preparation, and *bowtie2* to remove common contaminating reads (e.g., rRNAs and tRNAs, etc). Subsequently, we use *STAR* to map reads to the reference genome. Finally, we use *Ribo-seQC* to assess Ribo-seq reads for the necessary statistics required for downstream analysis (Fig. 2).

1. Download published data for practice from NCBI SRA using *sra-tools*. Alternatively, use your own Ribo-seq and RNA-seq data.

```
# prefetch <accession number> command will download the
<accession number> run file.
prefetch SRR15700008 SRR15700002 -O ./DATA/
# fasterq-dump converts .sra files (the Sequence Read Archive
format) into FASTQ files
fasterq-dump -3 ./DATA/SRR15700008/SRR15700008.sra -O ./DATA/
fasterq-dump -3 ./DATA/SRR15700002/SRR15700002.sra -O ./DATA/
```

2. Assess the reads using *FastQC*. Since Ribo-seq uses 50-bp sequencing and RPFs have a length of 21–30 bp, Ribo-seq reads should contain the adapter sequence used in the library preparation (*see* **Note 4**). Other common contaminant sequences, such as from rRNAs and tRNAs, are also expected in the library.

```
fastqc -o ./DATA/ ./DATA/*.fastq
```

3. Trim the adapter sequences and low-quality bases using *cutadapt*. Ribo-seq reads that do not contain the adapter sequence are discarded since they are unlikely to be an RPF.

```
# Trim adapter sequences from Ribo-seq reads
cutadapt -m 20 -M 40 -q 5 --discard-untrimmed -a AGATCGGAAGAG
CACACGTCT \
 -o ./DATA/SRR15700008_trim.fq ./DATA/SRR15700008.fastq

# Trim adapter sequences from paired-end RNA-seq reads
cutadapt -m 20 -q 5 -a AGATCGGAAGAGCACACGTCTGAACTCCAGTCA \
 -A AGATCGGAAGAGCGTCGTGTAGGGAAAGAGTGT --poly-a \
 -o ./DATA/SRR15700002_1_trim.fq -p ./DATA/SRR15700002_2_trim.
fq \
 ./DATA/SRR15700002_1.fastq ./DATA/SRR15700002_2.fastq
```

4. Repeat *FastQC* to ensure that adapters have been removed.

```
fastqc -o ./DATA/ ./DATA/*_trim.fq
```

5. Remove contaminant RNAs (rRNA, tRNA, snRNA, and snoRNA) using *bowtie2*.

```
# Build index of contaminant RNAs
bowtie2-build -f ./GENOME_FILES/CONTAM_RNAS.FASTA ./DATA/
BT2_INDEX

# Align reads to contaminant RNAs and output unaligned reads
bowtie2 -L 20 -x ./DATA/BT2_INDEX -S ./DATA/SRR15700008_a-
ligned.sam \
 ./DATA/SRR15700008_trim.fq --un ./DATA/SRR15700008_decon.fq
bowtie2 -L 20 -x ./DATA/BT2_INDEX \
 -1 ./DATA/SRR15700002_1_trim.fq -2 ./DATA/SRR15700002_2_trim.
fq \
 -S ./DATA/SRR15700002_alignments.sam \
 --un-conc ./DATA/SRR15700002_%_decon.fq
```

6. Repeat *FastQC* to check if most of the overrepresented sequences are removed (there may still be some). Any additional overrepresented sequence identified after *FastQC* here can be added to the `CONTAM_RNAS.FASTA` to remove those as well.

```
fastqc -o ./DATA/ ./DATA/*_decon.fq # repeat FastQC
```

7. Map reads to reference genome using splicing-aware mapper *STAR*:
   (a) Build a separate index for the Ribo-seq and RNA-seq.
   (b) Map the "decontaminated" reads from **step 6** above to the reference genome. The resulting BAM files that end in "_Aligned.sortedByCoord.out.bam" will be used for the next steps. Considerations about multi-mapping reads are discussed in *see* **Note 5**.

```
# Generate indices for Ribo-seq reads (from single-end sequen-
cing)
STAR --runMode genomeGenerate \
 --genomeDir ./DATA/RIBO_STAR_INDEX \
 --genomeFastaFiles ./GENOME_FILES/GENOME.FASTA \
 --sjdbGTFfile ${gtf_path} ./GENOME_FILES/GENOME.GTF \
 --sjdbOverhang 34 # we assume the max Ribo-seq reads after
adapter trimming is 35, 35-1=34. See STAR manual for details.

# Generate indices for RNA-seq reads (from paired-end sequen-
cing)
STAR --runMode genomeGenerate \
 --genomeDir ./DATA/RNA_STAR_INDEX \
 --genomeFastaFiles ./GENOME_FILES/GENOME.FASTA \
 --sjdbGTFfile ${gtf_path} ./GENOME_FILES/GENOME.GTF \
 --sjdbOverhang 99 # If the sequence is paired-end
```

```
100, 100-1=99

# Align Ribo-seq reads to genome
# See STAR manual the meaning of the parameters
STAR --genomeDir ./DATA/RIBO_STAR_INDEX \
 --readFilesIn ./DATA/SRR15700008_decon.fq \
 --alignIntronMax 5000 \
 --alignIntronMin 15 \
 --outFilterMismatchNmax 1 \
 --outFilterMultimapNmax 20 \
 --outFilterType BySJout \
 --alignSJoverhangMin 8 \
 --alignSJDBoverhangMin 2 \
 --outSAMtype BAM SortedByCoordinate \
 --quantMode TranscriptomeSAM \
 --outSAMmultNmax 1 \
 --outMultimapperOrder Random \
 --outFileNamePrefix "./DATA/SRR15700008_star_"

# Align RNA-seq reads to genome
STAR --genomeDir ./DATA/RNA_STAR_INDEX \
 --readFilesIn ./DATA/SRR15700002_1_decon.fq \
 ./DATA/SRR15700002_2_decon.fq \
 --alignIntronMax 5000 \
 --alignIntronMin 15 \
 --outFilterMismatchNmax 1 \
 --outFilterMultimapNmax 20 \
 --outFilterType BySJout \
 --alignSJoverhangMin 8 \
 --alignSJDBoverhangMin 2 \
   outSAMtype BAM SortedByCoordinate \
 --quantMode TranscriptomeSAM \
 --outSAMmultNmax 1 \
 --outMultimapperOrder Random \
 --outFileNamePrefix "./DATA/SRR15700002_star_"
```

8. Run *Ribo-seQC* in R. *Ribo-seQC* provides an HTML report with many useful statistics for Ribo-seq. *Within the Ribo-seQC report*, Subheading 4.3 shows a plot of 3-nt periodicity, and Subheading 4.2.2 summarizes the P-site offsets for different read lengths; this information is required for running *RiboTaper* discussed in Subheading 3.3.

```
# Set variables
IN_BAM <- "./DATA/SRR15700008_star_Aligned.sortedByCoord.out.
bam"
OUT_HTML <- "./DATA/SRR15700008_RiboseQC.html"
GTF_DIR <- "./GENOME_FILES/"
GTF_PATH <- "./GENOME_FILES/GENOME.GTF"
```

```
RANNOT <- "./GENOME_FILES/GENOME.GTF_Rannot"
FASTA_PATH <- "./GENOME_FILES/GENOME.FASTA"
CIRCLE_SEQS <- c("ChrM", "ChrC") # list of circular chromo-
somes

# Prepare annotation files for Ribo-seQC
prepare_annotation_files(annotation_directory = GTF_DIR,
 genome_seq = FASTA_PATH, gtf_file = GTF_PATH,
 export_bed_tables_TxDb = F, forge_BSgenome = F,
 create_TxDb = T, circ_chroms = CIRCLE_SEQS)

# Run analysis and generate HTML report
RiboseQC_analysis(annotation_file = RANNOT, bam_files = IN_-
BAM,
 report_file = OUT_HTML, write_tmp_files = T)
```

3.2 (Optional) Reference-Guided Transcriptome Assembly

To facilitate the identification of sORFs in the unannotated transcripts, it may be desirable to perform reference-guided de novo transcriptome assembly to capture transcripts that are missing from the annotation. Here, we use *Scallop2* to assemble novel transcripts and *Gffcompare* to identify new transcripts. New transcripts that are likely to contain a sORF are then added to the annotation file using *rtracklayer* for downstream analysis.

1. Assemble transcripts from RNA-seq data using *Scallop2*.

```
scallop2 --library_type second \
 -i ./DATA/SRR15700002_star_Aligned.sortedByCoord.out.bam \
 -o ./DATA/NEW_SCALLOP2_GTF.GTF
```

2. Compare assembled transcripts to annotated transcripts using *Gffcompare*. Each new transcript will be assigned a class code based on its structure relative to the reference transcripts.

```
gffcompare -V -r ./GENOME_FILES/GENOME.GTF -o ./DATA/MERGED \
 ./DATA/NEW_SCALLOP2_GTF.GTF
```

3. Add new transcripts with the class codes *i*, *x*, *y*, *o*, and *u*, which are likely to contain a novel ORF, to the original GTF annotation using *rtracklayer* in R.

```
ANNOT_GTF_PATH <- "./GENOME_FILES/GENOME.GTF"
SCALLOP_GTF_PATH <- "./DATA/MERGED.annotated.gtf"
FINAL_GTF_PATH <- "./DATA/FINAL_MERGED.GTF"

# import GTFs as GRanges objects
ANNOT_GTF <- rtracklayer::import(ANNOT_GTF_PATH, format =
"gtf")
SCALLOP_GTF <- rtracklayer::import(SCALLOP_GTF_PATH, format =
"gtf")
```

```
# get the subset of new transcripts from the i, x, y, o, u
classes
GENES_ixyou <- SCALLOP_GTF$gene_id[SCALLOP_GTF$class_code %in
% c("i", "x", "y", "o", "u")]
GTF_ixyou <- SCALLOP_GTF[SCALLOP_GTF$gene_id %in% GENES_ixyou]

# set new transcripts as noncoding since they do not contain an
annotated CDS
GTF_ixyou$biotype <- "noncoding"

# append the new transcripts to the end of the original GTF
annotation
FINAL_GTF <- c(ANNOT_GTF, GTF_ixyou)
rtracklayer::export(FINAL_GTF, FINAL_GTF_PATH, format =
"gtf")
```

3.3 ORF Discovery Using RiboTaper

RiboTaper identifies translated ORFs based on significant 3-nt periodicity [22]. The output file "ORFs_max_filt" contains a list of identified translated ORFs. The file "P_sites_all" contains the Ribo-seq P-site positions (*see* **Note 6**), which can be reformatted for *ggRibo* visualization [27] described in Subheading 3.4.

1. Prepare annotation files for *RiboTaper*. Here, we assume readers performed transcriptome assembly and obtained the FINAL_MERGED.GTF. If not, please use the original GTF for the following steps.

```
create_annotations_files.bash ./DATA/FINAL_MERGED.GTF \
./GENOME_FILES/GENOME.FASTA false false ./DATA/RIBOTAPER_AN-
NOT/
```

2. Run *RiboTaper* to identify translated ORFs using the alignments from *STAR* (Subheading 3.1, **step 7**) and P-site offsets from *Ribo-seQC* (Subheading 3.1, **step 8**).

```
Ribotaper.sh ./DATA/SRR15700008_star_Aligned.sortedByCoord.
out.bam \
 ./DATA/SRR15700002_star_Aligned.sortedByCoord.out.bam \
 ./DATA/RIBOTAPER_ANNOT/ \
 22,23,24,25,26,27,28,29 6,7,8,9,10,11,12,12 8
```

3. (For data visualization in Subheading 3.4, also *see* **Note 6**) reformat P-sites for running *ggRibo*.

```
# P_sites_all file is from RiboTaper output
cut -f 1,3,6 P_sites_all | sort | uniq -c \
 | sed -r 's/^( *[^ ]+) +/\1\t/' > ./DATA/RIBOTAPER_P_SITES.
GGRIBO
```

3.4 Visualization/Validation of Individual sORF Expression Using ggRibo

ggRibo is a *ggplot2*-based R package for visualizing Ribo-seq/RNA-seq expression of individual genes [27]. It shows Ribo-seq 3-nt periodicity and RNA-seq coverage simultaneously in the context of gene structure, enabling visualization of novel ORFs, isoform translation, and various translational features. Users can also zoom in to visualize DNA and amino acid sequences for the region of interest in parallel. Beyond Ribo-seq, *ggRibo* supports other single-nucleotide resolution datasets (e.g., degradome, CAGE-seq, and TI-seq), facilitating multi-omics comparisons and comprehensive gene expression analysis at multiple levels (*see* **Note 7**). Below, we plot three examples to visualize one well-studied sORF and one novel sORF discovered in a long ncRNAs in Arabidopsis. These examples use the complete dataset from [10] (*see* **Note 8**); users may download the processed data as described in Subheading (2.4, **item 2**) for practice.

3.4.1 Example of a Well-Studied, Annotated Small Protein, IRONMAN

The IRON MAN (IMA) peptides play an important role in plant iron (Fe) homeostasis and immune responses [35, 36]. IMA peptides are encoded by eight genes in Arabidopsis and are highly conserved across angiosperms. Here, we use *ggRibo* to visualize *IMA1* Ribo-seq/RNA-seq expression, highlighting the frame of Ribo-seq reads with three colors (Fig. 3). By default, *ggRibo* colors the annotated frame (frame 0) and the +1 and +2 frames with red, blue, and green, respectively.

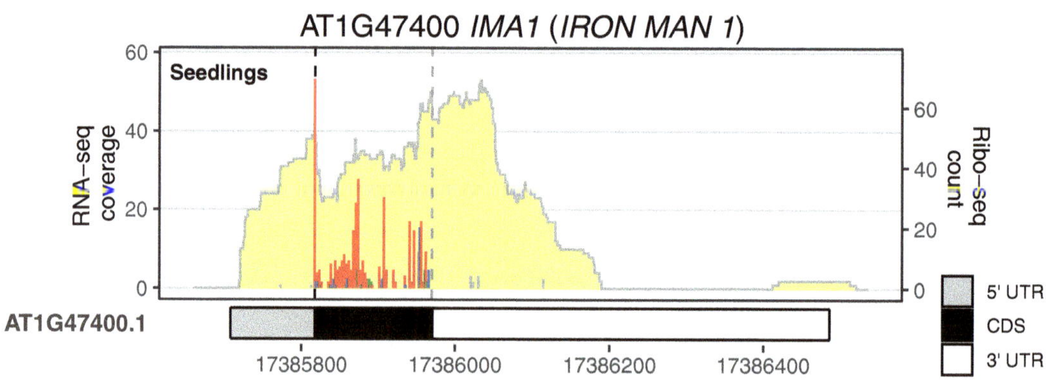

Fig. 3 Example *ggRibo* plot of a well-characterized sORF encoded by *IMA1*. Ribo-seq and RNA-seq expression profiles of *IMA1* in Arabidopsis seedlings are shown. In the gene model, the annotated coding sequence (CDS) is represented by a black box, and the 5′ UTR and 3′ UTR are depicted as grey and white boxes, respectively. RNA-seq coverage is displayed with a light-yellow background, and Ribo-seq reads are represented by three colored lines: red indicates the primary reading frame, while blue and green indicate the other two reading frames. Ribo-seq reads outside the coding sequence are colored grey. The start and stop codons of the coding sequence are marked with black and grey dashed lines, respectively. The numbers below the gene model indicate genomic coordinates. Ribo-seq and RNA-seq data were reported in [10]

1. Load the datasets and set up parameters. Here, we assume that you have downloaded the datasets from Mendeley Data and saved them in the "./Arabidopsis" folder.

```
library(ggRibo)
CTRL_RNA <- "./Arabidopsis/Arabidopsis/RNA_CTRL_merged.bam"
CTRL_ribo  <-  "./Arabidopsis/CTRL_expressed_P_sites_sort_-
count"
Samples = c("Seedlings")

RNA_files <- list(CTRL_RNA)
Ribo_files <- list(CTRL_ribo)
#Load data with create_seq_input
inputs_full <- create_seq_input(rna_files = RNA_files,

 ribo_files = Ribo_files, sample_names = Samples)
#Load gtf
gtf_import(annotation  =  "./Arabidopsis/Araport11+-
CTRL_20181206.gtf", format = "gtf",
 dataSource = "Araport",organism = "Arabidopsis thaliana")
```

2. Plot AT1G47400 (*IMA1*).

```
#Example for an annotated protein coding gene
ggRibo(tx_id = "AT1G47400.1",
 NAME = "IMA1 (IRON MAN 1)",
 Extend = 50)
```

3.4.2 Examples of a Novel sORF from a Long ncRNA

Next, we use *ggRibo* to visualize a sORF within an annotated long ncRNA. Since ncRNAs lack annotated coding sequences, *ggRibo* assigns the reading frame from the start of the RNA sequence rather than from the start of a coding sequence. Consequently, the enriched reading frame for the sORF could fall into the red, blue, or green frame (*see* **Note 9**). The Ribo-seq reads for the example sORF are enriched in the red frame (Fig. 4).

1. Plot lncRNA example.

```
#Use the same data loaded in Example 1
#Example for an annotated non-coding RNA
LNC1 <- ggRibo(tx_id = "AT3G03505.1",
 NAME = "lncRNA",
 Extend = 50)
LNC1
```

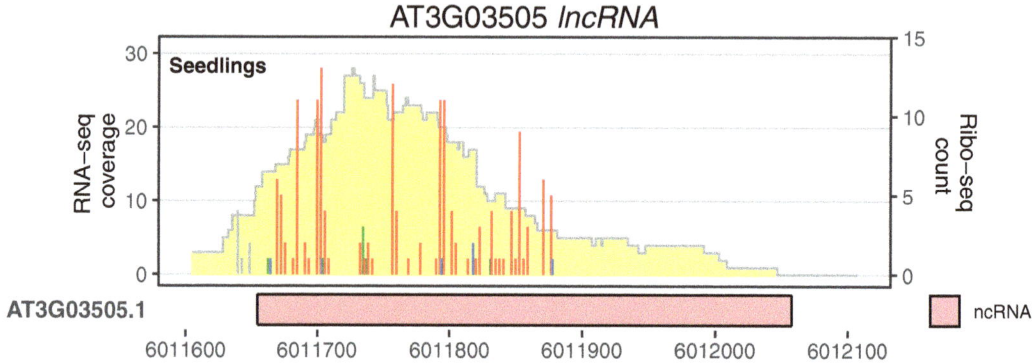

Fig. 4 Example of a novel sORF identified within an annotated long ncRNA using the "red" reading frame. (**a**) Ribo-seq and RNA-seq expression profiles of a novel sORF identified within an annotated lncRNA in Arabidopsis seedlings are shown. In the gene model, the ncRNA region is illustrated in pink. The data representations are similar to those described in Fig. 3, except for ncRNAs, no coding sequence range is indicated, as ncRNAs lack an annotated coding sequence by default. The strong 3-nt periodicity of Ribo-seq reads in the "red" reading frame suggests the presence of a translated sORF. (**b**) Users can zoom in to inspect the sequence and positions of the corresponding start and stop codons. In this case, the real start codon appears to be upstream of the annotated transcript start site due to the presence of RNA-seq reads and in-frame Ribo-seq reads (marked with red arrow). Ribo-seq and RNA-seq data were reported in [10]

4 Notes

1. Since this chapter focuses on data analysis, readers interested in Ribo-seq preparation can refer to our recent papers [10, 37] for the principles and steps required for high-quality Ribo-seq library preparation in plants. Due to the short length of RPFs, Illumina single-end 50 bp sequencing is well-suited for Ribo-seq.
2. Ideally, RNA-seq should be paired-end 100 bp or longer and prepared from the same biological samples as the Ribo-seq libraries. We recommend using RiboZero or other rRNA depletion kits to remove rRNA. Poly(A) selection may introduce a 3′ bias toward poly(A) tails, potentially interfering with downstream analyses such as reference-guided transcriptome assembly.
3. Other considerations for the sequencing data and analysis:
   (a) Is the depth of your Ribo-seq reads sufficient for the study? High sequencing depth is essential for identifying translated small ORFs.
   (b) Does the Ribo-seq show high coverage across transcripts? A high number of sequencing reads does not necessarily equate to greater depth, as over-amplification of a small number of RPFs does not improve coverage. The advantages of improved-coverage Ribo-seq reads were discussed in [10].

(c) Would reference-guided transcriptome assembly benefit your study? If your goal is to identify small translated ORFs in unannotated ncRNAs, we recommend reference-guided transcriptome assembly (*see* Subheading 3.2), which requires a paired-end 100 bp or longer RNA-seq library.

(d) Carefully consider which transcriptome annotation to use. Some annotations are highly complex and include more isoforms than other versions, which can lead to an overestimation of translated ORFs and make data visualization more challenging.

4. The adapter is added during library preparation, so its sequence may vary depending on the protocol used.
5. Multi-mapping reads (i.e., reads mapping to multiple locations) can be a significant concern when analyzing Ribo-seq and short RNA-seq reads, particularly in complex genomes. We use two key parameters in *STAR* alignment:

   (a) —outFilterMultimapNmax 20: This parameter sets the maximum number of locations a read is allowed to align to. If a read aligns to more than 20 loci, *STAR* treats it as unmapped. If your genome of interest is not highly complex or repetitive, you can lower this number (e.g., 5 or 10).

   (b) —outSAMmultNmax 1: This parameter determines how many of those multiple alignments are included in the final SAM/BAM file. Even if a read can map to up to 20 locations (as defined by —outFilterMultimapNmax 20), *STAR* will output only one alignment for that read.

   As a result, these two parameters effectively reduce the impact of reads mapping to highly repetitive regions in downstream analyses, since such reads are either discarded or only partially represented in the final alignments.

6. The P-site files can also be generated from Ribo-seQC output. Please see the GitHub link listed in Subheading 2. Materials for details.
7. For examples of multi-omic analysis using *ggRibo*, please refer to [27].
8. To ensure high-coverage reads for plotting, Ribo-seq and RNA-seq data from all six Arabidopsis seedling datasets were combined from [10].
9. The sORFs can also be plotted as protein-coding genes if users create a customized GTF file and provide the coding ranges for them.

Acknowledgments

This work used computational resources and services provided by the Institute for Cyber-Enabled Research at Michigan State University. This work was supported by a predoctoral training award under Grant Number T32-GM110523 from the National Institute of General Medical Sciences of the National Institutes of Health to IDK, and research grants from the National Science Foundation under Award Numbers 2051885 and 2425390, and the National Institute of General Medical Sciences of the National Institutes of Health under Award Number R35GM155375 to PYH. The content is solely the responsibility of the authors and does not necessarily represent the official views of the NSF or NIH.

References

1. Hellens RP, Brown CM, Chisnall MAW, Waterhouse PM, Macknight RC (2016) The emerging world of small ORFs. Trends Plant Sci 21:317–328
2. Hsu PY, Benfey PN (2018) Small but mighty: functional peptides encoded by small ORFs in plants. Proteomics 18:1700038
3. Bhati KK, Dolde U, Wenkel S (2021) Micro-Proteins: expanding functions and novel modes of regulation. Mol Plant 14:705–707
4. Wu H-YL, Jen J, Hsu PY (2024) What, where, and how: regulation of translation and the translational landscape in plants. Plant Cell 36:koad197
5. Hsu PY, Calviello L, Wu HYL, Li FW, Rothfels CJ, Ohler U et al (2016) Super-resolution ribosome profiling reveals unannotated translation events in Arabidopsis. Proc Natl Acad Sci USA 113:E7126–E7135
6. Li S, Le B, Ma X, Li S, You C, Yu Y et al (2016) Biogenesis of phased siRNAs on membrane-bound polysomes in Arabidopsis. elife 5: e22750
7. Yoshikawa M, Iki T, Numa H, Miyashita K, Meshi T, Ishikawa M (2016) A short open reading frame encompassing the micro-RNA173 target site plays a role in trans-acting small interfering RNA biogenesis. Plant Physiol 171:359–368
8. Bazin J, Baerenfaller K, Gosai SJ, Gregory BD, Crespi M, Bailey-Serres J (2017) Global analysis of ribosome-associated noncoding RNAs unveils new modes of translational regulation. Proc Natl Acad Sci USA 114:E10018–E10027
9. Iwakawa HO, Lam AYW, Mine A, Fujita T, Kiyokawa K, Yoshikawa M et al (2021) Ribosome stalling caused by the Argonaute-microRNA-SGS3 complex regulates the production of secondary siRNAs in plants. Cell Rep 35:109300
10. Wu H-YL, Ai Q, Teixeira RT, Nguyen PHT, Song G, Montes C et al (2024) Improved super-resolution ribosome profiling reveals prevalent translation of upstream ORFs and small ORFs in Arabidopsis. Plant Cell 36: 510–539
11. Basrai MA, Hieter P, Boeke JD (1997) Small open reading frames: beautiful needles in the haystack. Genome Res 7:768–771
12. Olsen AN, Mundy J, Skriver K (2002) Peptomics, identification of novel cationic Arabidopsis peptides with conserved sequence motifs. In Silico Biol 2:441–451
13. Lease KA, Walker JC (2006) The Arabidopsis unannotated secreted peptide database, a resource for plant peptidomics. Plant Physiol 142:831–838
14. Ruiz-Orera J, Messeguer X, Subirana JA, Alba MM (2014) Long non-coding RNAs as a source of new peptides. elife 3:e03523
15. Wu HYL, Song G, Walley JW, Hsu PY (2019) The tomato translational landscape revealed by transcriptome assembly and ribosome profiling. Plant Physiol 181:367–380
16. Sandmann C-L, Schulz JF, Ruiz-Orera J, Kirchner M, Ziehm M, Adami E et al (2023) Evolutionary origins and interactomes of human, young microproteins and small peptides translated from short open reading frames. Mol Cell 83:994–1011.e18
17. Feng Y, Jiang M, Yu W, Zhou J (2023) Identification of short open reading frames in plant genomes. Front Plant Sci [Internet] 14.

Available from: https://www.frontiersin.org/articles/10.3389/fpls.2023.1094715
18. Ingolia NT, Brar GA, Stern-Ginossar N, Harris MS, Talhouarne GJS, Jackson SE et al (2014) Ribosome profiling reveals pervasive translation outside of annotated protein-coding genes. Cell Rep 8:1365–1379
19. Brar GA, Weissman JS (2015) Ribosome profiling reveals the what, when, where and how of protein synthesis. Nat Rev Mol Cell Biol 16:651–664
20. Ingolia NT, Ghaemmaghami S, Newman JRS, Weissman JS (2009) Genome-wide analysis in vivo of translation with nucleotide resolution using ribosome profiling. Science 324:218–223
21. Ji Z, Song R, Regev A, Struhl K (2015) Many lncRNAs, 5′UTRs, and pseudogenes are translated and some are likely to express functional proteins. elife 4:e08890
22. Calviello L, Mukherjee N, Wyler E, Zauber H, Hirsekorn A, Selbach M et al (2016) Detecting actively translated open reading frames in ribosome profiling data. Nat Methods 13:165–170
23. Zhang P, He D, Xu Y, Hou J, Pan B-F, Wang Y et al (2017) Genome-wide identification and differential analysis of translational initiation. Nat Commun 8:1749
24. Xiao Z, Huang R, Xing X, Chen Y, Deng H, Yang X (2018) De novo annotation and characterization of the translatome with ribosome profiling data. Nucleic Acids Res 46:e61
25. Choudhary S, Li W, Smith DA (2020) Accurate detection of short and long active ORFs using Ribo-seq data. Bioinformatics 36:2053–2059
26. Harnett D, Meerdink E, Calviello L, Sydow D, Ohler U (2021) Genome-wide analysis of actively translated open reading frames using RiboTaper/ORFquant. Methods Mol Biol 2252:331–346
27. Wu H-YL, Kaufman ID, Hsu PY (2025) The *ggRibo* single-gene viewer reveals insights into translatome and other nucleotide-resolution omics data. Genome Research 35: 2130-2142
28. Martin M (2011) Cutadapt removes adapter sequences from high-throughput sequencing reads. EMBnet J 17:10–12
29. Langmead B, Salzberg SL (2012) Fast gapped-read alignment with bowtie 2. Nat Methods 9: 357–359
30. Dobin A, Davis CA, Schlesinger F, Drenkow J, Zaleski C, Jha S et al (2013) STAR: ultrafast universal RNA-seq aligner. Bioinformatics 29: 15–21
31. Calviello L, Sydow D, Harnett D, Ohler U (2019) Ribo-seQC: comprehensive analysis of cytoplasmic and organellar ribosome profiling data. bioRxiv:601468
32. Zhang Q, Shi Q, Shao M (2022) Accurate assembly of multi-end RNA-seq data with Scallop2. Nat Comput Sci 2:148–152
33. Pertea G, Pertea M (2020) GFF utilities: GffRead and GffCompare [Internet]. F1000Res. Available from: https://f1000research.com/articles/9-304
34. Lawrence M, Gentleman R, Carey V (2009) Rtracklayer: an R package for interfacing with genome browsers. Bioinformatics 25:1841–1842
35. Grillet L, Lan P, Li W, Mokkapati G, Schmidt W (2018) IRON MAN is a ubiquitous family of peptides that control iron transport in plants. Nat Plants 4:953–963
36. Cao M, Platre MP, Tsai H-H, Zhang L, Nobori T, Armengot L et al (2024) Spatial IMA1 regulation restricts root iron acquisition on MAMP perception. Nature 625:750–759
37. Wu HYL, Hsu PY (2022) A custom library construction method for super-resolution ribosome profiling in Arabidopsis. Plant Methods 18:1–8

Chapter 8

Translation Signature Scores: Data-Driven Approach to Assess Evidence for Active Translation

Devika Menon, Owen Rackham, and Sonia Chothani

Abstract

Microproteins encoded from small open reading frames (smORFs) in the human genome have long been hypothesized to play physiological and regulatory roles, but they have historically been excluded from genome annotations due to the challenges in their identification. The advent of ribosome profiling (Ribo-seq), a deep sequencing technology to capture genome-wide translation, has surfaced thousands of novel ORFs with the potential to encode previously unannotated microproteins. However, due to variability in data quality and sparseness in read coverage, distinguishing truly translated smORFs from background noise remains challenging. While there are many approaches to address these challenges, here we describe the translation signature approach. This approach utilizes large-scale pooled Ribo-seq data to enable visualization of translation in individual ORFs with unprecedented clarity as compared to individual samples. It then quantifies evidence of translation by using translation signature scores, which include three metrics, namely, P-sites in frame, uniformity, and drop-off. Lastly, annotated protein-coding ORFs are used as a reference to learn the expected range of the translation signature scores reflecting active translation and novel ORFs with such scores are prioritized. We summarize here the key data resources and methods, as well as essential considerations for conducting such an analysis. Additionally, we describe a web resource hosting the default set of smORFs generated using the described workflow. With an increasing volume of high-quality Ribo-seq datasets, the translation signature scores provide a robust framework to prioritize ORFs with strong evidence of translation.

Key words Translation, RNA, smORFs, Small open reading frames, Microproteins

1 Introduction

Proteins are building blocks of life with over 20,000 encoded in the human genome [1–3]. The expression of proteins determines the cell type, regulates cellular processes and is crucial in cell survival [4, 5]. Dysregulation of proteins results in cellular malfunctions and disease at the physiological level [5]. Understanding the full

Supplementary Information The online version contains supplementary material available at https://doi.org/10.1007/978-1-0716-5013-4_8.

Stephan Wenkel (ed.), *Microproteins: Methods and Protocols*, Methods in Molecular Biology, vol. 2992,
https://doi.org/10.1007/978-1-0716-5013-4_8,

spectrum of proteins encoded by the human genome is thus critical to advancing our knowledge of human health and disease, and for exploiting the full potential of therapeutic targets [6].

Protein synthesis occurs in two stages: Transcription and translation [7]. First, the DNA sequence is transcribed into a messenger RNA (mRNA) sequence and then typically an open reading frame (ORF) is translated into protein. The ORF is read three nucleotides (a codon) at a time, with each codon being translated into the corresponding amino acid, resulting in a chain of amino acids or polypeptide sequence [8]. This unique pattern of ribosome engagement with ORFs, periodic movement, and disengagement after translation is what we refer to as the translation signature. This can be quantified using the ribosome profiling (Ribo-seq) technique [9] and used to identify novel ORFs that show similar translation signatures to those of known proteins. Uncovering such previously missed translated ORFs can expand our understanding of the proteome and provide new targets for therapeutic intervention.

In this chapter, we focus on the following questions:

1. How do we quantify the translation signature for a given ORF?
2. How can we prioritize novel ORFs that show translation signatures similar to those of known proteins?

1.1 Mechanism of RNA Translation

Translation is the process of "translating" genetic information from an RNA sequence to a protein sequence. The RNA sequence is made of four bases: "Adenine" (A), "Guanine" (G), "Cytosine"(C) and "Uracil"(U) [10], whereas the protein sequences comprise 20 different amino acids [11]. In the 1960s, it was discovered that a set of three consecutive RNA bases corresponds to one amino acid [12]. Following the discovery of the first triplet-to-amino-acid relationship, AUG was identified as a start-codon that signals translation initiation and UAA, UAG, and UGA triplets were identified as stop-codons that signal translation termination [13]. The translation complex was understood to move a coding unit or codon (three nucleotides) at a time, translating each triplet of nucleotides across the length of the mRNA sequence [14]. This led to the definition of an open reading frame (ORF) as a stretch of triplet coding units between a start-codon and a stop-codon that is translated into a protein (Fig. 1a) [15].

In parallel, studies on the translation complex led to the discovery that a transfer RNA (tRNA) with a specific amino acid attached [16] interacts with the matching codon in the mRNA at the peptidyl-site (P-site) of the ribosome [17]. Translation initiates when a tRNA carrying methionine binds to the ribosome at the start codon (Fig. 1b). During elongation, a defining feature of translation is the 3-nucleotide periodic movement of the mRNA in relation to the ribosome [18], which persists throughout the

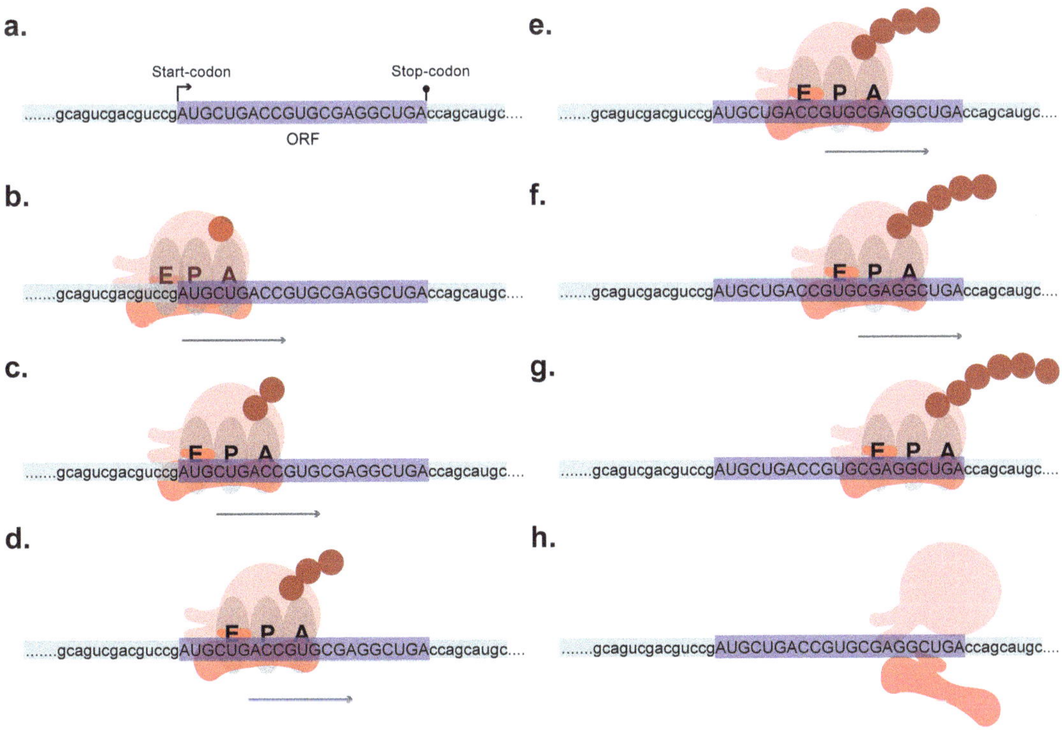

Fig. 1 Translation of an open reading frame (ORF). Schematic depicting process of translation. (**a**) Rectangular box showing ORF encoded within the exonic regions of a transcript. Gray box: Exon, Purple box: ORF. (**b**) Translation initiation shown with a ribosome (red) assembled at the start-codon (here AUG). Ribosome schematic shown with Exit-, Peptidyl- and Aminoacyl- sites (represented by E, P, and A) and polypeptide chain shown with filled red circles. (**c–f**) Translation Elongation shown across a given ORF, one codon at a time, with the growing polypeptide chain formation. (**g–h**) Translation termination shown as polypeptide chain is released, and translation machinery disassembled at the stop codon (here UGA)

length of the ORF, with the P-site hosting the growing polypeptide chain (Fig. 1c–g). Finally, recognition of the stop-codon by release factors leads to translation termination, releasing the peptide chain and disassembly of the ribosomes (Fig. 1h) [8].

1.2 Technological Advances in Studying RNA Translation

Early methods for studying translation focused on simplified experimental systems, wherein a synthetic RNA transcript was provided as the template, and in vitro translation of a single RNA could be measured in a controlled environment [19, 20]. In the 1960s, pulse-chase analysis was developed to study translation in living cells. Cells are first cultured with radioactively labeled amino acids for a brief time period (pulse) and then transferred to normal media (chase), enabling the quantification of newly synthesized proteins [21, 22]. Furthermore, in order to study the entire translatome, i.e., all actively translated transcripts in the cell, polysome profiling was developed [23]. This technique utilizes sucrose gradient ultracentrifugation to separate transcripts based on their ribosome associations. Transcripts bound to multiple ribosomes (polysomes) have a higher molecular weight and are distinguished from transcripts associated with monosomes or not associated with ribosomes. Thus, the technique can distinguish between highly translated and lowly translated transcripts, providing insights into the relative translation efficiency of different mRNAs. Toeprinting was developed to study translation initiation by utilizing reverse transcriptase to reverse transcribe mRNA into cDNA. This process would be stalled when it encountered a translation complex, thereby generating toeprints and allowing identification of ribosome positions on the mRNA [19]. A major breakthrough came in 2009, with the development of Ribosome profiling or Ribo-seq, which allows for the quantification of a cellular response to stimuli on the translational level by providing genome-wide information on protein synthesis in vivo [24]. It is a deep sequencing technique that maps genome-wide ribosome-protected mRNA fragments at single-nucleotide resolution [25]. Traditionally, RNA-sequencing (RNA-seq) has been used to capture RNA fragments from the entire transcripts, providing a genome-wide snapshot of transcript abundance. In contrast, Ribo-seq captures only RNA fragments protected by ribosomes, also known as RPFs (ribosome-protected fragments). Sequenced ribosome-protected mRNA fragments (RPFs) are expected to have a fixed length ranging between 28 and 30 bp. Furthermore, each RPF can be aligned to the transcriptome and used to computationally infer the location of the P-site during the snapshot, identifying the exact codon that is being translated (Fig. 2, see electronic Supplementary Video 1). Thus, Ribo-seq provides an additional layer of data with nucleotide resolution, highlighting regions with active translation across the transcriptome (Fig. 2a, b).

Since translation occurs one codon (or three nucleotides) at a time, in translated ORFs, a 3-nucleotide periodic signal is observed in mapped Ribo-seq reads [26]. Various approaches exploit the 3 nt-periodicity signal [27, 28] as well as length distributions [29] to identify novel regions of translation. Furthermore, in a true translation event, this periodic signal should be observed throughout the length of the ORF, i.e., reads from Ribo-seq will peak every three nucleotides of the translating frame [28, 30]. Lastly, since termination takes place at the stop codon, the abrupt absence of Ribo-seq reads can be observed after this position [31]. These distinct read-outs from Ribo-seq that correspond to the well-established biology of translation (Fig. 1b–h), form the foundation of the translation signature scores we describe in this chapter (details described in Subheadings 3.1, 3.2, 3.3 and 3.4).

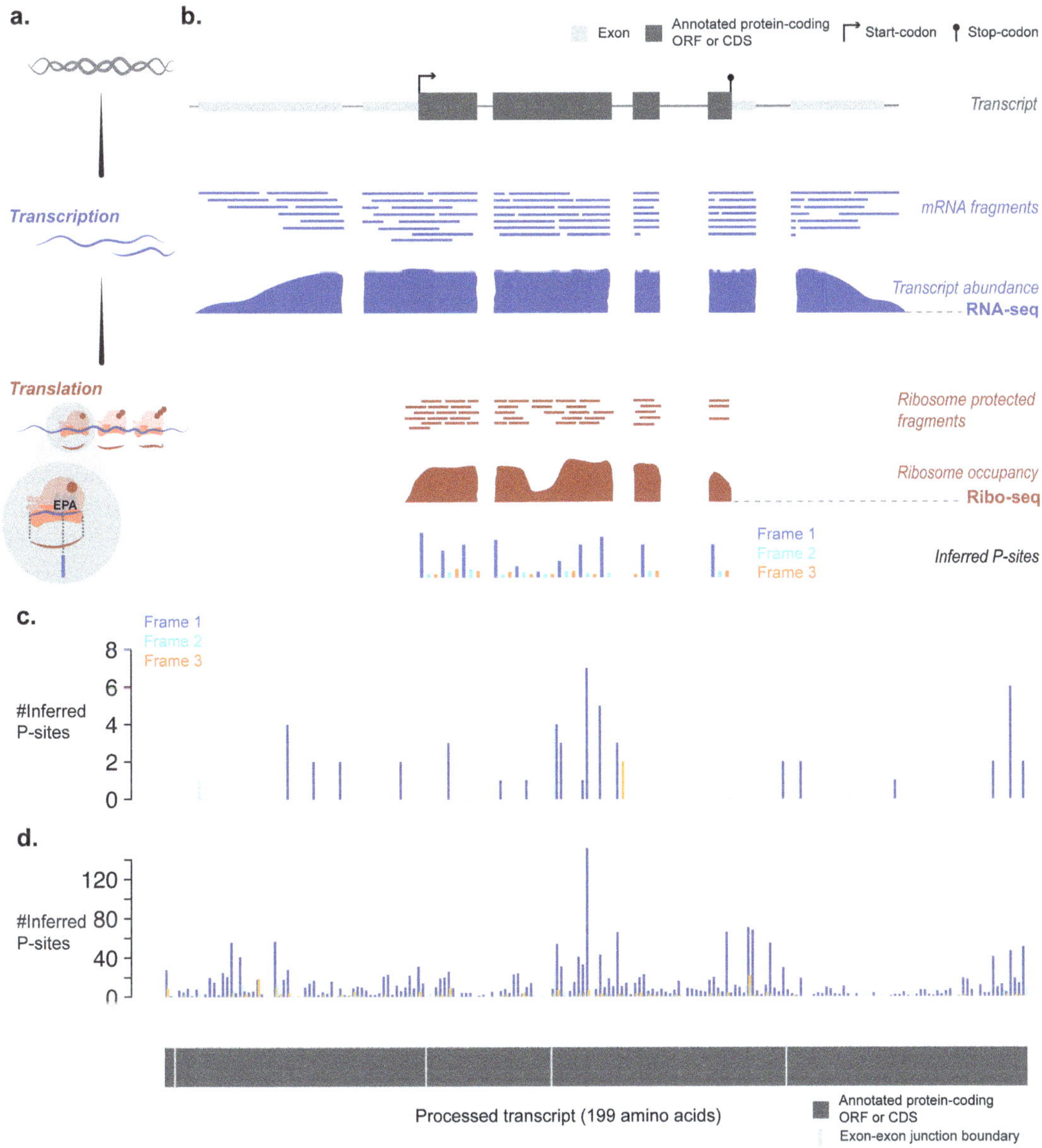

Fig. 2 RNA sequencing and ribosome profiling (Ribo-seq) techniques to measure transcription and translation. (**a**) Schematic showing simplified description of transcription and translation. Blue lines: mRNA transcript, Red lines: ribosome-protected mRNA fragments. Ribosomes shown with Exit-, Peptidyl- and Aminoacyl-sites marked as E, P, and A respectively. (**b**) An example transcript structure is shown with corresponding RNA-seq read coverage, Ribo-seq read coverage and inferred P-sites. RNA-seq captures the full-transcript expression, whereas Ribo-seq captures the expression of the coding sequences within the transcript. Light gray boxes: exons, Dark gray boxes: CDS, Blue lines: RNA-seq reads, blue area chart: Transcript abundance inferred from RNA-seq. Red lines: Sequenced reads covering ribosome-protected fragments (RPFs) obtained using Ribo-seq. Red area chart: Ribosome occupancy inferred from Ribo-seq. The bar chart (bottom) depicts inferred P-sites from Ribo-seq (y-axis) at every position (x-axis). Dark blue: P-sites in Frame 1, Light blue: P-sites in Frame 2 and Orange: Psites in Frame 3. Abbreviations: ORF, Open Reading Frame; CDS, coding sequence. (**c**, **d**) Bar plot of Inferred P-sites of the Interleukin-11 gene from single sample (**c**) and pooled human body-map (**d**) Dark blue: Frame 1, Cyan: Frame 2, Orange: Frame 3. Abbreviations: IL-11, Interleukin-11

1.3 Small Open Reading Frames: Missed Translation Events

In the 1990s, it was already established that there could be functional ORFs smaller than 100 amino acids [32], with several examples of microproteins characterized such as CAMP-dependent protein kinase inhibitor alpha or PKIA (76aa), an inhibitor of protein kinase A [33] and neurogranin (78aa), which regulates calmodulin's affinity for calcium [34]. Despite their importance, early genome annotations required an ORF length of at least 100 amino acids [35]. This was because not every combination of a start- and a stop-codon would translate into a functional protein, and given there are millions of ORFs in the human genome, there were experimental and computational challenges to distinguish a translated ORF from a random occurrence [29, 36]. With the development of ribosome profiling, thousands of previously unannotated ORFs that could be potentially translated were identified [25, 37]. Thus, leading to the identification of various functional microproteins such as ELABELA, an endogenous ligand of apelin receptor, which has been indicated in early cardiac development processes [38]. The microprotein Myomixer (MYMX) has been found to be crucial for skeletal muscle development [39], while Sarcolipin (Sln, 31aa) and Myoregulin (46aa) have been shown to regulate calcium uptake in muscle [40, 41]. Microproteins CYREN (Cell Cycle Regulator of Non-Homologous End Joining), APPLE and ASAP have been strongly associated with cancer [42–44]). MP31 is an upstream ORF of the tumor suppressor PTEN that inhibits the activity of lactate dehydrogenase [45]. Additionally, mitochondria-localized smORFs such as MOCCI [46] and BRAWNIN [47] play essential roles in the electron transport chain.

However, a major challenge remains in globally annotating and prioritizing functional smORFs from the tens of thousands of putative ORFs detected [27]. In this chapter, we will describe the data and input requirements and methodologies to quantify the translation signature scores and utilise them to prioritize genome-wide smORFs with strong evidence of translation.

2 Materials

The translation signature scores utilize a data-driven approach to prioritize ORFs with strong evidence of translation. Thus, the quantification of the scores requires various datasets and resources that are described here in the following subsections: Human translation body-map database (Subheading 2.1), ORF callers and databases (Subheading 2.2) and gene transcript structures (Subheading 2.3).

2.1 Human Translation Body-Map Database

The reliability of the identification of translated ORFs largely depends on the quality and depth of the underlying Ribo-seq data. Given the inherent sparseness and variability in individual Ribo-seq datasets [27], coverage across the full length of an ORF is often limited when using single samples. To overcome this limitation and enable robust quantification of translation signature scores, we utilized our recently published, pooled and stringently curated large-scale human body-map of translation [48] generated using the Ribo-seq technique. This body-map spans six primary cell types and five human tissues, comprising a total of 187 Ribo-seq samples with each sample exhibiting at least 60% periodicity and each cell-type data totalling to a minimum of 50 million inferred P-site reads (Table 1). On pooling, we obtained a resource of 1.3 billion inferred P-site reads with an overall high 3-nt periodicity of 85%. This pooled data is used for the quantification of translation signature scores.

2.1.1 Pooled P-Site Coverage File

A pooled P-site coverage file in bedGraph format is used to quantify the P-site coverage within a given ORF for the translation signature score calculation. The pooling approach maximises coverage across the full length of ORFs, allowing for the accurate identification of

Table 1
Summary of the human body-map translation database. Total P-sites for each cell-type or tissue within the human body-map translation dataset are shown with corresponding quality metric i.e., 3 nt-periodicity and number of samples

Tissue/Cell-type	Number of samples	Total no. of P-sites	Overall periodicity
Brain	8	87012832	84%
Embryonic stem cell	6	62871344	90%
Fat	6	80443194	86%
Fibroblast	52	317762905	82%
Heart	80	416978194	88%
Hepatocytes	5	145405984	85%
Human aortic endothelial cells (HA_EC)	6	19964429	75%
Human coronary artery endothelial cells (HCAEC)	5	94190525	81%
Human umbilical vein endothelial cells (HUVEC)	3	12363634	75%
Kidney	1	4494474	71%
Skeletal muscle	1	328464	85%
Vascular smooth muscle cells (VSMC)	11	59758852	84%

translated regions with robust ribosome occupancy. The bedGraph file can also be used for visualization of translation profiles across any ORF of interest.

For example, in the case of a known protein-coding gene such as *Interleukin-11* (Fig. 2c, d), translation signatures were not clearly visible in individual samples but became apparent upon pooling, illustrating the power of aggregated high-quality Ribo-seq data. Additional datasets can be incorporated into this resource to achieve even higher coverage across the length of ORFs, further enhancing its utility for identifying novel translated ORFs with high confidence (*see* **Note 1**).

2.2 Computational Tools and Databases to Obtain Translated ORFs

In the last decade, several computational algorithms utilizing Ribo-seq data have been developed [49] with the aim of identifying actively translated ORFs. Various statistical and signal-processing techniques have been applied to distinguish active translation from random occurrences of ORF sequences. In our published catalog, three such computational tools were utilized for identifying small open reading frames from Ribo-seq data.

1. PRICE (probabilistic inference of codon activities by an EM algorithm) models experimental noise to distinguish real smORFs from noise and infers actively translating codons using a binomial distribution [50].
2. Ribo-TISH uses a frame test to compare RPF counts of a frame against alternative reading frames to predict translating open reading frames [51].
3. RiboTaper identifies ORFs using the presence of a 3-nt periodic signal [52]. In our published resource, these tools were independently applied to the pooled Ribo-seq data from the health atlas to identify translated ORFs.
4. In addition to the ORFs called using these methods, a published database of small open reading frames (sorfs.org) was incorporated [53].

Finally, ORFs identified using different methods and databases were compiled into a file (Table 2) containing ORF genomic coordinates, which are then used as input for translation signature score calculation. Here we describe three tools and a database used in our published catalog, but other tools and databases can also be used as described below (*see* **Notes 2–4** for guidelines on converting tool output into the required format for translation signature score calculation and Subheading 3.1 for more details on how to generate the required input format file).

Table 2
Exemplar ORF exon coordinate file. Each line represents a given exonic part of the ORF. ORFs spanning across multiple exons will have multiple lines within this file. Each line includes information on chromosome number, feature, start and stop positions, strand, gene, and ORF ID

Chr	Feature	Start	Stop	.	str	.	Gene_id	ORF_id
2	exon	46660802	46660942	.	–	.	GENE_001	ORF_001
2	exon	46647204	46647269	.	–	.	GENE_001	ORF_002
8	exon	78723985	78724149	.	+	.	GENE_002	ORF_003
8	exon	78723861	78723905	.	+	.	GENE_002	ORF_003
8	exon	78805288	78805347	.	+	.	GENE_003	ORF_004

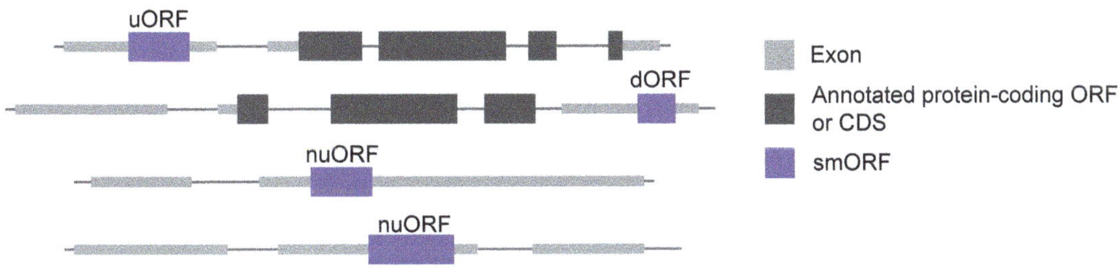

Fig. 3 ORF types and their host genes. Schematic showing positions of various types of smORFs relative to the regions within the host gene and transcript. Light gray box: Exon, Dark gray box: CDS, Purple box: ORF. Abbreviations: uORF upstream open reading frame, dORF downstream open reading frame, nuORF novel unannotated open reading frames i.e., ORFs found in previously annotated non-coding transcripts, ORF open reading frame, CDS coding sequence

2.3 Annotated Transcript Structures

Identifying the gene and transcript that encodes the ORF is required for the quantification of translation signature score, specifically the drop-off score (as detailed in Subheading 3.4).

2.3.1 Gene Transfer Format (GTF)

Gene transfer format (GTF) file of the human genome assembly hg38 obtained from any of the standardized databases, such as Ensembl [54], GENCODE [55], Refseq [3], provides exact genomic coordinates of exons within a transcript and can be used for identifying host gene and relative position of novel ORFs. *See* **Note 5**.

In our published catalog, we utilize the hg38 human genome GTF obtained from Ensembl v86, but either of the above databases or different versions can be used. Furthermore, this GTF, along with ORF coordinates obtained from Subheading 2.2, can also be used to identify the host gene and ORF biotype. For instance, ORFs found in the 5′ untranslated region (5′ UTR) of a gene are known as upstream ORFs (uORFs), and ORFs found in 3′ UTRs are known as downstream ORFs (dORFs) (Fig. 3).

3 Methods

Assessing ORFs for evidence of translation using the translation signature scores (Fig. 4) involves three key steps. First, the coordinates of potential translated ORFs are obtained using results from various tools and databases. Second, translation signature scores are quantified for each ORF using the pooled human body-map Ribo-seq data. Finally, translation signature scores are calculated for annotated protein-coding ORFs to establish thresholds for identifying novel ORFs with equivalent evidence. The following section details this process: processing of ORF coordinate file (Subheading 3.1), calculation of translation signature scores using three metrics (P-sites in frame (Subheading 3.2), uniformity (Subheading 3.3), drop-off (Subheading 3.4)), threshold determination of translation signature scores from annotated protein-coding ORFs and visualization of inferred P-sites in ORFs (Subheading 3.5). Lastly, we describe a web resource that shares the reference set of smORFs identified using the translation signature workflow on the human body-map data (Subheading 3.6).

3.1 Streamlining of ORF Coordinates for Translation Signature Quantification

To quantify translation signature scores for a given ORF, its genomic coordinate information is essential. While typically, this is a standard output of ORF-calling tools and databases, they are often provided in various formats, and splicing information is sometimes missing (*see* to **Notes 3** and **4** for variations in

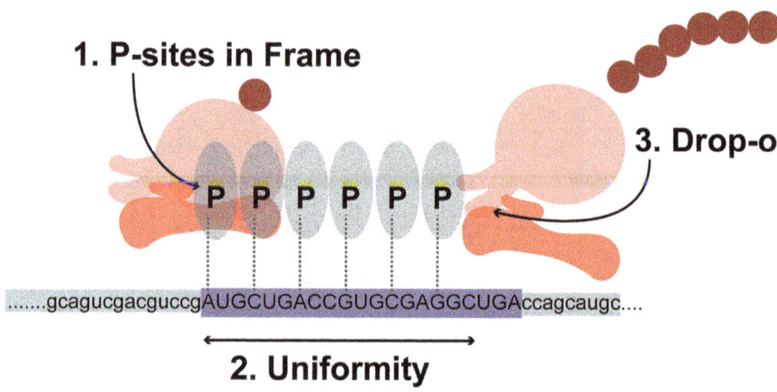

Fig. 4 Translation signature scores. Translation biology-derived computational metrics namely, P-sites in Frame (PIF), uniformity and drop-off (pink: ribosome; red: polypeptide chain; grey box: Exon; Purple box: ORF; Gray oval: P-site position). (1) PIF: Proportion of P-sites detected in translating frame over P-sites detected in all three frames. (2) Uniformity: Ratio of codons with >33.33% P-sites in-frame with respect to total number of codons within the ORF. This is indicative of uniform coverage through the length of the ORF. (3) Drop-off: Ratio of P-sites in frame (within a given flanking region) before the stop codon to all P-sites in frame before and after stop codon (within flank region). This is a measure of ribosome disengagement at the stop codon

processing). Additionally, multiple ORF callers and databases may identify identical or overlapping ORFs, leading to redundancy when collecting ORFs from multiple sources. Therefore, standardizing the results is necessary for consolidation and downstream analysis. Although different tools generate varied output formats, the general steps involved in preparing a standardized coordinate file are as follows: (1) Removing ORFs that overlap completely with annotated CDSs in the same reading frame, (2) Removing ORFs that lack essential information, such as chromosome and genomic coordinates (with missing spliced coordinates for ORFs spanning splice junctions), (3) Generate a BED file with chromosome, genomic coordinates of each part of the ORF and a unique ORF id (Refer Table 2). This standardized BED file is then used to quantify translation signature scores for all the smORFs retained after the above-mentioned filtering steps.

3.2 Quantification of P-Sites in Frame (PIF)

Once the ORF coordinates file is prepared, translation signature scores can be calculated for each ORF in the file, starting with the P-sites in frame or PIF metric. The PIF metric quantifies the proportion of total P-sites detected in the translating frame, effectively serving as an overall 3 nt-periodicity test for the ORF and thus an indicator of its translation. An actively translating smORF will exhibit a significantly high fraction of its total P-sites in Frame 1, suggesting that the observed reads likely represent true translation activity.

The PIF is quantified as follows:

$$\mathrm{PIF} = \frac{\mathrm{P-sites\ in\ Frame\ 1}}{\mathrm{P-sites\ in\ Frame\ 1} + \mathrm{P-sites\ in\ Frame\ 2} + \mathrm{P-sites\ in\ Frame\ 3}}$$

The detailed steps to quantify PIF are described here:

1. For each ORF, obtain the coordinates from the bed file generated in Subheading 3.1. This includes the chromosome, strand, and start and stop positions for each ORF. If the ORF spans multiple exons, sort the exons by their start positions.
2. For each ORF, extract the start and stop coordinates and make a continuous vector ranging from the start to stop of the ORF with each element as a given nucleotide position. If the ORF spans multiple exons, generate a vector for each ORF segment by extracting its start and stop coordinates, and then concatenate these vectors to form a continuous vector of positions in the full length of the ORF. Refer to **Note 6** for more details on preparing the coordinate vector for negative-strand genes.
3. For each position in the vector, the corresponding number of inferred P-sites is quantified using the P-site coverage bedGraph file from the pooled human body-map data described in Subheading 2.1.1.

4. Using the coverage vector obtained from Subheading 3.2, **step 3**, the PIF metric is calculated as follows. First, the inferred P-sites in the translating frame or Frame 1 are quantified by summing the coverage values at the first position (n) and every n + 3 position until the end of the vector. Second, the total P-sites are calculated by summing all values in the coverage vector. Finally, PIF is calculated as a ratio of the P-sites in Frame 1 (or the translating frame) to the total number of P-sites.

3.3 Quantification of Uniformity

The second metric is called uniformity, which is a measure of the presence of 3 nt-periodicity throughout the length of the ORF. This metric distinguishes ORFs with inferred P-sites concentrated at only a few positions from ORFs that have evidence of P-sites distributed across the majority of the ORF length, highlighting those with a higher likelihood of active translation or stronger evidence. Uniformity is calculated by quantifying the codons that have >33% periodicity in the translating frame, divided by the total number of codons.

$$\text{Uniformity} = \frac{\text{No.of codons with} > 33.33\%\ \text{P} - \text{sites in Frame 1}}{\text{Total no.of codons in ORF}}$$

The detailed steps to quantify uniformity are described here:

1. The steps described in Subheading 3.2, **step 1–3** derive a coverage vector for the given ORF that is used for the uniformity calculation as described in the next steps.
2. For each of the three positions (or codons) from the start of the coverage vector: The ratio of P-sites in the first position with respect to total P-sites within the codon is quantified. Repeat this calculation for all codons in the ORF to determine the 3 nt-periodicity % for each codon in the ORF.
3. Count the number of codons where the 3 nt-periodicity value exceeds 33%. This threshold is based on the principle that since there are three frames possible, if the value is 33% or less, the reads are either random or not indicative of active translation in the selected reading frame.
4. Finally, the uniformity score is calculated as the ratio of the number of codons that pass the threshold in Subheading 3.3 **step 3** to the total number of codons in the ORF.

3.4 Quantification of Drop-Off

The third metric, drop-off, measures ribosome disengagement after the stop-codon. It evaluates the evidence of ribosome drop-off at the stop codon of an identified ORF by comparing the P-sites in the translating frame before and after the identified stop codon. To calculate the drop-off, a limited flanking region is considered on both sides of the stop codon so as to avoid interference by signals from other potential downstream ORFs. Specifically, the drop-off is calculated by dividing the total number of P-sites in Frame 1 before the stop codon within the flanking region by the total number of reads in Frame 1 within the entire flanking region. A high drop-off score of 1 indicates that all P-sites are located before the stop codon, signifying a clear ribosome disengagement at the stop codon. Conversely, a lower drop-off, such as 0.5, suggests an equal number of P-sites in Frame 1 before and after the stop-codon, indicating either noise or the stop-codon is not being actively used.

$$\text{Drop} - \text{off} = \frac{\text{P} - \text{sites in Frame 1 within flanking region before stop} - \text{codon}}{\text{P} - \text{sites in Frame 1 within flanking region before and after stop} - \text{codon}}$$

The detailed steps to quantify the drop-off score are described here:

1. Identify the host transcript of the given ORF and obtain a processed transcript vector. To obtain the processed transcript vector, the exon coordinates for the transcript are obtained using the Ensembl hg38 human GTF file (described in Subheading 2.3). The exons are sorted based on the start coordinate of the exon. A continuous vector of coordinates is created from the start to the stop of a given exon. This is obtained for each exon, following which all the exon vectors are concatenated to obtain a processed transcript coordinate vector. Refer to **Note 6** for more details on this step, specifically for a negative-strand gene.
2. The location of the stop-codon of the smORF is identified within the processed transcript vector obtained in Subheading 3.4, **step 1**.
3. Determining Flank Regions Around the Stop Codon: As the drop-off score calculates ribosome occupancy before and after the stop codon, the regions flanking the left and right of the stop codon need to be identified. The left flank or a 5′ extension from the stop codon is within the coding sequence, and the right flank or a 3′ extension from the stop codon is beyond the coding sequence. Within the processed transcript vector, the start of the left flank is determined by subtracting 18 from the location of the stop codon obtained in Subheading 3.4, **step 2**. Similarly, the end of the right flank is calculated by adding 15 to the location of the stop-codon obtained in Subheading 3.4, **step 2**. This results in a region with five codons before and five codons after the stop codon of the smORF. Refer to **Notes 6–8** for other parameters and special considerations for this step.
4. Extracting Coverage Values for the Flanking Regions: Once the left and right flank regions are defined for calculation, the corresponding number of P-sites at these positions is obtained using the human body-map P-site coverage bedGraph file outlined in Subheading 2.1.1.

5. Obtaining a Coverage Vector for the Translating Frame in the Flanking Regions: Starting from the first position, every third position in the flanking-region coverage vector obtained in Subheading 3.4, **step 4** is identified as position in the translating frame, and corresponding coverages can be extracted from the human body-map alignment or bedGraph file. As the drop-off score tests whether there are reads in-frame before and after the stop codon, we only utilize read coverage in the translating frame.
6. Calculating Drop-Off Score: First, P-sites in the left-flank are determined by summing the coverage values in the positions 1–5. Second, the total P-sites in the vector obtained in Subheading 3.4, **step 5** are summed. Finally, the drop-off score is calculated as the ratio of the P-sites in the left-flank divided by the total P-sites in this flanking region around the stop codon.

3.5 Identifying ORFs with Translation Signatures Equivalent to Annotated Protein-Coding ORFs

This section demonstrates how the scores can be utilized to identify ORFs with strong evidence of translation. The PIF, uniformity, and drop-off score can be calculated for all novel identified ORFs as detailed in Subheadings 3.2, 3.3 and 3.4 (Fig. 5). In order to generate a reference set and determine a threshold for translation signature scores, the translation signature scores are also quantified for annotated protein-coding ORFs using the same method as described in Subheadings 3.2, 3.3 and 3.4 (*see* **Note 5**). A normal distribution was fitted to the PIF and uniformity score distributions for annotated protein-coding, and the threshold was determined by subtracting two standard deviations from the mean, thus encompassing 95% of the genes. As the drop-off did not follow a normal distribution, the mean value was set as the threshold for drop-off. Thus, indicating ORFs that pass these thresholds of scores as having equivalent evidence of translation as the annotated protein-coding ORFs. We show here a few functionalized smORFs, their observed translation signature and respective translation signature scores (Fig. 6). P-sites from annotated CDS, UTRs, and introns for a given gene can also be visualized using tools such as RiboPlotR—an R package which utilizes both the P-sites and the GTF file to visualize [56] or Trips-Viz—a transcriptome browser which displays triplet periodicity and various start- and stop-in all three frames in the given region [57] (*see* **Note 9**).

3.6 A Web Resource for smORFs with Clear Translation Signatures

We provide a set of ORFs identified using such a workflow at http://smorfs.ddnetbio.com as previously described in Chothani et al. [48]. Here, we used three ORF callers (namely PRICE [48, 50], Ribotaper [52], and RiboTISH [51] and a database (sorfs.org) [53] to generate a catalog of potential ORFs using the human body-map translation data (Subheading 2.1). Furthermore, we quantified translation signature scores for all these ORFs as well

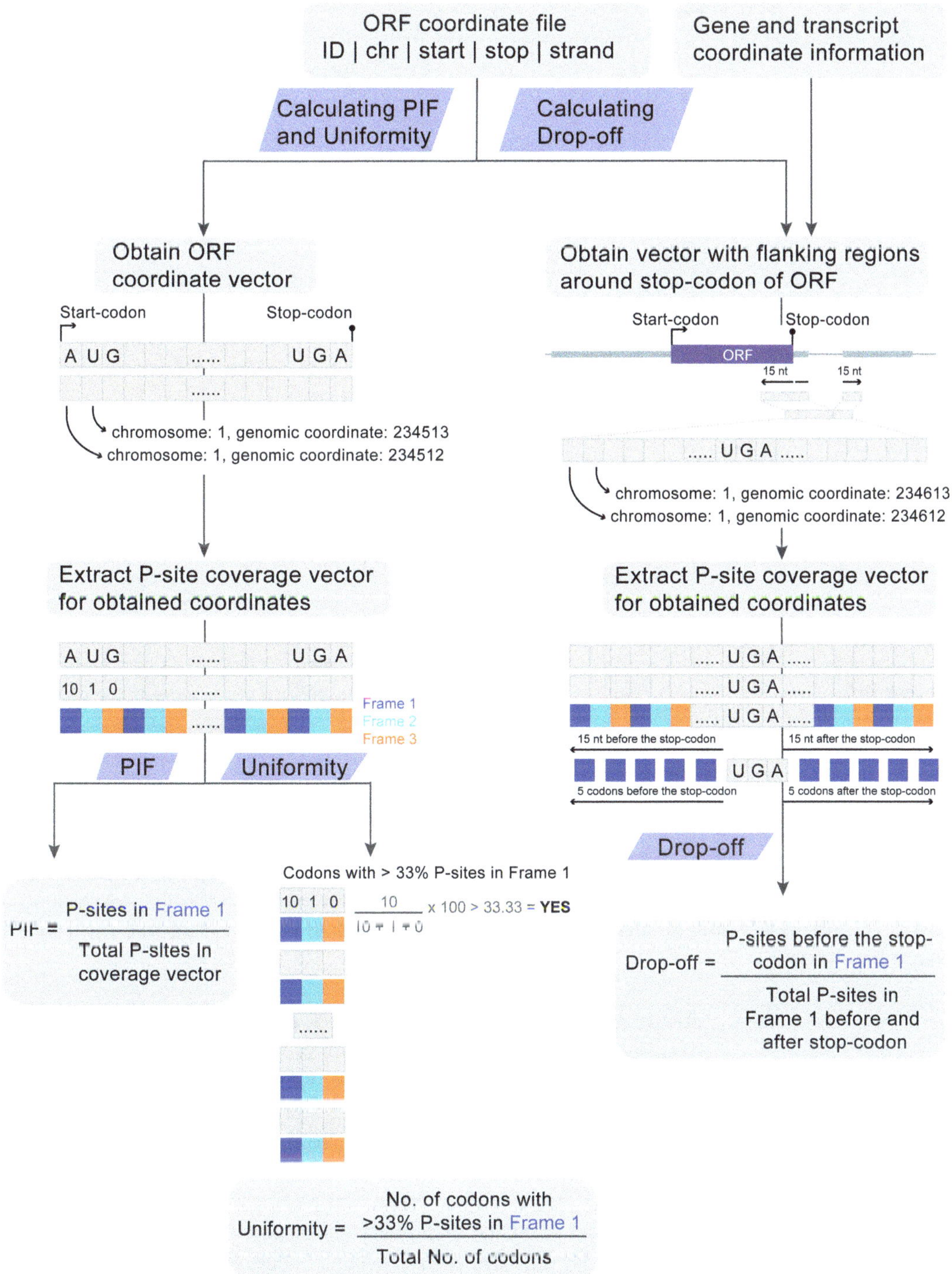

Fig. 5 Translation signature score quantification workflow. The translation signature scores require quantification of three metrics: P-sites in Frame (PIF), uniformity, and drop-off. The quantification of these scores requires ORF coordinate information as well as host transcript coordinate information. For PIF and Uniformity calculation, coordinates from start to stop of the ORF, including splice junction information, are required and used to derive an ORF position vector. For drop-off score, a position vector with flanking regions around the

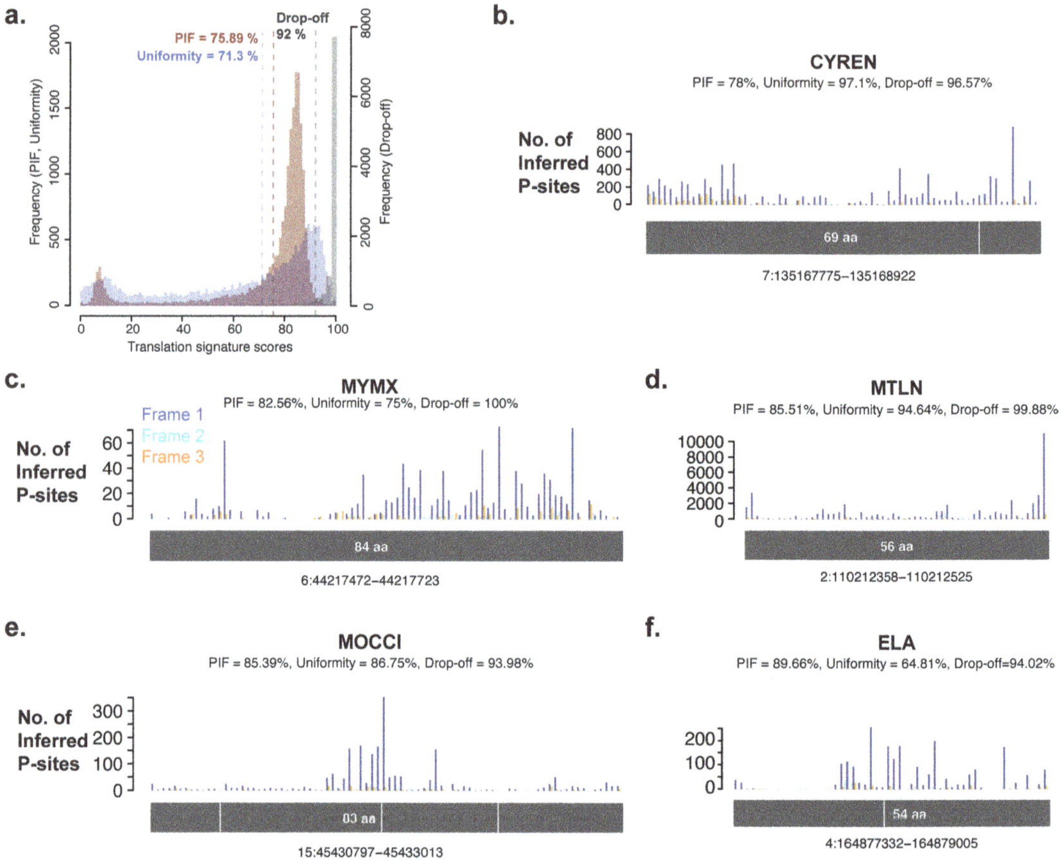

Fig. 6 Prioritization of novel ORFs with translation signatures similar to known proteins. (**a**) Histogram showing distribution of P-sites in Frame, uniformity, and drop-off scores for annotated protein-coding ORFs. (**b–g**) Barplots showing inferred P-sites observed (y-axis) at every nucleotide position (x-axis) within the ORFs: Cell cycle regulator of non-homologous end joining (CYREN) [42] (**b**), Myomixer (MYMX) [39] (**c**), Mitoregulin (MTLN) [71] (**d**), MOCCI [46] (**e**), and ELABELA (ELA) [38] (**f**). Blue bar: Frame 1; light blue bar: Frame 2; orange bar: Frame 3

as annotated protein-coding ORFs using the human body-map translation data. Using translation signature scores for annotated protein-coding ORFs, thresholds were determined as >75.89% PIF, >71.3% uniformity, and > 92% drop-off as described in Sect. 3.5. Using these determined thresholds, we identified 7,767 ORFs

Fig. 5 (continued) stop codon is derived using the stop-codon coordinates and host transcript exonic coordinate information. P-site coverage in respective position vectors is obtained using the human body-map data to obtain coverage vectors, followed by PIF, uniformity, and drop-off calculation. Dark blue: Frame 1, light blue: Frame 2, orange: Frame 3. Dark purple box: smORFs, light gray box: Exon. Abbreviations: chr chromosome, PIF P-sites in frame. Note: For illustration purposes, the figure only shows AUG as the start and UGA as the stop codon, but the scores can be quantified for any other alternative initiation codons and stop codons that have similar translation signatures as annotated protein-coding ORFs. The ORFs and their metadata are shared on the web application in three parts: (1) Annotation, (2) Expression, (3) Is it a smORF?. In the Annotation tab, the ORF coordinates, its host gene, ORF type (uORF, dORF, etc), ORF sequence, evolutionary conservation and other sequence features such as signal peptide are included. In the expression tab, the ORF's expression values based on Ribo-seq and RNA-seq are shown across various cell types and tissues within the human body-map translation database. Lastly, the web application also includes a feature "Is it a smORF?" wherein a query list with gene IDs can be submitted and whether any of those genes contain a smORF is displayed as output (Fig. 7).

4 Notes

1. We use the human body-map dataset to quantify translation signature scores. This can be carried out using any other dataset. There are several other large-scale Ribo-seq databases such as RiboSeq.Org [58], RPFdb [59] and TranslatomeDB [60]. It is important to incorporate high-quality data with high 3 nt-periodicity, as low-quality data will reduce the signal-to-noise ratio for this approach.
2. We utilized three independent ORF calling tools here, namely, PRICE [50], RiboTISH [51] and RiboTaper [52] to identify a list of potential ORFs. There are many other computational tools [61] that can be used to identify novel ORFs, and subsequently, translation signature scores can be used to test them. Some examples of such tools include RibORFv1.0 [62], Ribo-Code [63], ORFquant [64], RiboTIE [65] and iRibo [66]. Similarly, apart from the sorfs.org database that was used in the above workflow, there are several other smORF databases such as SmProt [67], MetaMORF [68], uORFdb [69], and OpenProt [70], which could also be incorporated into this workflow.
3. The output files from different smORF callers may have to be processed differently to generate the smORF information file, as described in Subheading 3.1. We outline here processing steps for the three tools that are used in the above workflow: (a) RiboTaper: The ORF coordinates can be found in the output file named "translated_ORFs_filtered_sorted.bed", and the corresponding gene annotations, such as gene ID and gene biotype, are stored in "ORFs_max_filt". The two files are combined based on ORF id, and ORFs that fall into the "ORFs_ccds" category are removed. (b) PRICE: Output from PRICE (*.orfs.tsv) was filtered to keep novel ORFs only by excluding lines that contain "CDS" or "Trunc" or "iORF".

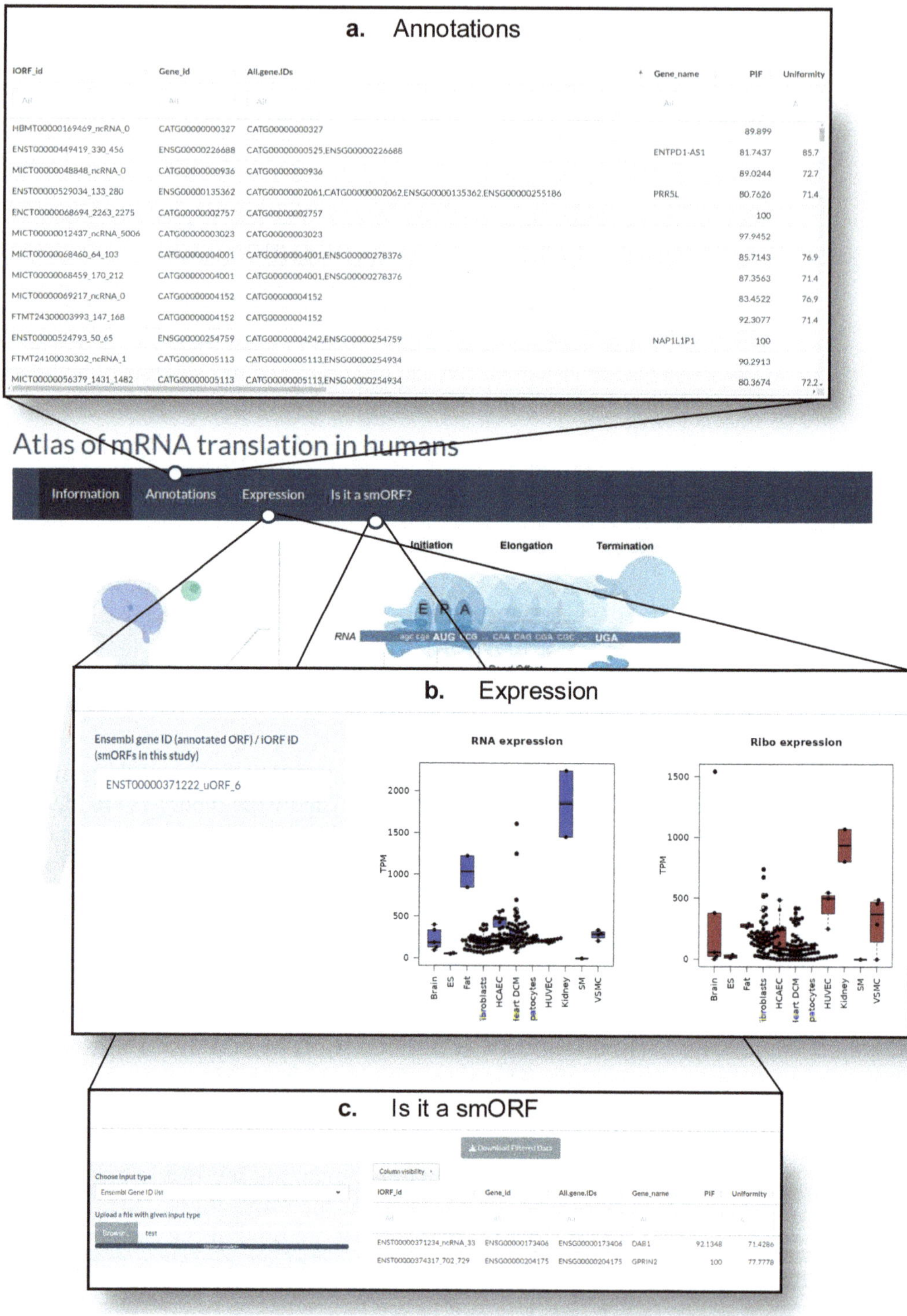

Fig. 7 Website for ORFs identified using the translation signature workflow. Screenshot of the landing page of the website showing human body-map and illustration of Ribo-seq. (**a**) Annotation tab showing a table including information for 7767 ORFs. This includes for each identified ORF their ORF ID, host gene information, translation signature scores, peptide sequence, evolutionary conservation, and other sequence features. (**b**) Expression tab showing boxplots with RNA-seq and Ribo-seq expression for each ORF across tissue and cell types in the human body-map data. (**c**) Lastly, "Is it a smORF?" tab offers a tool for users to provide an input list of Gene IDs that can be tested for the presence of ORFs. This results in a table showing ORFs within the host genes provided. Link: https:/smorfs.ddnetbio.com

Following this, p-values from the results were adjusted by BH correction such that only novel ORFs with a significant p-adj < 0.05 were selected. (c) RiboTISH: Output file is filtered to retain ORFs with "Novel|5'UTR|3'UTR" annotations and remove lines that contain "CDSFrameOverlap" and "Novel: Known". Coordinates from these files are then used to derive an exon coordinate file, such as in Table 2.

4. Different ORF callers may find the same ORFs, thus, for any downstream analysis, it would be important to collapse such redundancies in ORF lists. Furthermore, various ORF callers may identify slightly different initiation or termination, which should also be collapsed as isoforms of one ORF to avoid redundancies.
5. In order to apply translation signature scores to known ORFs we used the Ensembl hg38 genome build, but other gene annotations such as GENCODE [55], RefSeq [3] and UniprotKB [2] may be utilized as well. The distribution of translation signature scores applied to protein-coding genes from the preferred gene annotation database may be used to set an updated threshold to filter smORFs.
6. We have described translation signature score quantification for a positive-strand gene. For negative-strand genes, in Subheading 3.2, **step 1** (for PIF and uniformity), the coverage vector needs to be reversed before the calculation of PIF and uniformity scores. As well as in Subheading 3.4, **step 1** (i.e., Drop-off calculation), for negative-strand genes, the processed transcript is to be reversed before proceeding to Subheading 3.4, **step 2** and **3**, thus the stop codon location is identified and flanking regions are obtained within this reversed processed transcript vector.
7. As certain ORFs may span across multiple exons, the Subheading 3.4, **step 1** is very important, and the processed transcript vector must be created before proceeding to the identification of flanking regions. This is because in such cases, it is possible that the 15 base pair extension may not fall within one exon. Hence, a vector of all positions across exons must be created as described in Subheading 3.4, **step 1**, thus avoiding coordinates and regions between exons for Drop-off calculation.
8. If the flank region is not sufficient i.e., stop-codon is less than 15 bp away from the end of the transcript, the drop-off is not calculated. As well as, when there are no reads present throughout the flanking-region coverage vector, the drop-off score was not used for the evaluation of translation evidence of the given ORF.
9. To generate the P-site visualization shown in Fig. 6b for a given smORF, the start and end positions of the smORF must be

extracted from the smORF coordinate file. For each coordinate or position of the smORF, inferred P-sites are identified using the human body-map P-site coverage file. This file is strand-specific, and care needs to be taken to match it with the gene-strand. The extracted P-site coverages are then plotted as a barplot with three unique colors provided to color the barplot resulting in one color for each frame. For negative-strand ORFs, the ORF's P-site coverage vector is reversed.

Conflict of Interest

None.

Funding

SC is funded by Singapore Ministry of Health's National Medical Research Council under OF-YIRG (OFYIRG23jan-0016, MOH-001340).

References

1. Mishra GR, Suresh M, Kumaran K et al (2006) Human protein reference database–2006 update. Nucleic Acids Res 34:D411–D414
2. UniProt Consortium (2025) UniProt: the universal protein knowledgebase in 2025. Nucleic Acids Res 53:D609–D617
3. O'Leary NA, Wright MW, Brister JR et al (2016) Reference sequence (RefSeq) database at NCBI: current status, taxonomic expansion, and functional annotation. Nucleic Acids Res 44:D733–D745
4. Pontén F, Gry M, Fagerberg L et al (2009) A global view of protein expression in human cells, tissues, and organs. Mol Syst Biol 5:337
5. Jia X, He X, Huang C et al (2024) Protein translation: biological processes and therapeutic strategies for human diseases. Signal Transduct Target Ther 9:44
6. Smith LM, Agar JN, Chamot-Rooke J et al (2021) The human proteoform project: defining the human proteome. Sci Adv 7:eabk0734
7. Crick F (1970) Central dogma of molecular biology. Nature 227:561–563
8. Ramakrishnan V (2002) Ribosome structure and the mechanism of translation. Cell 108: 557–572
9. Ingolia NT, Ghaemmaghami S, Newman JRS et al (2009) Genome-wide analysis in vivo of translation with nucleotide resolution using ribosome profiling. Science 324:218–223
10. Alberts B, Johnson A, Lewis J et al (2002) From DNA to RNA. Garland Science
11. Lopez MJ, Mohiuddin SS (2024) Biochemistry, essential amino acids. In: StatPearls [Internet]. StatPearls Publishing
12. Matthaei JH, Jones OW, Martin RG et al (1962) Characteristics and composition of RNA coding units. Proc Natl Acad Sci USA 48:666–677
13. Stahl FW (1995) The amber mutants of phage T4. Genetics 141:439–442
14. Crick FH, Barnett L, Brenner S et al (1961) General nature of the genetic code for proteins. Nature 192:1227–1232
15. Sieber P, Platzer M, Schuster S (2018) The definition of open reading frame revisited. Trends Genet 34:167–170
16. Holley RW, Apgar J, Everett GA et al (1965) Structure of a ribonucleic acid. Science 147: 1462–1465
17. Monro RE, Cerná J, Marcker KA (1968) Ribosome-catalyzed peptidyl transfer: substrate specificity at the P-site. Proc Natl Acad Sci 61:1042–1049
18. Cooper GM (2000) Translation of mRNA. In: The cell: a molecular approach, 2nd edn. Sinauer Associates

19. Nilsen TW (2013) Toeprinting. Cold Spring Harb Protoc 2013:pdb.prot077180
20. Nakano H, Yamane T (1998) Cell-free protein synthesis systems. Biotechnol Adv 16:367–384
21. Jamieson JD, Palade GE (1967) Intracellular transport of secretory proteins in the pancreatic exocrine cell. I. Role of the peripheral elements of the Golgi complex. J Cell Biol 34:577–596
22. Jamieson JD, Palade GE (1967) Intracellular transport of secretory proteins in the pancreatic exocrine cell. II. Transport to condensing vacuoles and zymogen granules. J Cell Biol 34:597–615
23. Chassé H, Boulben S, Costache V et al (2016) Analysis of translation using polysome profiling. Nucleic Acids Res 45:e15
24. Michel AM, Baranov PV (2013) Ribosome profiling: a Hi-Def monitor for protein synthesis at the genome-wide scale. Wiley Interdiscip Rev RNA 4:473
25. Brar GA, Weissman JS (2015) Ribosome profiling reveals the what, when, where, and how of protein synthesis. Nat Rev Mol Cell Biol 16:651
26. Michel AM, Choudhury KR, Firth AE et al (2012) Observation of dually decoded regions of the human genome using ribosome profiling data. Genome Res 22:2219–2229
27. Chothani S, Ho L, Schafer S et al (2023) Discovering microproteins: making the most of ribosome profiling data. RNA Biol 20:943–954
28. Bazzini AA, Johnstone TG, Christiano R et al (2014) Identification of small ORFs in vertebrates using ribosome footprinting and evolutionary conservation. EMBO J 33:981–993
29. Ingolia NT, Brar GA, Stern-Ginossar N et al (2014) Ribosome profiling reveals pervasive translation outside of annotated protein-coding genes. Cell Rep 8:1365–1379
30. Fedorova AD, Kiniry SJ, Andreev DE et al (2022) Thousands of human non-AUG extended proteoforms lack evidence of evolutionary selection among mammals. Nat Commun 13:7910
31. Guttman M, Russell P, Ingolia NT et al (2013) Ribosome profiling provides evidence that large noncoding RNAs do not encode proteins. Cell 154:240–251
32. Basrai MA, Hieter P, Boeke JD (1997) Small open reading frames: beautiful needles in the haystack. Genome Res 7:768–771
33. Olsen SR, Uhler MD (1991) Inhibition of protein kinase-A by overexpression of the cloned human protein kinase inhibitor. Mol Endocrinol 5:1246–1256
34. Watson JB, Battenberg EF, Wong KK et al (1990) Subtractive cDNA cloning of RC3, a rodent cortex-enriched mRNA encoding a novel 78 residue protein. J Neurosci Res 26: 397–408
35. Mohsen JJ, Martel AA, Slavoff SA (2023) Microproteins-discovery, structure, and function. Proteomics 23:e2100211
36. Su M, Ling Y, Yu J et al (2013) Small proteins: untapped area of potential biological importance. Front Genet 4:286
37. Choudhary S, Li W, Smith AD (2019) Accurate detection of short and long active ORFs using Ribo-seq data. Bioinformatics 36:2053
38. Chng SC, Ho L, Tian J et al (2013) ELABELA: a hormone essential for heart development signals via the apelin receptor. Dev Cell 27:672–680
39. Bi P, Ramirez-Martinez A, Li H et al (2017) Control of muscle formation by the fusogenic micropeptide myomixer. Science 356:323–327
40. Anderson DM, Anderson KM, Chang C-L et al (2015) A micropeptide encoded by a putative long noncoding RNA regulates muscle performance. Cell 160:595–606
41. Magny EG, Pueyo JI, Pearl FMG et al (2013) Conserved regulation of cardiac calcium uptake by peptides encoded in small open reading frames. Science 341:1116–1120
42. Arnoult N, Correia A, Ma J et al (2017) Regulation of DNA repair pathway choice in S and G2 phases by the NHEJ inhibitor CYREN. Nature 549:548–552
43. Sun L, Wang W, Han C et al (2021) The oncomicropeptide APPLE promotes hematopoietic malignancy by enhancing translation initiation. Mol Cell 81:1193–1508.e9
44. Ge Q, Jia D, Cen D et al (2021) Micropeptide ASAP encoded by LINC00467 promotes colorectal cancer progression by directly modulating ATP synthase activity. J Clin Invest 131: e152911
45. Huang N, Li F, Zhang M et al (2021) An upstream open Reading frame in phosphatase and Tensin homolog encodes a circuit breaker of lactate metabolism. Cell Metab 33:128–144.e9
46. Lee CQE, Kerouanton B, Chothani S et al (2021) Coding and non-coding roles of MOCCI (C15ORF48) coordinate to regulate host inflammation and immunity. Nat Commun 12:2130
47. Zhang S, Reljić B, Liang C et al (2020) Mitochondrial peptide BRAWNIN is essential for vertebrate respiratory complex III assembly. Nat Commun 11:1312

48. Chothani SP, Adami E, Widjaja AA et al (2022) A high-resolution map of human RNA translation. Mol Cell 82:2885–2899.e8
49. Kiniry SJ, Michel AM, Baranov PV (2020) Computational methods for ribosome profiling data analysis. Wiley Interdiscip Rev RNA 11: e1577
50. Erhard F, Halenius A, Zimmermann C et al (2018) Improved Ribo-seq enables identification of cryptic translation events. Nat Methods 15:363–366
51. Zhang P, He D, Xu Y et al (2017) Genome-wide identification and differential analysis of translational initiation. Nat Commun 8:1749
52. Calviello L, Mukherjee N, Wyler E et al (2016) Detecting actively translated open reading frames in ribosome profiling data. Nat Methods 13:165–170
53. Olexiouk V, Van Criekinge W, Menschaert G (2018) An update on sORFs.org: a repository of small ORFs identified by ribosome profiling. Nucleic Acids Res 46:D497–D502
54. Dyer SC, Austine-Orimoloye O, Azov AG et al (2025) Ensembl 2025. Nucleic Acids Res 53: D948–D957
55. Mudge JM, Carbonell-Sala S, Diekhans M et al (2025) GENCODE 2025: reference gene annotation for human and mouse. Nucleic Acids Res 53:D966–D975
56. Wu H-YL, Hsu PY (2021) RiboPlotR: a visualization tool for periodic Ribo-seq reads. Plant Methods 17:124
57. Kiniry SJ, Judge CE, Michel AM et al (2021) Trips-Viz: an environment for the analysis of public and user-generated ribosome profiling data. Nucleic Acids Res 49:W662–W670
58. Tierney JAS, Świrski MI, Tjeldnes H et al (2025) RiboSeq.Org: an integrated suite of resources for ribosome profiling data analysis and visualization. Nucleic Acids Res 53:D268
59. Wang H, Yang L, Wang Y et al (2019) RPFdb v2.0: an updated database for genome-wide information of translated mRNA generated from ribosome profiling. Nucleic Acids Res 47:D230
60. Liu W, Xiang L, Zheng T et al (2018) TranslatomeDB: a comprehensive database and cloud-based analysis platform for translatome sequencing data. Nucleic Acids Res 46:D206
61. Tong G, Hah N, Martinez TF (2023) Comparison of software packages for detecting unannotated translated small open reading frames by Ribo-seq. bioRxiv. https://doi.org/10.1101/2023.12.30.573709
62. Yang H, Li Q, Stroup EK et al (2024) Widespread stable noncanonical peptides identified by integrated analyses of ribosome profiling and ORF features. Nat Commun 15:1–18
63. Xiao Z, Huang R, Xing X et al (2018) De novo annotation and characterization of the translatome with ribosome profiling data. Nucleic Acids Res 46:e61
64. Calviello L, Hirsekorn A, Ohler U (2020) Quantification of translation uncovers the functions of the alternative transcriptome. Nat Struct Mol Biol 27:717–725
65. Clauwaert J, McVey Z, Gupta R et al (2024) Deep learning to decode sites of RNA translation in normal and cancerous tissues. bioRxiv. https://doi.org/10.1101/2024.03.21.586110
66. Turcan A, Lee J, Wacholder A et al (2024) Integrative detection of genome-wide translation using iRibo. STAR Protoc 5:102826
67. Hao Y, Zhang L, Niu Y et al (2018) SmProt: a database of small proteins encoded by annotated coding and non-coding RNA loci. Brief Bioinform 19:636–643
68. Choteau SA, Wagner A, Pierre P et al (2021) MetamORF: a repository of unique short open reading frames identified by both experimental and computational approaches for gene and metagene analyses. Database (Oxford) 2021: baab032
69. Wethmar K, Barbosa-Silva A, Andrade-Navarro MA et al (2014) uORFdb–a comprehensive literature database on eukaryotic uORF biology. Nucleic Acids Res 42:D60–D67
70. Leblanc S, Yala F, Provencher N et al (2024) OpenProt 2.0 builds a path to the functional characterization of alternative proteins. Nucleic Acids Res 52:D522–D528
71. Stein CS, Jadiya P, Zhang X et al (2018) Mitoregulin: a lncRNA-encoded microprotein that supports mitochondrial supercomplexes and respiratory efficiency. Cell Rep 23:3710–3720.e8

Chapter 9

Identification of Potentially Coding Small ORFs in Plant Transcriptomes

Daria Ganaeva, Irina Lyapina, Anna Mamaeva, and Igor Fesenko

Abstract

Small open reading frames (smORFs, <100 amino acids) are a ubiquitous feature of eukaryotic and prokaryotic genomes. However, only a fraction of these smORFs encode functional proteins, known as microproteins. Functional microproteins have been found in all three domains of life, although due to limitations in their prediction and characterization, only a few have been described in plants. Therefore, the catalogization and functional analysis of plant smORF-encoded microproteins is far from complete. In this chapter, we present a bioinformatic pipeline designed to predict and validate coding smORFs in plant transcriptomes. The workflow comprises several key steps: transcriptome assembly, prediction of potentially coding smORFs, filtration of candidates, and validation of their translation through analysis of mass-spectrometry data.

Key words smORFs, Transcriptomes, Plant genomes, Mass spectrometry, Microproteins

1 Introduction

Genomes contain millions of small open reading frames (smORFs; <100 codons). Even though most of them appear randomly, functional smORF-encoded microproteins have been identified in a range of organisms, including drosophila, mice, yeast, and plants [1–5]. They were shown to play important roles mainly associated with the modulation activity of other proteins and protein complexes [6–10]. However, due to the limitations in computational prediction and further biochemical validation, smORFs are often overlooked during gene annotations [9–12]. Therefore, the diversity and roles of functional microproteins in plants remain poorly understood.

The majority of microproteins are lineage-specific and can originate from noncoding sequences during the process of de novo gene birth. For example, a recent study identified microprotein microRPG1 that regulates kernel dehydration in *Zea mays*

Stephan Wenkel (ed.), *Microproteins: Methods and Protocols*, Methods in Molecular Biology, vol. 2992,
https://doi.org/10.1007/978-1-0716-5013-4_9,

[13]. Phylogenetic analysis confirmed that microRPG1 appeared de novo from a noncoding sequence during *Zea* evolution. Because most smORFs are not evolutionarily conserved, identification of novel microproteins based on homology search is hindered. Thus, a variety of approaches, including ribosomal profiling (Ribo-seq) or mass-spectrometry (MS) analysis, can be used for the identification of coding smORFs [14–18].

Depending on position, the different types of smORFs can be found on eukaryotic transcripts. The smORFs located in the upstream and downstream regions of mRNAs primarily regulate the translation of the main ORF. In addition, long noncoding RNAs (lncRNAs) or unidentified intergenic transcripts can be a source of functional microproteins [5, 19–25]. In plants, several studies identified functional microproteins in *Physcomitrium patens*, *Populus deltoides*, *Phaseolus vulgaris*, *Medicago truncatula*, *Glycine max*, and *Lotus japonicus* [4, 26].

Here, we propose an integrated pipeline for the identification of coding smORFs using transcriptomic data analysis. Application of this pipeline resulted in the identification of 31,946 transcribed smORFs with high coding potential in the model plant *Physcomitrium patens*. Additionally, we described an approach for identification of translatable, functional smORFs based on MS data analysis.

2 Materials

This section presents the tools and datasets that are required for the analysis. We expect readers to have some experience working with the command line and applications in Unix-based operating systems. In our pipeline, we will use Python and Bash scripts that can be downloaded from GitHub: https://github.com/GanaevaDR/plant_smORFs.

2.1 Datasets

This protocol requires FASTQ files from the RNA-sequencing experiment as input. These data will be used for transcriptome assembly, followed by searching for smORFs with high coding potential and verification by proteomics. The readers can use their own RNA-seq data or download FASTQ files from databases, e.g., SRA (https://www.ncbi.nlm.nih.gov/sra) and ENA (https://www.ebi.ac.uk/ena/browser/home).

Here, we used published RNA-seq data of moss *Physcomitrium patens* infected by the *Botrytis cinerea* pathogen at different time points ([27]; SRA project SRP274010). For the demonstration of our workflow, wild-type control samples were used (samples SRR12341095, SRR12341104, and SRR12341105). For the validation step, peptidomic datasets available in the PRIDE database (https://www.ebi.ac.uk/pride/), which were previously obtained in our laboratory, were used (project PXD009532). The annotated proteome of *P. patens* (available at https://data.jgi.doe.gov/, version V3.3) was used to generate the database for reanalysis of mass-spectrometry (MS) data.

2.2 Software

This section presents the tools that are required at each step of the analysis. Programs used in this protocol can be installed using package managers such as conda (https://docs.conda.io/projects/conda/en/latest/user-guide/install/index.html). Detailed installation guidelines can be found on our GitHub: https://github.com/GanaevaDR/plant_smORFs.

2.2.1 Pre-processing of FASTQ Reads

Open-access web-service SRA explorer (https://sra-explorer.info/#) was used for the download of raw FASTQ files by SRA accession. FastQC v0.12.1 ([28]; https://www.bioinformatics.babraham.ac.uk/projects/fastqc/) and Multiqc v1.25.2 ([29]; https://seqera.io/multiqc/) are used for quality control of FASTQ files. Fastp 0.24.0 ([30]; https://github.com/OpenGene/fastp) is used for FASTQ files trimming. Hisat2-2.2.1 ([31]; https://daehwankimlab.github.io/hisat2/manual/) and samtools 1.13 ([32]; https://www.htslib.org/doc/samtools.html) are used for mapping reads to the reference genome and further processing of mapping output files.

2.2.2 Transcriptome Assembly

StringTie-2.2.3 ([33]; https://ccb.jhu.edu/software/stringtie/index.shtml?t=manual) is used for transcriptome assembly. Generating a FASTA file with assembled transcripts is done using gffread 0.9.12 ([34]; https://ccb.jhu.edu/software/stringtie/gff.shtml#gffread). Gffcompare v0.12.6 ([34]; https://ccb.jhu.edu/software/stringtie/gffcompare.shtml) can be applied for the comparison of StringTie and reference annotation.

2.2.3 smORFs Prediction

MiPepid ([35]; https://github.com/MindAI/MiPepid) is used for the prediction of small open reading frames (smORFs) on assembled transcripts. Search for predicted smORFs among annotated proteins can be done using MMseqs2 ([36]; https://github.com/soedinglab/MMseqs2).

2.2.4 MS/MS Data Analysis

An annotated proteome can be downloaded from the Phytozome data archive (https://data.jgi.doe.gov/). PRIDE database (https://www.ebi.ac.uk/pride/) serves as the source of LC–MS/MS. LC–MS/MS data can be analyzed using FragPipe v21.1 (https://fragpipe.nesvilab.org/).

3 Methods

We provide a step-by-step pipeline for the identification of potentially coding smORFs in transcriptomic data. Our pipeline contains

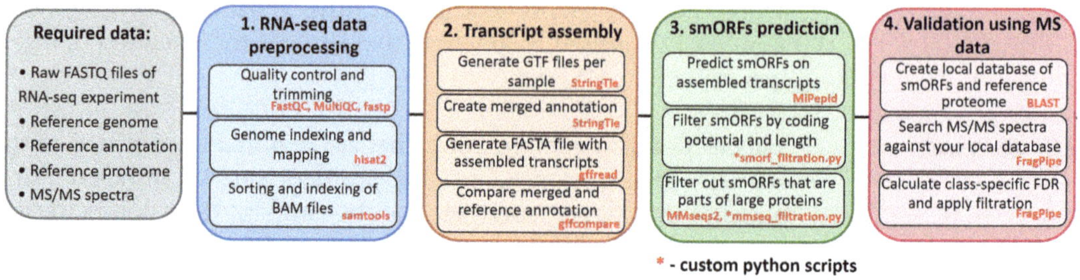

Fig. 1 Scheme of data analysis pipeline

four major steps (Fig. 1), which are preprocessing of raw transcriptomic data (Subheading 3.1), transcriptome assembly (Subheading 3.2), prediction and filtration of smORFs (Subheading 3.3), and validation with mass-spectrometry data (Subheading 3.4). Commands that should be executed in the Linux terminal start with > sign and are written in Courier new format on a grey background. The reverse slash symbol (\) is used for line breaks when writing long commands. Scripts that are used in this analysis can be downloaded from GitHub: https://github.com/GanaevaDR/plant_smORFs.

3.1 Preprocessing of RNA-Seq Data

1. First of all, we should create a directory for our project where all steps of the analysis will be performed:

```
> mkdir smORF_project
> cd smORF_project
```

2. Before starting the analysis, the following reference data should be downloaded: genome assembly FASTA file, GFF annotation, and FASTA file with annotated proteins (*see* **Note 1**). Here, we used the model plant *Physcomitrium patens* v3.3 for the demonstration of the pipeline. The reference data were downloaded from Phytozome (https://phytozome-next.jgi.doe.gov). In order to download reference data from Phytozome, select your species of interest and press the "get standard data files" button. In the open tab, select the desired file and add it to your cart. Next, in the drop-down menu under the "download" button, select "command line download" and use the generated URL to download the data.

```
> mkdir reference
> wget -P ./reference <URL_generated_by_database>
```

3. In order to download RNA-seq datasets to your machine, we suggest using web server SRA Explorer (https://sra-explorer.info/#). The reader needs to input the desired dataset accession, press "search," and add selected samples to the collection.

Then, the reader would get a URL for downloading raw FASTQ. Run the following commands (*see* **Note 2**):

```
> mkdir raw_data
> cd raw_data
> wget -P ./raw_data <URL_generated_by_SRA_Explorer>
> cd ..
```

4. Then, we are going to perform quality control of raw reads using the FastQC program [28]. This tool allows analyzing the quality of reads, GC content, and checking for duplicate sequences and adapters. The FastQC tool can process both raw and archived (e.g., .gz) FASTQ files.

```
> mkdir qc
> mkdir qc/fastqc_raw
> fastqc -o ./qc/fastqc_raw ./raw_data/*
```

5. The FastQC generates an HTML file that can be opened in a web browser. MultiQC [29] is a convenient tool for merging reports from multiple FASTQ files into a single report. To use MultiQC, run the following:

```
> multiqc ./qc/fastqc_raw -o ./qc/fastqc_raw
```

6. Trimmomatic [37], Cutadapt [38], and fastp [30] can be used to perform read trimming. Depending on the quality of RNA-seq data, a custom trimming command could be generated. Here, we provide an example of how to trim paired-end reads of low quality and filter out reads of length less than 50 nucleotides using fastp [30]:

```
> mkdir trimmed_data
>fastp --in1 ./raw_data/<FASTQ_forward_reads> \
--in2 ./raw_data/<FASTQ_reverse_reads> \
--out1 ./trimmed_data/<FASTQ_forward_trimmed_reads> \
--out2 ./trimmed_data/<FASTQ_reverse_trimmed_reads> \
-l 50 --cut_front --cut_right
```

7. We are going to perform quality control after trimming by FastQC [28]. It is important to compare FastQC reports before and after trimming to verify that the procedure successfully fixed the issues. Run the following to generate FastQC and MultiQC reports for trimmed reads:

```
> mkdir qc/fastqc_trimmed
> fastqc -o ./qc/fastqc_trimmed ./trimmed_data/*
> multiqc ./qc/fastqc_trimmed -o ./qc/fastqc_trimmed
```

8. Next, we are going to map trimmed reads to our reference genome using HISAT2 [31]. Before mapping, we will build a genome index using HISAT2 [31]. To do this, run the following commands:

```
> mkdir mapping
> hisat2-build ./reference/<genome_FASTA_file> ./reference/
index
```

9. Next, we should map trimmed reads to the reference genome. By default, HISAT2 [31] provides mapping output in SAM format. Conversion of SAM files to a compressed binary BAM format can be further performed using samtools [32]. These two steps can be executed in one command using a pipe (|). Here, we provide a command example for mapping one file with FASTQ paired-end Illumina short reads. This step can be automated for processing several files using a bash script. An example of such a script is presented on our GitHub: https://github.com/GanaevaDR/plant_smORFs/blob/main/scripts/mapping.sh. This script can be downloaded and executed using **`bash mapping.sh`** command.

```
> hisat2 --no-softclip --summary-file ./mapping/<log_file> \
-x ./reference/index \
-1 ./trimmed_data/<FASTQ_forward_trimmed_reads> \
-2 ./trimmed_data/<FASTQ_reverse_trimmed_reads> \
| samtools view -Sb - > ./mapping/<BAM_mapping_file>
```

10. The mapping procedure should be followed by sorting and indexing the obtained BAM files. To do this, run the following:

```
> samtools sort -o ./mapping/<BAM_mapping_file> \
./mapping/<BAM_mapping_file>
> samtools index ./mapping/<BAM_mapping_file>
```

3.2 De Novo Transcriptome Assembly

Since coding smORFs are often overlooked during gene annotation, we propose an annotation-free (*see* **Note 3**) reference-guided approach for transcript assembly. This approach allows us to assemble not only annotated but also novel transcripts. In this protocol, we are going to use StringTie ([33]; *see* **Note 4**).

1. In the first step, we are going to generate GTF annotation files for each sample using the StringTie tool. For this, run the following command for each BAM file. We should also create a **list.txt** in the stringtie directory. This file should contain names of GTF files obtained in the previous step, one file name per line, and save it to a stringtie directory. Alternatively, you can use the bash script provided on our GitHub (https://github.com/GanaevaDR/plant_smORFs/blob/main/scripts/stringtie_assembly.sh) for performing this step for multiple samples and generation of list.txt file. This script can be downloaded and executed using **`bash stringtie_assembly.sh`** command.

```
> mkdir stringtie
> stringtie -o ./stringtie/<GTF_file> ./mapping/<BAM_file>
```

2. Then, we will combine the GTF files obtained at the previous step to create a nonredundant set of transcripts. This command merges GTF files mentioned in **list.txt** into one GTF annotation:

```
> stringtie --merge -o ./stringtie/<merged_GTF_file> \
./stringtie/list.txt
```

3. Finally, we generate a FASTA file with nucleotide sequences of transcripts using gffread software [34]:

```
> gffread -w ./stringtie/<transcripts_FASTA_file> \
-g ./reference/<genome_FASTA> ./stringtie/<merged_GTF_file>
```

4. To compare assembled transcripts with a reference genome annotation, the reader can use the gffcompare software [34]. The output is a table with annotated genes and assembled transcripts with class-codes (*see* **Note 5**) as well as a new GTF file containing information on both annotated and novel transcripts. The quality of transcriptome assembly can further be estimated (*see* **Note 6**).

```
> cd stringtie
> gffcompare -r ../reference/<reference_annotation> \
./<merged_GTF_file>
> cd ..
```

3.3 smORF Prediction

Because plant genomes contain millions of smORFs, it is important to distinguish between spurious and potentially translated smORFs. For this analysis, one can use proteogenomic approaches or computational predictions that are based on sequence features.

Here, we will use the MiPepid tool to predict smORFs with high coding potential using the obtained transcripts (*see* **Note 7**).

1. For proper execution of the MiPepid script, the environment can be set using the conda package manager. The following program versions can be used: Python 3.6.15, conda 24.5.0, pandas 1.1.5, numpy 1.19.5, scikit-learn 0.19.1; conda channels: conda-forge, bioconda, defaults.
2. At the second step, download MiPepid from GitHub repository (*see* **Note 8**):

```
> git clone https://github.com/MindAI/MiPepid/
```

3. To predict smORFs based on assembled transcripts, we are going to use the following command:

```
> python ./MiPepid/src/mipepid.py \
./stringtie/<transcripts_FASTA_file> \
./MiPepid/mipepid_res.csv
```

4. At this step, we are going to filter the results of MiPepid prediction. For further analysis, the predicted smORFs will be selected based on coding potential and the length (>=30 aa). To get files with smORF coordinates and FASTA files with nucleotide and amino acid sequences, run the following command in the terminal:

```
> python <path_to_smorf_filtration.py> \
--table ./MiPepid/mipepid_res.csv \
--out_bed ./MiPepid/filtered_smORFs.bed \
--out_table ./MiPepid/filtered_smORFs.csv \
--out_nucl ./MiPepid/smORFs_nucleotide.fasta \
--out_prot ./MiPepid/smORFs_protein.fasta
```

5. After prediction of potentially coding smORFs, we are going to filter out those that are just a part of large proteins. We will use the MMSeqs2 search tool [36] for identification of such smORFs. Create a new directory named MMSEQ_SEARCH and run the following:

```
> mkdir MMSEQ_SEARCH
> mmseqs createdb ./MiPepid/smORFs_protein.fasta MMSEQ_-
SEARCH/SMORFDB
> mmseqs createdb ./reference/<protein_fasta> MMSEQ_SEARCH/
PROTDB
> mmseqs search ./MMSEQ_SEARCH/SMORFDB ./MMSEQ_SEARCH/PROTDB
\
./MMSEQ_SEARCH/SMORF_PROT tmp --cov-mode 0 -s 7.5 --min-seq-
```

```
id 0.95
 > mmseqs convertalis ./MMSEQ_SEARCH/SMORFDB ./MMSEQ_SEARCH/
PROTDB \
 ./MMSEQ_SEARCH/SMORF_PROT ./MMSEQ_SEARCH/SMORF_PROT_results.
tsv
```

6. This will generate a SMORF_PROT_results.tsv file that contains the results of MMSeqs2 search. This file will be used for excluding smORFs that are part of annotated proteins. Run the following command to get a novel FASTA file with smORF amino acid sequences:

```
> python <path_to_mmseq_filtration.py> \
--mmseq ./MMSEQ_SEARCH/SMORF_PROT_results.tsv \
--fasta ./MiPepid/smORFs_protein.fasta \
--out_smorfs ./MMSEQ_SEARCH/filtered_smORFs.fasta
```

3.4 Mass-Spectrometry Validation of Predicted smORFs

For validation of translation of predicted smORFs, the reader can use either proteomic or peptidomic datasets. It is required to modify the analysis parameters according to the chosen data.

1. The reader can generate their own MS data or, alternatively, download MS data from a public database, such as PRIDE (https://www.ebi.ac.uk/pride/). Using keywords, find desired datasets, click the "FTP" button above the list of files on the chosen project page, and select the raw data files. Then the reader can download it directly to a local computer or copy the link and download it to the server via the following command:

```
> mkdir MS_data
> wget -P ./MS_data <URL_generated_by_PRIDE>
```

The MS analysis itself will be executed on a Windows machine, so this data does not need to be downloaded to the Linux server.

2. Create a local protein database from translated predicted sequences of smORFs, obtained in 3.3.6 step, and an annotated reference proteome. Next, obtain the concatenated FASTA file (*see* **Note 9**) by running this command:

```
> cat ./reference/<reference_proteome_fasta> \
./MMSEQ_SEARCH/filtered_smORFs.fasta > ./MS_data/mydb.fasta
```

Then you can transfer the generated database to your local computer using the following command:

```
> scp -P <port> login@address:<path_to_database>/mydb.fasta \
<path_to_destination_on_computer>
```

3. Search MS raw data against a combined database using the proper software, such as FragPipe v21.1 (https://fragpipe.nesvilab.org/). The reader can find the tutorial for running FragPipe at https://fragpipe.nesvilab.org/docs/tutorial_fragpipe.html.
4. In the "Workflow" tab, select a workflow according to the type of your MS/MS data (peptidomic or proteomic) from the dropdown menu and click "Load workflow." The reader can use a customized workflow if your data requires specific settings. For example, for a standard peptidomic dataset, one should select the "Nonspecific-peptidome" option. With these settings, the output data would be filtered to 1% FDR, and the protein FDR is disabled. For a standard proteomic dataset, the reader would need to load a "Default" workflow. We also recommend checking with the PRIDE description of your MS data if you have no information on how these data were obtained.
5. Then, in the "Workflow" tab, add your MS data files and load the workflow.
6. In the "Database" tab, select the database FASTA generated in step 3.4.2, then click "Add decoys" and select "Add decoys and contaminants."
7. To start the analysis, on the "Run" tab, select or create a target directory on the local computer, click the "RUN" button, and wait for the completion message to appear in the Console. The reader can now analyze the resulting .tsv files with peptide, protein and PSM data.
8. After FragPipe peptide search, for identifying smORF-encoded microproteins, compute the class-specific FDR to perform FDR estimation by target-decoy strategy separately for each class of peptides [39]. Calculate a class-specific FDR as the number of decoy smORF hits divided by the number of target smORF hits. Use the 1% class-specific FDR to select the smORF PSMs.

4 Notes

1. *Source of Reference Data*. Possible sources for plant reference genome could be Phytozome (https://phytozome-next.jgi.doe.gov/), Ensembl Plants (https://plants.ensembl.org/index.html) or a specific database for your organism.

2. *Downloading FASTQ Files.* Alternatively, programs such as fastq-dump, fasterq-dump can be used for command line file download.
3. *StringTte Annotation-Free Approach.* Here, we do not provide any reference annotation to the software. Therefore, transcripts are assembled directly from the RNA-sequencing reads alignment to the reference genome.
4. *StringTie Input Requirements.* Stringite takes as input BAM, SAM, or CRAM files obtained from such mapping programs as HISAT2, STAR, or TopHat. Short, long reads or both can be provided as input.
5. *Analysing gffcompare Class-Codes.* Output flags can be analyzed using this webpage, section «Transcripts classification codes»: https://ccb.jhu.edu/software/stringtie/gffcompare.shtml.
6. *Quality Control of Transcriptome Assembly.* In order to estimate the quality of the assembled transcriptome, such statistics as the number and length of assembled transcripts, the N50 metric, and GC content can be collected using the seqkit stats program. Additionally, you can use BUSCO in order to estimate the completeness of your assembly by discovering gene content of one-copy orthologs of your target lineage (e.g., eudicots).
7. The coding potential of predicted smORFs can be estimated using different tools [40, 41]. The resulting FASTA file that contains predicted microproteins can be used for downstream analysis.
8. GitHub has removed support for password-based authentication for Git operations. Instead, you need to use a Personal Access Token (PAT) or SSH key to authenticate.
9. Sequences of different origins should be differentiated by name in the final database file.

Acknowledgments

This work was supported by the Russian Science Foundation (project No. 23-74-10048).

References

1. Lease KA, Walker JC (2006) The Arabidopsis unannotated secreted peptide database, a resource for plant peptidomics. Plant Physiol 142:831. https://doi.org/10.1104/pp.106.086041
2. Hanada K, Zhang X, Borevitz JO, Li W-H, Shiu S-H (2007) A large number of novel coding small open reading frames in the intergenic regions of the Arabidopsis thaliana genome are transcribed and/or under purifying selection. Genome Res 17:632–640. https://doi.org/10.1101/gr.5836207
3. Hanada K, Higuchi-Takeuchi M, Okamoto M, Yoshizumi T, Shimizu M, Nakaminami K, Nishi R, Ohashi C, Iida K, Tanaka M, Horii Y, Kawashima M, Matsui K, Toyoda T,

Shinozaki K, Seki M, Matsui M (2013) Small open reading frames associated with morphogenesis are hidden in plant genomes. PNAS 110:2395–2400. https://doi.org/10.1073/pnas.1213958110

4. Fesenko I, Kirov I, Knyazev A, Khazigaleeva R, Lazarev V, Kharlampieva D, Grafskaia E, Zgoda V, Butenko I, Arapidi G, Mamaeva A, Ivanov V, Govorun V (2019) Distinct types of short open reading frames are translated in plant cells. Genome Res 29(9):1464–1477. https://doi.org/10.1101/gr.253302.119
5. Fesenko I, Shabalina SA, Mamaeva A, Knyazev A, Glushkevich A, Lyapina I, Ziganshin R, Kovalchuk S, Kharlampieva D, Lazarev V, Taliansky M, Koonin EV (2021) A vast pool of lineage-specific microproteins encoded by long non-coding RNAs in plants. Nucleic Acids Res 49:10328–10346. https://doi.org/10.1093/nar/gkab816
6. Magny EG, Pueyo JI, Pearl FMG, Cespedes MA, Niven JE, Bishop SA, Couso JP (2013) Conserved regulation of cardiac calcium uptake by peptides encoded in small open reading frames. Science 341:1116–1120. https://doi.org/10.1126/science.1238802
7. Nelson BR, Makarewich CA, Anderson DM, Winders BR, Troupes CD, Wu F, Reese AL, McAnally JR, Chen X, Kavalali ET, Cannon SC, Houser SR, Bassel-Duby R, Olson EN (2016) A peptide encoded by a transcript annotated as long noncoding RNA enhances SERCA activity in muscle. Science 351:271–275. https://doi.org/10.1126/science.aad4076
8. Kondo T, Plaza S, Zanet J, Benrabah E, Valenti P, Hashimoto Y, Kobayashi S, Payre F, Kageyama Y (2010) Small peptides switch the transcriptional activity of Shavenbaby during Drosophila embryogenesis. Science 329:336–339. https://doi.org/10.1126/science.1188158
9. Hassel KR, Brito-Estrada O, Makarewich CA (2023) Microproteins: overlooked regulators of physiology and disease. iScience 26:106781. https://doi.org/10.1016/j.isci.2023.106781
10. Chanut-Delalande H, Zanet J (2024) Small ORFs, big insights: drosophila as a model to unraveling microprotein functions. Cells 13:1645. https://doi.org/10.3390/cells13191645
11. Ladoukakis E, Pereira V, Magny EG, Eyre-Walker A, Couso JP (2011) Hundreds of putatively functional small open reading frames in Drosophila. Genome Biol 12:R118. https://doi.org/10.1186/gb-2011-12-11-r118
12. Dong X, Zhang K, Xun C, Chu T, Liang S, Zeng Y, Liu Z (2023) Small open reading frame-encoded micro-peptides: an emerging protein world. Int J Mol Sci 24:10562. https://doi.org/10.3390/ijms241310562
13. Yu Y, Li W, Liu Y, Liu Y, Zhang Q, Ouyang Y, Ding W, Xue Y, Zou Y, Yan J, Jia A, Yan J, Hao X, Gou Y, Zhai Z, Liu L, Zheng Y, Zhang B, Xu J, Yang N, Xiao Y, Zhuo L, Lai Z, Yin P, Liu H-J, Fernie AR, Jackson D, Yan J (2025) A Zea genus-specific micropeptide controls kernel dehydration in maize. Cell 1888(1):44-59.e21. https://doi.org/10.1016/j.cell.2024.10.030
14. Ingolia NT, Lareau LF, Weissman JS (2011) Ribosome profiling of mouse embryonic stem cells reveals the complexity of mammalian proteomes. Cell 147:789–802. https://doi.org/10.1016/j.cell.2011.10.002
15. Guttman M, Russell P, Ingolia NT, Weissman JS, Lander ES (2013) Ribosome profiling provides evidence that large non-coding RNAs do not encode proteins. Cell 154:240–251. https://doi.org/10.1016/j.cell.2013.06.009
16. Aspden JL, Eyre-Walker YC, Phillips RJ, Amin U, Mumtaz MAS, Brocard M, Couso J-P (2014) Extensive translation of small Open Reading Frames revealed by Poly-Ribo-Seq. eLife 3:e03528. https://doi.org/10.7554/eLife.03528
17. Bazzini AA, Johnstone TG, Christiano R, Mackowiak SD, Obermayer B, Fleming ES, Vejnar CE, Lee MT, Rajewsky N, Walther TC, Giraldez AJ (2014) Identification of small ORFs in vertebrates using ribosome footprinting and evolutionary conservation. EMBO J 33:981–993. https://doi.org/10.1002/embj.201488411
18. Verheggen K, Volders P-J, Mestdagh P, Menschaert G, Van Damme P, Gevaert K, Martens L, Vandesompele J (2017) Noncoding after all: biases in proteomics data do not explain observed absence of lncRNA translation products. J Proteome Res 16:2508–2515. https://doi.org/10.1021/acs.jproteome.7b00085
19. Andrews SJ, Rothnagel JA (2014) Emerging evidence for functional peptides encoded by short open reading frames. Nat Rev Genet 15:193–204. https://doi.org/10.1038/nrg3520
20. Laing WA, Martínez-Sánchez M, Wright MA, Bulley SM, Brewster D, Dare AP, Rassam M, Wang D, Storey R, Macknight RC, Hellens RP (2015) An upstream open reading frame is essential for feedback regulation of ascorbate biosynthesis in Arabidopsis. Plant Cell 27:772–786. https://doi.org/10.1105/tpc.114.133777

21. Lauressergues D, Couzigou J-M, Clemente HS, Martinez Y, Dunand C, Bécard G, Combier J-P (2015) Primary transcripts of microRNAs encode regulatory peptides. Nature 520:90–93. https://doi.org/10.1038/nature14346
22. Couso J-P, Patraquim P (2017) Classification and function of small open reading frames. Nat Rev Mol Cell Biol 18:575–589. https://doi.org/10.1038/nrm.2017.58
23. Brunet MA, Brunelle M, Lucier J-F, Delcourt V, Levesque M, Grenier F, Samandi S, Leblanc S, Aguilar J-D, Dufour P, Jacques J-F, Fournier I, Ouangraoua A, Scott MS, Boisvert F-M, Roucou X (2019) OpenProt: a more comprehensive guide to explore eukaryotic coding potential and proteomes. Nucleic Acids Res 47:D403–D410. https://doi.org/10.1093/nar/gky936
24. Ruiz-Orera J, Villanueva-Cañas JL, Albà MM (2020) Evolution of new proteins from translated sORFs in long non-coding RNAs. Exp Cell Res 391:111940. https://doi.org/10.1016/j.yexcr.2020.111940
25. Zhao S, Meng J, Wekesa JS, Luan Y (2023) Identification of small open reading frames in plant lncRNA using class-imbalance learning. Comput Biol Med 157:106773. https://doi.org/10.1016/j.compbiomed.2023.106773
26. Guillén G, Díaz-Camino C, Loyola-Torres CA, Aparicio-Fabre R, Hernández-López A, Díaz-Sánchez M, Sanchez F (2013) Detailed analysis of putative genes encoding small proteins in legume genomes. Front Plant Sci 4:208. https://doi.org/10.3389/fpls.2013.00208
27. Reboledo G, Agorio AD, Vignale L, Batista-García RA, Ponce De León I (2021) Transcriptional profiling reveals conserved and species-specific plant defense responses during the interaction of Physcomitrium patens with Botrytis cinerea. Plant Mol Biol 107:365–385. https://doi.org/10.1007/s11103-021-01116-0
28. Andrews S (2010) FastQC: a quality control tool for high throughput sequence data. https://www.bioinformatics.babraham.ac.uk/projects/fastqc/. Accessed 6 May 2024
29. Ewels P, Magnusson M, Lundin S, Käller M (2016) MultiQC: summarize analysis results for multiple tools and samples in a single report. Bioinformatics 32:3047–3048. https://doi.org/10.1093/bioinformatics/btw354
30. Chen S, Zhou Y, Chen Y, Gu J (2018) Fastp: an ultra-fast all-in-one FASTQ preprocessor. Bioinformatics 34:i884–i890. https://doi.org/10.1093/bioinformatics/bty560
31. Kim D, Langmead B, Salzberg SL (2015) HISAT: a fast spliced aligner with low memory requirements. Nat Methods 12:357–360. https://doi.org/10.1038/nmeth.3317
32. Li H, Handsaker B, Wysoker A, Fennell T, Ruan J, Homer N, Marth G, Abecasis G, Durbin R, 1000 Genome Project Data Processing Subgroup (2009) The sequence alignment/map format and SAMtools. Bioinformatics 25:2078–2079. https://doi.org/10.1093/bioinformatics/btp352
33. Pertea M, Pertea GM, Antonescu CM, Chang T-C, Mendell JT, Salzberg SL (2015) StringTie enables improved reconstruction of a transcriptome from RNA-seq reads. Nat Biotechnol 33:290–295. https://doi.org/10.1038/nbt.3122
34. Pertea G, Pertea M (2020) GFF utilities: GffRead and GffCompare. F1000Res 9:ISCB Comm J-304. https://doi.org/10.12688/f1000research.23297.2
35. Zhu M, Gribskov M (2019) MiPepid: MicroPeptide identification tool using machine learning. BMC Bioinformatics 20:559. https://doi.org/10.1186/s12859-019-3033-9
36. Steinegger M, Söding J (2017) MMseqs2 enables sensitive protein sequence searching for the analysis of massive data sets. Nat Biotechnol 35:1026–1028. https://doi.org/10.1038/nbt.3988
37. Bolger AM, Lohse M, Usadel B (2014) Trimmomatic: a flexible trimmer for Illumina sequence data. Bioinformatics 30:2114–2120. https://doi.org/10.1093/bioinformatics/btu170
38. Martin M (2011) Cutadapt removes adapter sequences from high-throughput sequencing reads. EMBnetjournal 17:10–12. https://doi.org/10.14806/ej.17.1.200
39. Nesvizhskii AI (2014) Proteogenomics: concepts, applications, and computational strategies. Nat Methods 11:1114–1125. https://doi.org/10.1038/nmeth.3144
40. Hanada K, Akiyama K, Sakurai T, Toyoda T, Shinozaki K, Shiu S-H (2010) sORF finder: a program package to identify small open reading frames with high coding potential. Bioinformatics 26:399–400. https://doi.org/10.1093/bioinformatics/btp688
41. Zhang M, Zhao J, Li C, Ge F, Wu J, Jiang B, Song J, Song X (2022) csORF-finder: an effective ensemble learning framework for accurate identification of multi-species coding short open reading frames. Brief Bioinform 23:bbac392. https://doi.org/10.1093/bib/bbac392

Part III

Computational, Structural, and Evolutionary Insights

Chapter 10

Structure and Disorder Predictions of Microproteins: Usage, Applications, and Pitfalls

Lars A. Eicholt

Abstract

In the past decade, computational predictions of protein structure and disorder have become widely accessible, achieving accuracy that, in some cases, rivals experimental results. These advances have been instrumental in identifying structural homologies, guiding protein design, and enhancing functional annotation. However, most prediction models are trained on "classical proteins," favoring specific sequence lengths and patterns, leading to biases. Microproteins—small proteins with emerging roles in development and disease—function similarly across the tree of life and stand to benefit significantly from structure and disorder predictions. Yet, their short length and molecular interactions present unique challenges, making homology detection more difficult and requiring careful methodological considerations. Here, I outline workflows for predicting, analyzing, and refining microprotein structures and disorders, emphasizing key precautions to ensure reliable insights.

Key words Protein structure prediction, Disorder prediction, Structural annotation, Microproteins, Molecular dynamics simulations, Sequence feature annotation

1 Introduction

Building upon decades of experimental structural biology and advancements in computational biology, at the 2020 edition of Critical Assessment of Techniques for Protein Structure Prediction (CASP) [1], AlphaFold2 [2] achieved for the first time computational protein structure prediction at an accuracy rivaling experimental structure determination. This opened protein structure prediction to a wider field of biology and chemistry and spurred a revolution in the development of further protein structure and disorder predictors based on varying machine learning approaches [3–5]. Predicting a protein's structure reliably can have many use cases, ranging from protein design [6], structural homology inference [7], and binder design [8] to functional annotation [9, 10].

Despite the advancement of the last years, these tools will be limited to a certain extent to the training data consisting of known

Stephan Wenkel (ed.), *Microproteins: Methods and Protocols*, Methods in Molecular Biology, vol. 2992,
https://doi.org/10.1007/978-1-0716-5013-4_10,

protein sequences and solved protein structures. Microproteins (sometimes also referred to as "micropeptides" or "small ORF-encoded peptides") are significantly shorter than average long proteins [11–13], and are underrepresented in publicly available structural databases [14–17]. Their limited presence often impedes template-based or evolutionary-coupling approaches, which rely on multiple sequence alignments (MSA) as used in AlphaFold2 and many other tools. Sequence alignments of short sequences, undergoing fast divergence, can yield sparse alignments with fewer matched residues or false positives [18–24]. These limitations can mask true homologs and confound tertiary structure prediction, especially when template libraries lack close structural analogs below the 50–60 residue range. In turn, structure predictions based on MSAs and templates may yield overfitted or erroneous models for microproteins that are far out of range of available training sets. These limitations can mask true homologs and confound tertiary structure. Protocols have been established to identify homology between microproteins more reliably and construct MSAs of them [24–26].

Moreover, there is an ongoing debate about the minimal length required to form a stable tertiary fold; some studies suggest that most autonomous, stable folds require at least 30–40 amino acids, while others argue for longer minimum [27–30]. Many microproteins lie at or below this threshold, with stable structures frequently relying on disulfide bridges, metal cofactors, or specialized motifs [31, 32]. These idiosyncrasies do not always generalize from larger proteins, further complicating predictions for microproteins.

Consequently, all these tools can be considered biased against shorter microproteins, owing to their lower abundance in many databases and a limited number of solved structures, as well as challenges in homology detection for shorter sequences. Special caution must therefore be taken when using these tools for microproteins.

Additionally, not only structure but also disorder predictors are often trained on well-characterized proteins, relying on MSAs and energy functions derived from larger, well-studied proteins. This, combined with the common use of sliding, window approaches that may be either too short or too long for microproteins, can lead to inaccuracies. Microproteins, due to their small size and unique structural features, frequently deviate from typical disorder patterns. Many exhibit hybrid characteristics, containing both disordered and structured regions, and may fold spontaneously without chaperones [33], interact with membranes [34–36], nucleic acids [37], or other proteins to form multimers [32, 38]. This results in a diverse folding and structural landscape that cannot be fully captured by single predictors, whether for structure or disorder.

Nevertheless, microproteins play crucial roles in diverse biological processes, including development, immune modulation, and disease [39–43]. Their functional versatility has also made them a growing focus of biotechnological research [44, 45]. Thus, predicting and analyzing their structures is of high importance. In the following sections, I will first describe (Methods) how to use various in silico tools—such as tertiary structure predictors (e.g., AlphaFold), disorder predictors (e.g., IUPred), and molecular dynamics simulations—efficiently for microproteins (and, by extension, other proteins), and will then discuss the critical pitfalls to consider in structural predictions of microproteins (**Note 4**).

2 Materials

All the tools and databases described are freely available. Below is a list, along with a brief description of their purpose and local or server-based usage.

2.1 Databases

- *RCSB protein data bank (PDB)*

  https://www.rcsb.org/ | Usage: Server or local (data download).

  The PDB is the central repository of experimentally determined protein structures. It enables both direct web-based browsing and bulk downloads for offline use.
- *AlphaFold DB (AFDB)*

  https://alphafold.ebi.ac.uk/ | Usage: Server or local (bulk download).

  This database hosts >200 Mio. predicted structures from AlphaFold2 for most known proteins. Users can explore the web interface or download predictions for offline analysis.
- *UniProt*

  https://www.uniprot.org/ | Usage: Server or local (bulk download).

  UniProt is a comprehensive database for protein sequences and annotations. It provides an online interface and downloadable datasets for large-scale analyses.
- *CAID*

  https://idpcentral.org/caid/ | Usage: Database & Benchmark.

  Critical assessment of protein intrinsic disorder (CAID) evaluates the performance of disorder prediction methods. It offers benchmark data and results for developers and users. The latest leading disorder predictor is PUNCH2 [46]. Future advancements in disorder prediction should be monitored for further improvements in accuracy and performance.

- *CASP*

  http://predictioncenter.org/ | Usage: Database & Benchmark.

  Critical assessment of protein structure prediction (CASP) is a community experiment for testing computational structure prediction methods every 2 years. It provides datasets and comparative results from multiple prediction rounds. In the future, it is worth checking which structure prediction tool performs best. However, since the AlphaFold2 breakthrough, performance differences among top tools have been relatively small. Most high-ranking tools are complex to install and use, as they are largely built on the AlphaFold2 architecture.

2.2 Tools & Software

- *Custom Scripts*

  The custom scripts to be used are deposited on GitHub: https://github.com/ArsLeicholt/structure_predictions.

  All scripts are annotated, explaining the functions of the code.

- *AlphaFold2*

  https://github.com/deepmind/alphafold | Usage: Local or cluster. | Structure Prediction.

  AlphaFold2 [2] predicts protein 3D structures using a deep learning approach based on MSAs. It requires a one-time installation of dependencies and can leverage GPUs for faster predictions.

- *AlphaFold3*

  https://github.com/google-deepmind/alphafold3 | Usage: Local or cluster. | Structure Prediction.

  AlphaFold3 [47] is based on a diffusion architecture and is especially suited for predicting interactions between proteins, with ligands and nucleic acids.

- *ESMfold*

  https://github.com/facebookresearch/esm | Usage: Local or cluster. | Structure Prediction.

  ESMfold [48] is a protein structure predictor from FAIR (Meta AI), relying on large language models, not requiring an MSA.

- *STRIDE*

  https://github.com/cmbi/dssp | Usage: Local. | Secondary Structure Annotation.

  STRIDE [49] is based on DSSP [50] and assigns secondary structure elements to protein residue coordinates in a PDB file. It runs from the command line and processes PDB files to identify helices, sheets, and other structural features.

- *PyMol & ChimeraX* [51, 52]

  https://www.cgl.ucsf.edu/chimera/ / https://pymol.org/2/ | Usage: Local. | Structure Visualization.

  These are two of many interactive visualization suites for molecular structures. They allow detailed inspection, annotation, and publication-quality rendering of proteins.
- *Foldseek*

  https://foldseek.com/ | Usage: Server or local. | Structural Homology Detection.

  Foldseek [53] is a fast tool for structure-based comparison and search. It can run via its web interface or be installed locally for high-throughput analyses.
- *PUNCH2 & PUNCH2-Light*

  https://github.com/deemeng/punch2?tab=readme-ov-file| Usage: local. | Disorder Prediction.

  PUNCH2 [46] and its lightweight version applies machine learning to identify intrinsically disordered regions, with an emphasis on short peptides and microproteins. They can be used locally. The light version does not use MSAs.
- *IUPred*

  https://iupred3.elte.hu/| Usage: local & web. | Disorder Prediction.

  IUPred [54] predicts intrinsically disordered regions based on pairwise energy estimation, without relying on MSAs. It offers three modes: short disorder, long disorder, and structured binding regions. The mode *long* is recommended to be used on microproteins since it predicts the global disorder, while the *short* predicts patches [55]. The most recent version is IUPred3.
- *Gromacs*

  http://www.gromacs.org/ | Usage: Local or cluster. | Molecular Dynamics.

  Gromacs [56] is a widely used molecular dynamics software for biomolecular simulations. It is optimized for parallel computing, making it suitable for HPC clusters.

3 Methods

Which tools to use—and how—ultimately depends on the research question and the size of the dataset. For predicting the structure or disorder of individual microproteins or small datasets, web-based servers are often the most convenient. However, for large-scale projects involving hundreds to thousands of microproteins, local installations, and especially high-performance computing clusters become indispensable (see Fig. 1).

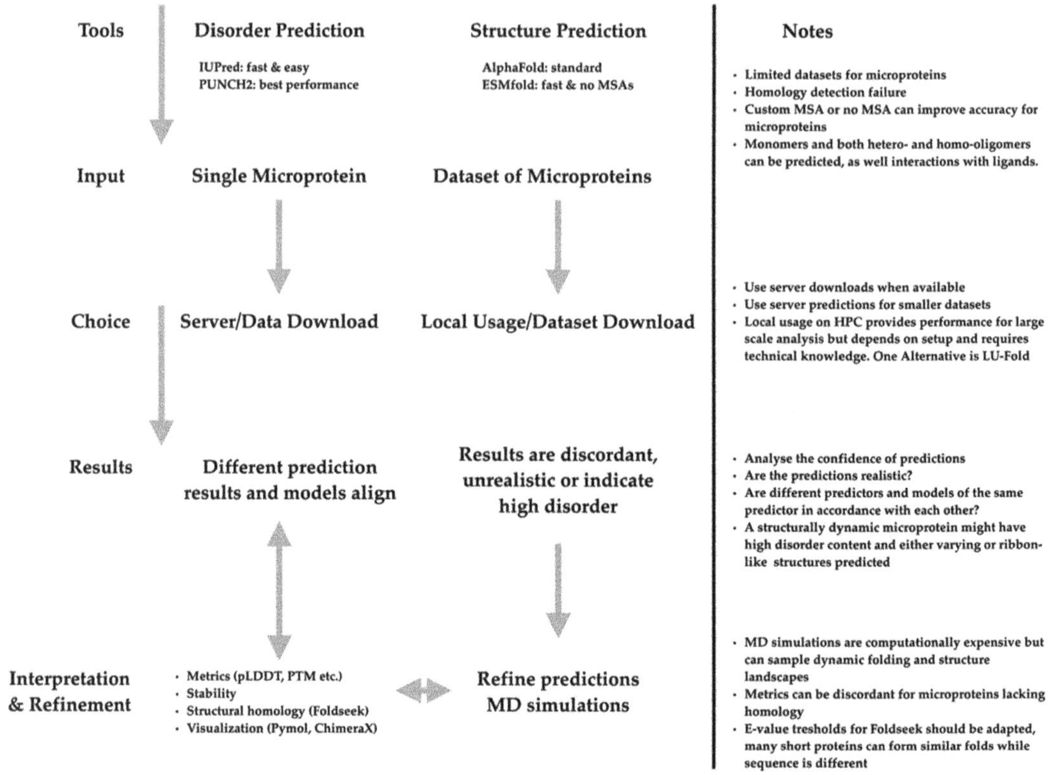

Fig. 1 The choice of tools, their usage and parameters depend largely on the research questions and quality of results

3.1 Tertiary Structure Predictions and Analysis

Before performing predictions, Uniprot, AFDB, and RCSPDB should be checked for an experimental or predicted structure of the microprotein(s) of interest is already deposited (see Materials). AlphaFold and most other structure prediction tools provide four key confidence metrics: predicted local distance difference test (pLDDT), which estimates the per-residue confidence in the model from zero (low) to 100 (high), and Predicted aligned error (PAE), which assesses the overall reliability of the global fold, particularly useful for evaluating domain packing, multimer confidence and protein–protein interactions. In contrast, ipTM evaluates the accuracy of the predicted relative positions of subunits in a protein–protein complex. Scores above 0.8 indicate high-confidence, high-quality predictions, while values below 0.6 suggest a likely failure. Predictions with ipTM scores between 0.6 and 0.8 fall into a gray area, where reliability is uncertain and should be interpreted with caution. Another metric in AlphaFold-Multimer is pTM, which is a global confidence metric that assesses how accurately AlphaFold-Multimer predicts the overall structure of a protein complex. A pTM score above 0.5 suggests that the predicted complex may resemble the true structure, whereas a score below

0.5 indicates a likely incorrect prediction. All these metrics are provided as output files with the prediction, except pLDDT which can be found in the PDB directly to be extracted and visualized (see Subheading 3.2).

Predictions of single microproteins can be performed quickly via AlphaFold3 webserver, for AlphaFold2 or ESMfold using their respective Google Colab:

- AlphaFold3: https://alphafoldserver.com/welcome.
- AlphaFold2 (Colab): https://colab.research.google.com/github/sokrypton/ColabFold/blob/main/AlphaFold2.ipynb.
- ESMfold (Colab): https://colab.research.google.com/github/sokrypton/ColabFold/blob/main/ESMFold.ipynb.

Google Colab allows for the provision of custom multiple sequence alignments (MSAs) or to perform predictions without an alignment at all (MSA options), which can, in some cases, improve accuracy for microproteins by reducing false-positive alignments (*see* Subheading 1 and **Notes 4**). Initially, default parameters are recommended to be used before custom alignments are used. It is also recommended to test AMBER relaxation on different top-ranked models (the output includes the top five ranked models, controlled by the num_relax parameter) to assess whether the microprotein of interest exhibits variations in predicted structures across different models. Note that a Google account is required, and Colab offers paid options for increased computational resources.

For larger-scale predictions or more extensive workflows, it is often preferable to run these tools on high-end workstations or high performance computing (HPC) clusters with robust GPUs and adequate storage. Installation instructions for local setups can be found in each tool's GitHub repository and mainly depend on the used workstation or cluster setup (see Subheading 2). We will focus here on HPC usage with the SLURM workload manager, which is commonly employed in academic HPC environments.

If setting up and maintaining these tools locally is not feasible, LU-Fold (https://atkinson-lab.com/lu-fold/) offers services to run predictions and analyses, serving as an external solution for researchers who require large-scale computations but lack HPC resources.

3.1.1 Running AlphaFold2 Using SLURM on HPC

Script: run_alphafold2.sh.

Link: https://github.com/ArsLeicholt/structure_predictions/blob/main/run_alphafold2.sh.

Run command: sbatch.

In general, required are access to an HPC environment with AlphaFold2 installed, its databases (which can total around three TB) and a GPU available. The example below demonstrates how to run multiple single-sequence FASTA files as a job array of 2000 with two jobs running simultaneously for AlphaFold2 predictions. Both monomeric and multimeric predictions are supported; for multimeric predictions, individual chains should be separated by a ":" in the FASTA input. This is particularly important for microproteins, as it allows testing for potential interactions within protein complexes.

Finally, ensure that the input FASTA files are correctly named and numbered for consistent job submission. Modules, paths and other parameters might differ from this template and will have to be adapted to the local setup. To run a single prediction, remove *#SBATCH --array=* and other parts of it in the script, and the input needs to be a single fasta file. The text on the right, starting with #, provides an explanation of the corresponding line.

3.1.2 Running AlphaFold3 Using SLURM on HPC

Script: run_alphafold3.sh.

Link: https://github.com/ArsLeicholt/structure_predictions/blob/main/run_alphafold3.sh.

Run command: sbatch.

Unlike AlphaFold2, AlphaFold3 can read JSON-formatted input to predict binding interactions with RNA or small molecules. This makes it especially interesting for microproteins that are interacting with ligands and nucleic acids.

Available on GitHub, there is (alphafold3_input.json: https://github.com/ArsLeicholt/structure_predictions/blob/main/alphafold3_input.json) an example JSON file demonstrating how to specify protein sequences, ligands, and RNA, followed by a sample SLURM submission script for HPC clusters. In this example, a dimer of two proteins (Chains A & B) is predicted and their interaction with the ligand ATP. It also provides paths to custom-paired MSA files.

- *name*

  A human-readable label that identifies the modeling project or dataset.
- *sequences*

  An array of definitions for each protein or ligand to be modeled. Each entry contains either a "protein" or a "ligand" object.
- *protein*

  Specifies details for one or more chains. Within this object:
  - *id*: One or more chain identifiers (e.g., ["A"]).
  - *sequence*: The complete amino acid sequence for the chain.

  - *unpairedMsa*: *(Optional)* A file path to an unpaired multiple sequence alignment (in A3M or similar format).
  - *pairedMsa*: *(Optional)* A file path to a paired MSA (useful in multimeric scenarios or advanced setups).
  - *templates*: *(Optional)* A list of template references for homology-based modeling. Leave empty if no templates are used.
- *ligand*

  Defines a small molecule or cofactor.
  - *id*: Chain-like identifiers that label the ligand in the final structure.
  - *ccdCodes*: One or more chemical component dictionary (CCD) codes that specify the ligand(s).
- *modelSeeds*

  A list of integer values for random seeds used in stochastic parts of the modeling process. Changing or adding seeds can yield slightly different predictions.
- *dialect*

  Specifies that this JSON file is structured for the "alpha-fold3" pipeline or a compatible parser.
- *version*

  Indicates the schema version for the input format, helping maintain compatibility across releases.

AlphaFold3 makes it easy to run predictions with custom MSAs or without, which makes it especially interesting for microprotein structure predictions. Some explanations of the details below. It has to be kept in mind that pairedMSA is especially relevant for multimer prediction, and customization of pairedMSAs is complicated. UnpairedMSA is relevant for monomer and multimer prediction. When AlphaFold3 performs the alignment for pairedMSA, it removes all insertions to ensure all sequences have the same length as the query, forming a rectangular matrix. It also ensures sequences from the same organism align in the same row. Naive concatenation can mix sequences from different organisms. AlphaFold3 uses UniProt organism IDs to pair sequences in pairedMsa.

- If unpairedMsa and pairedMsa are unset or null, AlphaFold 3 builds both MSAs automatically (recommended).
- If unpairedMsa is a non-empty A3M string and pairedMsa is an empty string (""), AlphaFold 3 uses unpairedMsa as is and runs pairedMSA-free.
- If pairedMsa is a non-empty A3M string and unpairedMsa is an empty string (""), AlphaFold 3 uses pairedMsa as is and runs unpairedMSA-free (not recommended).

- If both unpairedMsa and pairedMsa are empty strings (""), AlphaFold 3 runs completely MSA-free, using only the query sequence.
- If both unpairedMsa and pairedMsa are custom non-empty A3M strings, AlphaFold3 uses them instead of building an MSA (expert option).

Note: Both fields must be either set (non-null) or unset (null). Typically, pairedMsa is set to an empty string when using unpairedMsa.

The MSA must be A3M format, which follows FASTA but allows lowercase letters for insertions and hyphens (-) for gaps. The first sequence matches the query exactly.

AlphaFold3 can then be run on HPC using SLURM, giving a correct installation with this example script. To provide multiple input JSON files from one directory, change --json_path= to --input_dir=/path/to/your/input/directory.

3.1.3 Running ESMfold Using SLURM on HPC

Script: run_esmfold.sh.

Link: https://github.com/ArsLeicholt/structure_predictions/blob/main/run_esmfold.sh.

Run command: sbatch.

ESMfold can be used on HPC in a similar form as AlphaFold2, but requires different flags. ESMfold is based on a Protein Language Model (PLM) and, therefore not dependent on an MSA. However, its accuracy is still dependent on the training set for the PLM and how close the training set sequences are to the query sequences. While ESMfold performs worse than MSA-based structure predictors, it has been shown to be especially suitable for the prediction of orphan sequences—which microproteins are partially. Additionally, not needing to find an alignment improves speed and makes it an especially good choice when a lot of sequences must be predicted.

3.2 Analysis Structure Predictions

3.2.1 pLDDT and Secondary Element Annotation Extraction

Script: analysis_plddt_stride.py.

Link: https://github.com/ArsLeicholt/structure_predictions/blob/main/analysis_plddt_stride.py.

Run command: python3.

When predicting protein structures, analysis should go beyond visual inspection and the metric outputs to include quantification of secondary structure elements and pLDDT. The following script scans a chosen directory (pdb_dir) for all files ending in .pdb. The required python3 packages (pandas, numpy, and biopython) [57–59] will have to be installed beforehand using e. g. pip or conda. For pLDDT extraction, it uses Biopython's PDBParser to read each structure and gather B-factors. AlphaFold's pLDDT values are stored in the PDB B-factor field, so we average these values to get a global confidence score per structure. Secondary Structure Assignment is performed using the external program, STRIDE, which assigns secondary structure codes (H, G, I, E, B, T, C, and S). We parse STRIDE's output to count how many residues fall into each category. STRIDE needs to be installed beforehand.

For each protein (PDB file), the script stores:

- The protein's identifier (filename without extension).
- The average pLDDT.
- The count of residues in each STRIDE secondary structure class (helix, sheet, turn, coil, etc.).

All results are collected into a Pandas DataFrame and saved to a CSV file (output_csv) for downstream analysis. By generalizing paths and adding all major STRIDE codes, this script can be easily adapted to different systems and use cases with minimal changes.

3.2.2 Graphical Visualization of Prediction Confidence

PyMOL and ChimeraX allow users to open and visualize predicted protein structures (e.g., AlphaFold models) and apply color schemes based on confidence metrics such as pLDDT values (often stored in the B-factor column). In PyMOL, various plugins and scripts are available—see https://github.com/BobSchiffrin/pymol_scripts for a popular option—while in ChimeraX, simply enter *color bfactor palette alphafold* at the command line to highlight regions according to pLDDT. For further tricks on how to use PyMol and ChimeraX be referred to their respective manuals.

3.2.3 Foldseek to Identify Structural Homology

Structural homology of the predicted structures of microproteins can help to assign evolutionary and functional relationships beyond sequence similarity. However, earlier studies have suggested that the structural space for microproteins is compact and different sequences can attain the same fold [60]. Foldseek offers an online server at search.foldseek.com where one can quickly run structural searches without any local installation. The typical workflow is as follows:

1. *Prepare the Structure*: Provide the 3D structure of the query protein (or microprotein). This often comes in the form of a PDB file (e.g., my_microprotein.pdb).
2. *Access the Server & Upload the Query*: Upload the PDB file, paste a PDB identifier or paste in a structure in an accepted format.
3. *Adjust Parameters (Optional)*: Keep defaults unless there are specific requirements. If the input is very short, consider adjusting the E-value thresholds to capture more subtle matches.
4. *Submit the Job*: The server will process the structure and compare it to its database of known protein structures.

5. *Review Results*: When the results are ready, there will appear a list of potential matches ranked by structural similarity. Download the alignments, evaluate superpositions, and check various statistical parameters (e.g., TM-score and RMSD).

For large-scale projects, or if complete control over data and analyses is preferred, it is advantageous to install Foldseek locally. Below is a general guide:

1. *Download Foldseek*: Visit the Foldseek GitHub or the official website. Download a precompiled binary or build from source if needed.
2. *Extract or Install*: Using a precompiled binary, simply extract it. If building from source, ensure the required compilers and dependencies are installed and follow the build instructions.
3. *Prepare Database*

   Obtain a reference structure database: Foldseek can search against large databases, such as the PDB or custom sets of structures. Download the structural datasets of choice. These databases can be very large.
4. *Index the Database*: Before search, run a database indexing step.

   For example:

```
foldseek createdb path/to/structures/*.pdb mydb
```

   This command collects and indexes the structures into a database named mydb.
5. *Search the Query*

   Use foldseek search or foldseek easy-search: Run a structural search against the newly created database.

```
foldseek easy-search query.pdb mydb result.m8 tmp_folder
```

   - query.pdb is the microprotein structure file.
   - mydb is the name of the database.
   - result.m8 is the output file containing search results in tabular format.
   - tmp_folder is a working directory for intermediate files.
   - If more fine-grained control over the search is required, use the lower-level foldseek search command with additional parameters.
6. *Interpretation of Results*

- The output will typically be a tabular file listing matches, along with statistics such as TM-score, alignment coverage, and E-values.
- To analyze or visualize the alignments, use other software (e.g., PyMOL and ChimeraX) to load superimposed PDB files and inspect how well they match in 3D space.

7. *Optimize Parameters for Microproteins*

   For microproteins, adjusting sensitivity or E-value thresholds might be necessary. This is because shorter structures can produce higher rates of false positives or miss borderline matches. Experiment with parameters such as —min-alignment-coverage or other advanced options to tailor the search to the dataset.

3.3 Disorder Predictions

Disorder Predictions Using PUNCH

PUNCH and PUNCH-light rank as the top-performing protein disorder predictors according to the CAID benchmark. Although multiple disorder predictors exist (such as IUPred), we recommend using PUNCH/PUNCH-light for the best results.

Before running PUNCH, generate embeddings (One-Hot, ProtTrans, and MSA Transformer) for the query sequences. Punch-Light is installed and runs the same way as PUNCH2 but does not use MSAs, making it faster and particularly suitable for microproteins.

1. *Pull the Embedding Docker Image*`docker pull dimeng851/embedding:v3`
2. *Run the Embedding Script*

   By default, all three embeddings (One-Hot, ProtTrans, and MSA transformer) are needed:

```
CMD python /embedding/main.py --embeddingType onehot,prot-
Trans,msaTrans
```

3. *Start the Embedding Container*`docker run -d \`

```
 -it \
 --name [CONTAINER_NAME] \
 --mount type=bind,source=[PATH_TO_INPUT_FASTA_FILE],tar-
get=/embedding/data/input.fasta \
 --mount type=bind,source=[PATH_TO_INPUT_A3M_FOLDER],tar-
get=/embedding/data/msa \
 --mount type=bind,source=[PATH_TO_INPUT_OUTPUT_FOLDER],
target=/embedding/data/output \
 --mount type=bind,source=[PATH_TO_INPUT_TORCH_CHECK-
POINT],target=/root/.cache/torch/hub/checkpoints/ \
 --mount type=bind,source=
```

```
[PATH_TO_INPUT_HUGGINGFACE_HUB],target=/root/.cache/hug-
gingface/hub/ \
 dimeng851/embedding:v3
```

Replace the placeholders in brackets with:

[CONTAINER_NAME]: Any container name.

[PATH_TO_INPUT_FASTA_FILE]: Path to input FASTA file.

[PATH_TO_INPUT_A3M_FOLDER]: Folder containing .a3m files from HHblits results.

[PATH_TO_INPUT_OUTPUT_FOLDER]: Folder for saving the output embeddings.

[PATH_TO_INPUT_TORCH_CHECKPOINT]: Folder containing the MSA Transformer models (esm_msa1b_t12_100M_UR50S-contact-regression.pt. and esm_msa1b_t12_100M_UR50S.pt), or an empty folder where they will be downloaded if not present.

[PATH_TO_INPUT_HUGGINGFACE_HUB]: Folder containing the ProtTrans models (e.g., models–Rostlab–prot_t5_xl_uniref50) or an empty folder where they will be downloaded if not present.

4. *Pull the PUNCH Docker Image*`docker pull dimeng851/punch2:v1`
5. *Start the PUNCH Container*

```
docker run -d \
 -it \
 --name [CONTAINER_NAME] \
 --mount type=bind,source=[PATH_TO_INPUT_FASTA],target=/
punch2/data/input.fasta \
 --mount type=bind,source=[PATH_TO_ONEHOT],target=/
punch2/data/onehot \
 --mount type=bind,source=[PATH_TO_MSATRANS],target=/
punch2/data/msaTrans \
 --mount type=bind,source=[PATH_TO_PROTTRANS],target=/
punch2/data/protTrans \
 --mount type=bind,source=[PATH_OUTPUT],target=/punch2/
output \
 dimeng851/punch2:v1
```

Replace the placeholders in brackets with:

[CONTAINER_NAME]: Any container name.

[PATH_TO_INPUT_FASTA]: Path to the single FASTA file containing *all* query sequences.

[PATH_TO_ONEHOT]: Folder containing all one-hot embeddings.

[PATH_TO_MSATRANS]: Folder containing all MSA transformer embeddings.

[PATH_TO_PROTTRANS]: Folder containing all ProtTrans embeddings.

[PATH_OUTPUT]: Folder where PUNCH output will be saved. This includes:

1. *timings.csv.*
2. *disorder* (a folder containing all prediction results).

Disorder Predictions Using IUPred3

Script: loop_iupred3.sh.

Link: https://github.com/ArsLeicholt/structure_predictions/blob/main/loop_iupred3.sh.

Run command: ./

While the multiple versions of IUPred are not within the top range in CAID [61], IUPred has been one of the most popular disorder predictors due to its easy usability. It can be used on a server (https://iupred3.elte.hu/) or locally installed for academic use. In any case, the long mode should be used on microproteins, since this mode predicts global disorder of a protein, despite some studies on small proteins using the short mode in the past [55]. When reporting the disorder level prediction of a protein, both median disorder and fraction of disorder-promoting residues should be provided to allow comparability. However, IUpred only prints the single residue scores in the terminal. I provide here a simple bash script that loops over the fasta files of a specified directory and runs Iupred3 on each single sequence fasta file, providing a csv file for each with the input file name as the output filename.csv. Smoothing is automatically turned off for sequences under 19 residues, since the sequence must be larger than the default window size (19) for smoothing.

Additionally, I provide here a python3 script to aggregate the output csv files to calculate the median and fraction of disorder-promoting residues (>=0.5).

Script: aggregate_iupred3.py.

Link: https://github.com/ArsLeicholt/structure_predictions/blob/main/aggregate_iupred3.py.

Run command: python3.

3.4 Molecular Dynamics Simulations

GROMACS

Predicted structures can be further refined using molecular dynamics (MD) simulations. However, MD simulations are computationally intensive, often requiring several days to complete, though they can be shorter for microproteins. MD is particularly valuable for microproteins with dynamic structural behavior, such as intrinsically disordered regions, structural variations across different ranked predictions, or interactions within multimers and ligand-bound states to assess binding stability. A key factor in MD simulations is the choice of force field. OPLS-AA/M [62] provides improved accuracy for both folded and unfolded small proteins. CHARMM36m [63] is recommended for simulations involving protein-DNA/RNA complexes. Additionally, running MD simulations in triplicate is considered best practice to ensure reproducibility and account for variations in trajectory sampling. I will here explain the steps to perform MD simulations on a PDB file containing either a single-chain protein or a protein complex using Gromacs on an HPC using SLURM. Multichain proteins can be simulated as well with the following script, but will require some changes.

1. *File Setup*

   First, generate three directories, containing each the start PDB file and the following files:

   - ions.mdp: http://www.mdtutorials.com/gmx/lysozyme/Files/ions.mdp.
   - minim.mdp: http://www.mdtutorials.com/gmx/lysozyme/Files/minim.mdp.
   - nvt.mdp: http://www.mdtutorials.com/gmx/lysozyme/Files/nvt.mdp.
   - npt.mdp: http://www.mdtutorials.com/gmx/lysozyme/Files/npt.mdp.
   - md.mdp: http://www.mdtutorials.com/gmx/lysozyme/Files/md.mdp (change nsteps to 125000000).

2. *Prepare MD Run*

   In each directory save the following script and run it on the HPC.

   Script: prep_gromacs_monomer.sh.

   Link: https://github.com/ArsLeicholt/structure_predictions/blob/main/prep_gromacs_monomer.sh.

   Run command: sbatch.

   For a multichain protein complex, the preparation script must be adapted.

   Script: prep_gromacs_multimer.sh.

   Link: https://github.com/ArsLeicholt/structure_predictions/blob/main/prep_gromacs_multimer.sh.

   Run command: sbatch.

3. Start MD Simulation

   After preparation, save the following script in the prepared directories and start the MD run.

   Script: start_gromacs.sh.

   Link: https://github.com/ArsLeicholt/structure_predictions/blob/main/start_gromacs.sh.

   Run command: sbatch.

4. Post-Processing with trjconv

   During the simulation, the protein may diffuse, appear fragmented, or jump across the periodic box. To correct for periodic boundary conditions (PBC) and center the protein in the box, use:

```
gmx trjconv -s final_md.tpr -f final_md.xtc -o final_md_noPBC.
xtc -pbc mol -center
```

   - Select 1 ("Protein") for centering.
   - Select 0 ("System") for output.

     All further analyses will use this PBC-corrected trajectory.

5. Root Mean Square Deviation (RMSD)

   RMSD measures the structural stability of the protein over time:

```
gmx rms -s final_md.tpr -f final_md_noPBC.xtc -o rmsd_runX.xvg
-tu ns
```

   - Choose 4 ("Backbone") for both fitting and RMSD calculation.
   - Use xmgrace rmsd_runX.xvg or convert .xvg to .csv for further analysis.

6. Root Mean Square Fluctuation (RMSF)

   RMSF identifies flexible and rigid regions in the protein:

```
gmx rmsf -s final_md.tpr -f final_md_noPBC.xtc -o rmsf_runX.
xvg -res
```

   - Choose 3 ("C-alpha") to assess per-residue flexibility.
   - Visualize with xmgrace or other plotting tools.

7. Radius of Gyration (Rg)

   Rg quantifies protein compactness and helps monitor unfolding:

```
gmx gyrate -s final_md.tpr -f final_md_noPBC.xtc -o gyrate_r-
unX.xvg
```

- Select 1 (“Protein”).
- Plot with xmgrace or other analysis tools.

8. Extracting Structural Snapshots

   To extract a PDB structure at a specific simulation time (e.g., 200 ns):

```
gmx trjconv -s final_md.tpr -f final_md_noPBC.xtc -o snap-
shot_200ns.pdb -b 200000 -e 200001
```

- Choose 1 (“Protein”).
- View in PyMOL alongside the starting structure for comparison.

  For different time points, adjust -b (start) and -e (end) (in ps, not ns).

  To generate a trajectory movie spanning 10 ns (e.g., 190–200 ns):

```
gmx trjconv -s final_md.tpr -f final_md_noPBC.xtc -o trajec-
tory_190_200ns.pdb -b 190000 -e 200001
```

4 Notes

1. Limited Benchmarking

There is a general scarcity of well-curated, experimentally characterized microproteins for large-scale benchmarking of structure prediction tools. Without well-defined validation sets containing diverse small proteins, it is difficult to assess the accuracy of new algorithms or refine existing ones. This impedes method development and can reinforce biases, since developers often optimize algorithms using datasets where microproteins are underrepresented or absent.

2. Are the Structure Predictions Realistic?

When evaluating a predicted structure, start by examining its confidence score (e.g., pLDDT and PTM). For microproteins lacking homology, pLDDT has been shown to behave discordantly [64]. If the protein sequence is rich in structure-breaking residues like prolines but still yields a stable fold, consider whether this is biologically plausible. Repetitive sequences often produce solenoid-like architectures in AlphaFold2 but not AlphaFold3

[65], so check whether the predicted structure aligns with known repeat patterns. Inspect lower-ranked or alternative models—if they differ substantially, the structure might be dynamic. Reviewing MSA or HMM-based alignments of the AlphaFold prediction, provided in the output files, can reveal whether there is sufficient evolutionary support for the proposed fold. In the case of microproteins, where alignments can often include false positives, perform again with custom MSA alignments created or run predictions without MSA. Also, be cautious if only a single beta sheet is predicted, as stand-alone beta sheets are uncommon. Lastly, compare results with other structure predictors to see if they converge on a similar outcome.

3. Use Multiple Predictors

Relying on a single method can overemphasize biases inherent in its training data or algorithmic design. Combining orthogonal predictors for structure (e.g., AlphaFold and ESMfold) and disorder (e.g., IUPred and PUNCH) can mitigate individual weaknesses [64]. For instance, a consensus approach can highlight regions where different algorithms consistently predict order or disorder, providing greater confidence in the result. When the structure prediction results vary over different ranked models and the protein is predicted to be disordered, attempt MD simulations.

4. Molecular Dynamics Simulations

Structure predictions, even when providing multiple models, can only capture a steric structure, most likely at the energy minima. MD simulations can sample structures and folding over time and capture dynamics. Microproteins often present challenges in molecular dynamics simulations due to their small size and inherent instability [36, 66]. Many lack well-defined hydrophobic cores, making them more prone to unfolding in simulations, which can lead to an overestimation of disorder. Additionally, most molecular dynamics force fields are optimized for well-folded, globular proteins and may not accurately capture the unique folding constraints of microproteins, such as disulfide bonds or specialized motifs. The accuracy of simulations also depends heavily on the initial predicted structure, often derived from predictive models like AlphaFold, which may contain inherent biases or errors. However, MD simulations provide unique opportunities to probe the conformational landscape of microprotein structures and their interactions with other molecules.

References

1. Jumper J, Evans R, Pritzel A, Green T, Figurnov M, Ronneberger O et al (2021) Applying and improving AlphaFold at CASP14. Proteins 89:1711–1721. https://doi.org/10.1002/prot.26257
2. Jumper J, Evans R, Pritzel A, Green T, Figurnov M, Ronneberger O et al (2021) Highly accurate protein structure prediction with AlphaFold. Nature 596:583–589. https://doi.org/10.1038/s41586-021-03819-2
3. Cramer P (2021) AlphaFold2 and the future of structural biology. Nat Struct Mol Biol 28: 704–705. https://doi.org/10.1038/s41594-021-00650-1
4. Lutomski CA, El-Baba TJ, Robinson CV, Riek R, Scheres SHW, Yan N et al (2022) The next decade of protein structure. Cell 185: 2617–2620. https://doi.org/10.1016/j.cell.2022.06.011
5. Fersht AR (2021) AlphaFold—a personal perspective on the impact of machine learning. J Mol Biol 433:167088. https://doi.org/10.1016/j.jmb.2021.167088
6. Woolfson DN (2021) A brief history of de novo protein design: minimal, rational, and computational. J Mol Biol 433:167160. https://doi.org/10.1016/j.jmb.2021.167160
7. Hamamsy T, Morton JT, Blackwell R, Berenberg D, Carriero N, Gligorijevic V et al (2024) Protein remote homology detection and structural alignment using deep learning. Nat Biotechnol 42(6):975–985. https://doi.org/10.1038/s41587-023-01917-2
8. Chang L, Perez A (2023) Ranking Peptide Binders by Affinity with AlphaFold. Angew. Chem. Int. Ed. 2023, 62, e202213362 https://doi.org/10.1002/anie.202213362
9. Nomburg, J., Doherty, E.E., Price, N. et al. Birth of protein folds and functions in the virome. Nature 633, 710–717 (2024). https://doi.org/10.1038/s41586-024-07809-y
10. Kustatscher G, Collins T, Gingras A-C, Guo T, Hermjakob H, Ideker T et al (2022) Understudied proteins: opportunities and challenges for functional proteomics. Nat Methods 19: 774–779. https://doi.org/10.1038/s41592-022-01454-x
11. Davids W, Fuxelius H-H, Andersson SGE (2003) The journey to smORFland. Comp Funct Genomics 4:537–541. https://doi.org/10.1002/cfg.325
12. Bhati KK, Dolde U, Wenkel S (2021) Micro-Proteins: expanding functions and novel modes of regulation. Mol Plant 14:705–707. https://doi.org/10.1016/j.molp.2021.01.006
13. Hassel KR, Brito-Estrada O, Makarewich CA (2023) Microproteins: overlooked regulators of physiology and disease. iScience 26: 106781. https://doi.org/10.1016/j.isci.2023.106781
14. wwPDB consortium (2019) Protein Data Bank: the single global archive for 3D macromolecular structure data. Nucleic Acids Res 47:D520–D528. https://doi.org/10.1093/nar/gky949
15. David A, Islam S, Tankhilevich E, Sternberg MJE (2022) The AlphaFold database of protein structures: a biologist's guide. J Mol Biol 434:167336. https://doi.org/10.1016/j.jmb.2021.167336
16. Varadi M, Anyango S, Deshpande MS, Nair SS, Natassia C, Yordanova G et al (2021) AlphaFold protein structure database: massively expanding the structural coverage of protein-sequence space with high-accuracy models. Nucleic Acids Res 50:D439–D444. https://doi.org/10/gqss2p
17. The UniProt Consortium, Bateman A, Martin M-J, Orchard S, Magrane M, Adesina A, et al. UniProt: the universal protein knowledgebase in 2025. Nucleic Acids Res 2025;53: D609–17.. https://doi.org/10/g83tzc
18. Altschul S (1997) Gapped BLAST and PSI-BLAST: a new generation of protein database search programs. Nucleic Acids Res 25: 3389–3402. https://doi.org/10/d8t8f7
19. Michaud JM, Madani A, Fraser JS (2022) A language model beats alphafold2 on orphans. Nat Biotechnol 40:1576
20. Moyers BA, Zhang J (2016) Evaluating phylostratigraphic evidence for widespread de novo gene birth in genome evolution. Mol Biol Evol 33:1245–1256. https://doi.org/10/f8j5vq
21. Moyers BA, Zhang J (2017) Further simulations and analyses demonstrate open problems of phylostratigraphy. Genome Biol Evol 9: 1519–1527. https://doi.org/10/gbs6mk
22. Moyers BA, Zhang J (2015) Phylostratigraphic bias creates spurious patterns of genome evolution. Mol Biol Evol 32:258–267. https://doi.org/10/f63gzr
23. Ladoukakis E, Pereira V, Magny EG, Eyre-Walker A, Couso JP (2011) Hundreds of putatively functional small open reading frames in drosophila. Genome Biol 12:R118. https://doi.org/10/b2ncxf

24. Leong AZ-X, Lee PY, Mohtar MA, Syafruddin SE, Pung Y-F, Low TY (2022) Short open reading frames (sORFs) and microproteins: an update on their identification and validation measures. J Biomed Sci 29:19. https://doi.org/10/gskxmj
25. Cheng H, Soon Chan W, Li Z, Wang D, Liu S, Zhou Y (2011) Small open reading frames: current prediction techniques and future prospect. CPPS 12:503–507. https://doi.org/10/d4twj7
26. Brito-Estrada O, Hassel KR, Makarewich CA (2022) An integrated approach for microprotein identification and sequence analysis. JoVE:63841. https://doi.org/10/g83v65
27. Alva V, Remmert M, Biegert A, Lupas AN, Söding J (2010) A galaxy of folds. Protein Sci 19:124–130. https://doi.org/10/b4zdxp
28. Alva V, Söding J, Lupas AN (2015) A vocabulary of ancient peptides at the origin of folded proteins. elife 4:e09410. https://doi.org/10/gf3q7g
29. Lupas AN, Ponting CP, Russell RB (2001) On the evolution of protein folds: are similar motifs in different protein folds the result of convergence, insertion, or relics of an ancient peptide world? J Struct Biol 134:191–203. https://doi.org/10/bjw8gg
30. Agozzino L, Dill KA (2018) Protein evolution speed depends on its stability and abundance and on chaperone concentrations. Proc Natl Acad Sci USA 115:9092–9097. https://doi.org/10/gd9gf8
31. Mohsen J, Martel A, Slavoff S (2023) Microproteins – discovery, structure and function. https://doi.org/10.22541/au.168854416.62840611/v1
32. Staudt A, Wenkel S (2011) Regulation of protein function by 'microProteins'. EMBO Rep 12:35–42. https://doi.org/10/dg6bjb
33. Vabulas RM, Raychaudhuri S, Hayer-Hartl M, Hartl FU (2010) Protein folding in the cytoplasm and the heat shock response. Cold Spring Harb Perspect Biol 2:a004390. https://doi.org/10/fw7kwt
34. Makarewich CA (2020) The hidden world of membrane microproteins. Exp Cell Res 388: 111853. https://doi.org/10/g83txq
35. Vakirlis N, Acar O, Hsu B, Castilho Coelho N, Van Oss SB, Wacholder A et al (2020) De novo emergence of adaptive membrane proteins from thymine-rich genomic sequences. Nat Commun 11:781. https://doi.org/10/gj3vfm
36. Peng J, Zhao L (2024) The origin and structural evolution of de novo genes in drosophila. Nat Commun 15:810. https://doi.org/10/gtg9gg
37. D'Lima NG, Ma J, Winkler L, Chu Q, Loh KH, Corpuz EO et al (2017) A human microprotein that interacts with the mRNA decapping complex. Nat Chem Biol 13:174–180. https://doi.org/10/f9gt2f
38. Chen J, Li Q, Xia S, Arsala D, Sosa D, Wang D et al (2024) The rapid evolution of de novo proteins in structure and complex. Genome Biol Evol 16:evae107. https://doi.org/10/g83v66
39. Schlesinger D, Dirks C, Navarro C, Lafranchi L, Spinner A, Raja GL et al (2025) A large-scale sORF screen identifies putative microproteins involved in cancer cell fitness. iScience:111884. https://doi.org/10/g83txv
40. Vakirlis N, Vance Z, Duggan KM, McLysaght A (2022) De novo birth of functional microproteins in the human lineage. Cell Rep 41: 111808. https://doi.org/10/grhxff
41. Sandmann C-L, Schulz JF, Ruiz-Orera J, Kirchner M, Ziehm M, Adami E et al (2023) Evolutionary origins and interactomes of human, young microproteins and small peptides translated from short open reading frames. Mol Cell 83:994–1011.e18. https://doi.org/10/grsqcx
42. Camarena ME, Theunissen P, Ruiz M, Ruiz-Orera J, Calvo-Serra B, Castelo R et al (2024) Microproteins encoded by noncanonical ORFs are a major source of tumor-specific antigens in a liver cancer patient meta-cohort. Sci Adv 10: eadn3628. https://doi.org/10/gvczrf
43. Hofman DA, Prensner JR, Van Heesch S (2024) Microproteins in cancer: identification, biological functions, and clinical implications. Trends Genet:S0168952524002117. https://doi.org/10/g8sb34
44. Qin Q, Jiang X, Huo L, Qian J, Yu H, Zhu H et al (2024) Computational design and engineering of self-assembling multivalent microproteins with therapeutic potential against SARS-CoV-2. J Nanobiotechnol 22:58. https://doi.org/10/g83txx
45. Petri L, Van Humbeeck A, Niu H, Ter Waarbeek C, Edwards A, Chiurazzi MJ et al (2024) Exploring the world of small proteins in plant biology and bioengineering. Trends Genet:S0168952524002129. https://doi.org/10/g8nrms
46. Meng D, Pollastri G (2024) PUNCH2: explore the strategy for intrinsically disordered protein predictor. https://doi.org/10.1101/2024.10.03.616458
47. Abramson J, Adler J, Dunger J, Evans R, Green T, Pritzel A et al (2024) Accurate

structure prediction of biomolecular interactions with AlphaFold 3. Nature 630:493–500. https://doi.org/10/gttgxm

48. Lin Z, Akin H, Rao R, Hie B, Zhu Z, Lu W et al (2023) Evolutionary-scale prediction of atomic-level protein structure with a language model. Science 379:1123–1130. https://doi.org/10/grzgts
49. Heinig M, Frishman D (2004) STRIDE: a web server for secondary structure assignment from known atomic coordinates of proteins. Nucleic Acids Res 32:W500–W502. https://doi.org/10/cmc2qg
50. Kabsch W, Sander C (1983) Dictionary of protein secondary structure: pattern recognition of hydrogen-bonded and geometrical features. Biopolymers 22:2577–2637. https://doi.org/10.1002/bip.360221211
51. Pettersen EF, Goddard TD, Huang CC, Meng EC, Couch GS, Croll TI et al (2021) UCSF ChimeraX: structure visualization for researchers, educators, and developers. Protein Sci 30:70–82. https://doi.org/10/ghr6mn
52. Delano WL. The PyMOL molecular graphics system, 2002
53. van Kempen M, Kim SS, Tumescheit C, Mirdita M, Gilchrist CLM, Söding J et al (2022) Foldseek: fast and accurate protein structure search. Bioinformatics. https://doi.org/10.1101/2022.02.07.479398
54. Erdős G, Pajkos M, Dosztányi Z (2021) IUPred3: prediction of protein disorder enhanced with unambiguous experimental annotation and visualization of evolutionary conservation. Nucleic Acids Res 49:W297–W303. https://doi.org/10.1093/nar/gkab408
55. Aubel M, Eicholt L, Bornberg-Bauer E (2023) Assessing structure and disorder prediction tools for de novo emerged proteins in the age of machine learning. F1000Res 12:347. https://doi.org/10.12688/f1000research.130443.1
56. Abraham MJ, Murtola T, Schulz R, Páll S, Smith JC, Hess B et al (2015) GROMACS: high performance molecular simulations through multi-level parallelism from laptops to supercomputers. SoftwareX 1–2:19–25. https://doi.org/10/f3pj2n
57. Reback J, jbrockmendel, McKinney W, den Bossche JV, Roeschke M, Augspurger T et al (2022) pandas-dev/pandas: Pandas 1.4.4. https://doi.org/10.5281/zenodo.7037953
58. Harris CR, Millman KJ, van der Walt SJ, Gommers R, Virtanen P, Cournapeau D et al (2020) Array programming with NumPy. Nature 585:357–362. https://doi.org/10/ghbzf2
59. Cock PJ, Antao T, Chang JT, Chapman BA, Cox CJ, Dalke A et al (2009) Biopython: freely available Python tools for computational molecular biology and bioinformatics. Bioinformatics 25:1422–1423. https://doi.org/10/d7zwd2
60. Kihara D, Skolnick J (2003) The PDB is a covering set of small protein structures. J Mol Biol 334:793–802. https://doi.org/10/bkcxxn
61. Predictors CAID, Curators DP, Necci M, Piovesan D, Tosatto SCE (2021) Critical assessment of protein intrinsic disorder prediction. Nat Methods 18:472–481. https://doi.org/10.1038/s41592-021-01117-3
62. Robertson MJ, Tirado-Rives J, Jorgensen WL (2015) Improved peptide and protein torsional energetics with the OPLS-AA force field. J Chem Theory Comput 11:3499–3509. https://doi.org/10/f7km22
63. Huang J, Rauscher S, Nawrocki G, Ran T, Feig M, De Groot BL et al (2017) CHARMM36m: an improved force field for folded and intrinsically disordered proteins. Nat Methods 14:71–73. https://doi.org/10/gfxkj3
64. Middendorf L, Eicholt LA (2024) Random, de novo, and conserved proteins: how structure and disorder predictors perform differently. Proteins:prot.26652. https://doi.org/10/gtfswb
65. Pratt OS, Elliott LG, Haon M, Mesdaghi S, Price RM, Simpkin AJ et al (2025) AlphaFold 2, but not AlphaFold 3, predicts confident but unrealistic β-solenoid structures for repeat proteins. Comput Struct Biotechnol J 27:467–477. https://doi.org/10/g83qvd
66. Middendorf L, Iyengar BR, Eicholt LA (2024) Sequence, structure, and functional space of drosophila de novo proteins. GBE 16(8). https://doi.org/10.1093/gbe/evae176

Chapter 11

Computational Methods for Tracing the Evolutionary History of Human Microproteins Encoded by Intronless Genes

Katia Aviña-Padilla and Maribel Hernández-Rosales

Abstract

MicroProteins (miPs), small functional proteins derived from short open reading frames (sORFs), play pivotal roles in posttranslational regulation by interacting with multidomain proteins. Acting as dominant-negative regulators, miPs modulate transcriptional activity and protein complex assembly. Their compact structure and functional versatility make them indispensable in cellular regulation, with significant implications in diseases such as cancer and neurodegenerative disorders. Intronless genes (IGs, characterized by the absence of introns in their coding sequences, are linked to key regulatory functions in development, cell proliferation, and disease pathways. While the roles of many IG-encoded proteins are well established, the potential of these genes to encode miPs, as well as their evolutionary history, remain largely unexplored. This chapter introduces a computational framework integrating three bioinformatics tools to investigate the evolution in vertebrates and the functional roles of miPs encoded by IGs. The framework begins with miPFinder2, which identifies potential miPs by annotating small peptides. Next, IGFinder classifies IGs based on genomic and UTR features. Finally, REvolutionH-tl reconstructs evolutionary histories that comprehend orthogroups, orthologs, paralogs, gene trees, species trees, and species tree/gene tree reconciliations. This integrative approach provides comprehensive insights into the structural, functional, and evolutionary reconstruction of miPs encoded by IGs, contributing to advancements in functional genomics, evolutionary biology, and disease research.

Key words Microproteins, Intronless genes, Short open reading frames, Functional genomics, Orthology, Gene duplication, Phylogenetic analysis, Evolutionary biology

1 Introduction

MicroProteins (miPs), defined as functional proteins smaller than 140 amino acids, present unique challenges in evolutionary and functional analysis due to their distinct origins and small size [1–4]. Unlike conventional proteins, miPs often arise from short open reading frames (sORFs) located in noncoding regions or through

Supplementary Information The online version contains supplementary material available at https://doi.org/10.1007/978-1-0716-5013-4_11.

Stephan Wenkel (ed.), *Microproteins: Methods and Protocols*, Methods in Molecular Biology, vol. 2992,
https://doi.org/10.1007/978-1-0716-5013-4_11,

mechanisms such as gene duplication and de novo emergence [5]. These characteristics, coupled with their high sequence divergence and structural simplicity, convolute their study using traditional genomic approaches [6]. Despite these difficulties, miPs are increasingly recognized for their pivotal roles in cellular processes, including transcriptional regulation, signaling pathways, and stress responses [7, 8]. Furthermore, their functional versatility and adaptability have linked miPs and other small proteins to a growing list of diseases, such as cancer, cardiovascular disorders, and neurodegenerative conditions [9–11].

A compelling connection exists between intronless genes (IGs) and miPs and other small proteins. IGs, characterized by the absence of introns in their coding sequences, enable streamlined transcriptional processes, bypassing the need for splicing. This structural simplicity facilitates the rapid and high-fidelity production of proteins, a property particularly advantageous for regulatory proteins like miPs [12]. IGs are frequently associated with specialized and critical cellular functions, encoding proteins such as transcription factors, histones, and stress response regulators [13]. These attributes suggest that IGs could act as genomic niches for the evolution of miPs. Although the number of miPs encoded by IGs identified thus far is limited, their overlap in functional domains highlights their potential biological significance.

For instance, the basic helix–loop–helix (bHLH) transcription factors (TFs), such as Ferd3l and Bhlhe23, encoded by IGs, display microprotein-like characteristics and are pivotal in neural development and brain function, with links to neurodevelopmental disorders such as autism and schizophrenia [14, 15]. Likewise, IG-encoded histones, including H1.2 and H1.4, are essential for chromatin structure regulation and are associated with tumorigenesis and epigenetic reprogramming [16, 17]. The IG-encoded miPs we have identified, such as those bHLH-TFs and histone variants, underscore their critical roles in modulating transcriptional networks and chromatin remodeling. These findings highlight the regulatory importance and therapeutic potential of IG-encoded miPs across diverse biological contexts, including neurodevelopment, oncology, and regenerative medicine. This chapter presents a computational framework for investigating the evolutionary trajectories and functional roles of miPs encoded by IGs. The pipeline integrates three specialized bioinformatics tools: miPFinder2 for the identification of miPs [18], IGFinder for classifying IGs through genome annotation and structural analysis [13], and REvolutionH-tl [19] for reconstructing evolutionary histories of gene families, which includes the identification of orthology and paralogy relationships among genes. This integrative approach enables a comprehensive analysis of miP function and evolution, bridging gaps in functional genomics and evolutionary studies, and providing valuable insights into their roles in health and disease.

2 Materials

2.1 Equipment

Computer Specifications: A computer with at least 8GB RAM and 4 CPU cores is recommended for running the pipeline efficiently. A high-memory server is recommended for better performance, but the pipeline can be run on standard desktop computers or laptops.

Operating System: Compatible with Linux, macOS, or Windows with Python 3.7 or higher installed.

Internet Access: Required for fetching gene data via the Ensembl API during the execution of IGFinder.

2.2 Software

The pipeline incorporates three specialized bioinformatics tools: miPFinder2 for the identification and annotation of miPs [18], IGFinder for IG classification through genome annotation and structural analysis [13], and REvolutionH-tl [19] for reconstructing event-labeled gene trees and identifying orthology and paralogy relationships. These steps are summarized in Fig. 1.

2.2.1 miPFinder2

Description: miPFinder2 is a specialized tool designed to identify potential miPs in protein sequence datasets from the Uniprot database.

Repository: https://github.com/ku-mip/mipfinder2.

Installation: Run the following bash command line.

% brew install gcc cmake hmmer clustal-omega.

Clone and compile miPFinder2: Run the following bash command lines.

% git clone https://github.com/ku-mip/mipfinder2.git

% cd mipfinder2.

% mkdir build && cd build.

% cmake.

% make.

Prepare Configuration File. Create a configuration.ini file specifying paths for the required input files. Example: Electronic Supplemental File 1.

2.2.2 IGFinder

Description: A pipeline designed to identify IGs in a specified organism. The tool employs the Ensembl API to retrieve genome annotation data and classify genes based on intronic and UTR information, Fig. 2.

Repository: https://github.com/GEmilioHO/IGFinder.

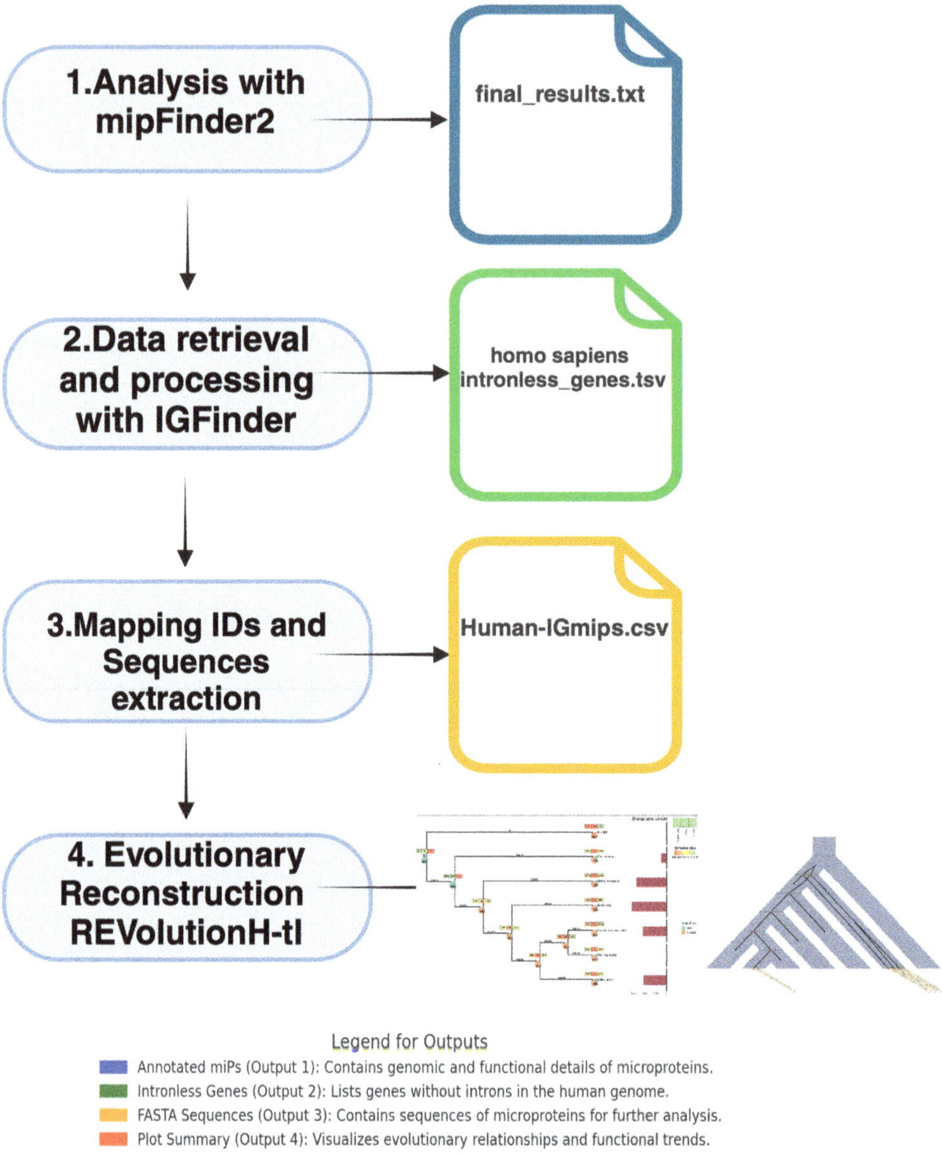

Fig. 1 Bioinformatics pipeline for analyzing the evolutionary history of human-IG encoded miPs. The workflow integrates multiple steps (1) miPFinder2 identifies human-miPs, producing functional annotations; (2) IGFinder retrieves and processes intronless gene data for further analysis; (3) sequence mapping and extraction prepare the data for evolutionary analysis, generating the Human-IGmips.csv file; and (4) REvolutionH-tl explores evolutionary reconstruction, including orthology/paralogy relationships, event-labeled gene tree reconciliation, and gene content evolution visualizations. The outputs of these tools include annotated human microproteins, human intronless gene datasets, sequence files, and comprehensive evolutionary plots of IG-miP encoded proteins in the human genome. (Image created using BioRender.com)

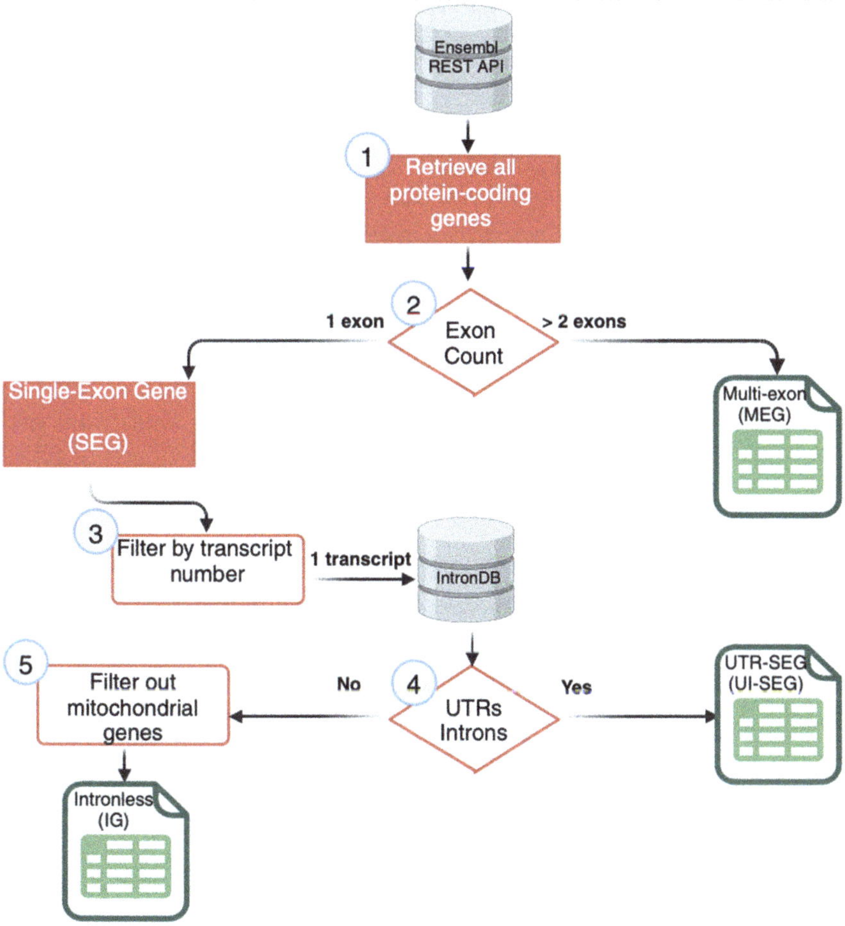

Fig. 2 Bioinformatics pipeline for generating IG, uiSEG and MEG datasets. This pipeline retrieves all protein-coding genes from the Ensembl REST API and categorizes them based on exon count. (automated steps 1:4) Genes with a single exon (single-exon genes, SEGs) are filtered by transcript number to identify those with exactly one transcript. These are further classified into Intronless Genes (IGs) if no introns are present in the untranslated regions (UTRs) or UTR-SEG (UI-SEGs) if introns are found in the UTRs. Genes with more than one exon are categorized as Multi-Exon Genes (MEGs). The process employs data integration with IntronDB to ensure precise characterization of intronic content within UTRs. Then, manual curation (step 5) excludes mitochondrial genes. Outputs are represented as green file squares. (Image created using BioRender.com)

Installation: Ensure Python 3.13.0 (latest) is installed: https://www.python.org/downloads/. Then, install the required libraries in bash:

```
% pip install requests pandas
```

Download fetch_intronless_genes.py and fetch_genes.py from the repository: https://github.com/GEmilioHO/IGFinder/tree/v1.0.0.

2.2.3 Material Required to Run R Codes to Identify the Human IG-Encoded miPs for Evolutionary Analysis

Description: This script demonstrates how to map Ensembl Gene IDs to Ensembl Peptide ID versions using the biomaRt package. The results are saved as a CSV file for further evolutionary analyses.

Repository: https://rpubs.com/kap84/1264360.

Installation: Ensure R is installed following the instructions on the CRAN website https://cran.r-project.org. Optionally, install RStudio for an enhanced coding experience.

Download the R script from the repository or copy it from the following link: https://rpubs.com/kap84/1264360.

2.2.4 REvolutionH-tl

Description: A graph-based tool for identifying orthogroups, orthology and paralogy relationships, inferring gene trees, species trees, and their reconciliations, while performing evolutionary analyses.

Repository: https://pypi.org/project/revolutionhtl/.

Installation: Ensure *REvolutionH-tl* is installed using the following command in bash:

% pip install revolutionhtl.

3 Methods

3.1 Identification of Microproteins with miPFinder2

miPFinder2 analyzes the human proteome from the Uniprot DB to identify, classify, and annotate potential microproteins. It leverages specific criteria, such as sequence length and functional domains, to distinguish microproteins from other peptides, enabling insights into their biological roles.

Input files:

(a) UP000005640_9606.fasta: Uniprot fasta file containing the protein sequences of human genes to search for miPs. Download from https://ftp.uniprot.org/pub/databases/uniprot/current_release/knowledgebase/reference_proteomes/Eukaryota/.

(b) Known_microproteins.fasta: A fasta file of known microproteins for the organism (*can be empty but must exist*).

(c) interpro_database.tsv: A three-column TSV file with a list of InterPro domains and their types. Download from InterPro: InterPro Entry List. https://www.ebi.ac.uk/interpro/download/. Alternatively, download using bash:

% curl -o interpro_database.tsv https://ftp.ebi.ac.uk/pub/databases/interpro/current_release/entry.list. An example is found in Electronic Supplemental file 2.

(d) go.obo: A gene ontology annotation file. Download from the Gene Ontology (GO) Consortium: GO Annotation File: http://purl.obolibrary.org/obo/go.obo

(e) uniprot_to_interpro.tsv: A tab-separated file mapping UniProt accessions to InterPro identifiers. An example is found in Electronic Supplemental file 3. It must contain the following three columns:

1. UniProt Accession: Unique identifier for a UniProt protein entry.
2. Sequence Version: The specific version of the protein sequence (important for isoforms).
3. InterPro Identifiers: A semicolon-separated list of InterPro domain identifiers associated with the UniProt protein. Download from UniProt API (UniProt to InterPro Mapping) or use bash with the following command line:

% curl -o uniprot_to_interpro.tsv "https://rest.uniprot.org/uniprotkb/stream?fields=accession%2Csequence_version%2Cxref_interpro&format=tsv&query=%28%28taxonomy_id%3A9606%29%29"

(f) uniprot_to_go.tsv: UniProt to GO Mapping. A tab-separated file mapping UniProt accessions to GO terms. This file includes:

1. UniProt Accession: Unique identifier for a UniProt protein entry.
2. Sequence Version: The specific version of the protein sequence (important for isoforms).
3. GO Terms: A semicolon-separated list of associated GO terms.

This file is generated mapping via the UniProt API or via Bash as follows:

% curl -o uniprot_to_go.tsv.gz "https://rest.uniprot.org/uniprotkb/stream?compressed=true&fields=accession%2Csequence_version%2Cgo&format=tsv&query=%28%28taxonomy_id%3A9606%29%2"

% gunzip uniprot_to_go.tsv.gz.

(g) configuration.ini. A file specifying the paths to all required input files for running the miPFinder2 tool. Download and modify from https://github.com/ku-mip/mipfinder2/blob/master/configuration.ini. In this file, set the organism_identifier to whatever name you want to provide to the results.

Run miPFinder2. Execute the tool using bash:

% ./mipfinder ../configuration.ini

Outputs: Results will be saved in results_folder_organism_identifier/containing the following outputs: cmip_results.fasta and final_results.txt with the potential miPs.

These outputs could offer critical insights for advanced analyses. For instance, the final_results.txt provides detailed annotations of the potential human microproteins, such as their stability, functional domains, evolutionary lineage, and homologues, making it a valuable resource for studying their biological roles and evolutionary significance. While the cmip_results.fasta, containing the microproteins amino acid sequences, could enable their experimental validation, structural modeling, and integration into multi-omics pipelines, facilitating a deeper understanding of microprotein function and regulation.

3.2 Generate a Comprehensive List of Ensembl IDs for the REvolutionH-tl Analyses

To perform the evolutionary analysis for the list of human miPs, Ensembl IDs will be required. Extract the UniProt isoform IDs from the final_results.txt file and map them to Ensembl IDs using the UniProt ID mapping tool. Follow these steps:

Extract ACCESSION_NAME: Use the following command to extract "*column 1*" from the final_results.txt:

% cut -f1 final_results.txt > accession_names.txt.

Map UniProt IDs to Ensembl IDs: Navigate to https://www.uniprot.org/id-mapping. Paste the content of accession_names.txt into the "Your identifiers" field. Select the following options:

From: UniProtKB AC/ID.

To: Ensembl Protein.

Click "Start" and download the resulting file containing the mapped Ensembl protein version IDs. Example: *"ENSP00000373254.5"*.

Output: The final output is a list of UniProtKB AC/ID corresponding to Ensembl protein version IDs saved in the downloaded file as uniprot_mapping_results.tsv.

3.3 Identification of IGs Using IGFinder

IGFinder identifies human IGs by analyzing Ensembl genomics data to detect genes lacking introns. It classifies these genes based on structural features such as their biotype, genomic location (start and end coordinates), and chromosomal. Additionally, it provides the gene symbol and a detailed description for each gene. For example, identified human IGs include SRY-box transcription factor 12 (SOX12) on chromosome 20 and zinc finger CCHC-type containing 3 (ZCCHC3), also located on chromosome 20. This classification facilitates functional insights into IGs, providing their genomic context and annotations to explore their roles in biological processes.

Input File: intron_UTR_db.txt: This file contains Ensembl information about genomic coordinates, introns, and UTRs for genes in the specified organism. Download this file from http://www.nextgenbioinformatics.org/IntronDB. Also available at https://drive.google.com/file/d/1c1t4eMQZ4SSHD0-iUiHph8eivTDYUq0l/view?usp=share_link.

Run IGFinder. Fetch genome annotations and identify IGs using the following bash command line:

% python fetch_intronless_genes.py homo sapiens "/path/to/intron_UTR_db.txt"

Outputs:

(a) homo_sapiens-ui_single_exon_genes.tsv: UTR intron-containing single-exon genes.

(b) homo_sapiens-multi_exon_genes.tsv: Multi-exon genes.

(c) homo_sapiens-intronless_genes.tsv: Intronless genes.

The outputs from IGFinder categorize eukaryotic genes based on their structural features, supporting targeted genomic and evolutionary studies. The homo_sapiens-ui_single_exon_genes.tsv file identifies human single-exon genes with untranslated regions (UTRs) and introns, highlighting unique regulatory elements and potential functional differences. The homo_sapiens-multi_exon_genes.tsv file lists genes with multiple exons, serving as a reference for comparative analyses against IGs and providing insights into gene structure complexity. Finally, the homo_sapiens-intronless_genes.tsv file focuses on completely human IGs, which are essential for understanding streamlined gene organization, rapid transcriptional responses, and their evolutionary significance. In this chapter, we will use this file to identify human IG-encoded miPs.

3.4 Mapping and Matching Peptide IDs Using miPFinder2 and IGFinder Outputs

This workflow describes the computational steps for mapping Ensembl Gene IDs to Ensembl Peptide ID Versions using the biomaRt package in RStudio. It merges data from miPFinder2 (potential human miPs) and IGFinder (human IGs), extracting the corresponding matching Ensembl peptide IDs. The process generates a CSV file, facilitating downstream evolutionary analyses.

Input Files:

(a) homo_sapiens-intronless_genes.tsv.

(b) uniprot_mapping_results.tsv.

Step 1: Load the Required Library. To interact with the Ensembl database, the biomaRt R package is used. This library provides a powerful interface to retrieve biological data.

R code: % library(biomaRt).

Step 2: Input Data Preparation. The workflow begins by loading the file (homo_sapiens-intronless_genes.tsv) containing Ensembl Gene IDs in the first column. Run the following commands in the R console or Rstudio:

R code: % input_file <- "homo_sapiens-intronless_genes.tsv"

R code: % data <- read.delim(input_file, header = TRUE)

R code: % ensembl_gene_ids <- data[[1]]

Step 3: Mapping Gene IDs to Peptide ID Versions. Convert Ensembl Gene IDs to Ensembl Peptide ID Versions using biomaRt, and we map the Ensembl Gene IDs to their corresponding Ensembl Peptide ID Versions.

R code: % mart <- useMart("ensembl", dataset = "hsapiens_-gene_ensembl") gene_peptide_versions <- getBM(attributes = c("ensembl_gene_id", "ensembl_peptide_id_version"), filters = "ensembl_gen-e_id", values = ensembl_gene_ids, mart = mart)

Step 4: Integrating mipFinder2 Data. The miPFinder2 output (uniprot_mapping_results.tsv) contains peptide IDs from UniProt mapped to Ensembl Gene IDs. This data is read into R and filtered to extract the Ensembl Peptide ID Versions for matching.

Example file:

To	From
Q7Z2F6	ENSP00000318635.4
P12345	ENSP00000488995.1

We load UniProt mapping results and extract Ensembl Peptide ID Versions for matching.

R code: % uniprot_file <- "/Users/path/to/uniprot_mappin-g_results.tsv"

R code: % uniprot_file <- "/Users/path/to/uniprot_mappin-g_results.tsv"

Extract the column with Ensembl Peptide ID Versions.

R code: % ensembl_peptide_versions <- uniprot_data[[2]]

Step 5: Matching Peptide IDs. The mapped Ensembl Peptide ID Versions from biomaRt are compared against the UniProt-derived data. Matching IDs are extracted and stored in a new dataset.

R code: % matched_versions <- gene_peptide_versions $ensembl_peptide_id_version[gene_peptide_versions $ensembl_peptide_id_version %in% ensembl_peptide_versions]

Step 6: Exporting the Results. The final dataset containing the matched Ensembl Peptide IDs will be saved as a CSV file Human-IGmips.csv for further analysis. This file contains the list of human IG-encoded miPs:

R code: % write.csv(matched_versions, "Human-IGmips.csv", row.names = FALSE)

3.5 IG-Encoded miPs Evolutionary Analysis with REvolutionH-tl1.3.0

The evolutionary analysis of IG-encoded miPs with REvolutionH-tl unravels their evolutionary trajectories, and could explore functional conservation, as well as uncover lineage-specific adaptations. This approach provides information on the biological significance of miPs across diverse species.

REvolutionH-tl runs five different steps for evolutionary analysis: (1) sequence alignments, (2) best hits computation and orthogroup detection, (3) best matches, orthology, and gene tree inference, (4) species tree inference, and (5) species tree and gene tree reconciliation. Although all steps can be run by default, in this analysis it is important to run the first steps 1 and 2, filter some results and then run steps 3 and 5. Step 4 is not necessary for this study.

Input Files:

(a) *Proteomes*: Complete peptide sequences of the following seven vertebrate species in FASTA format, downloaded from Ensembl:

1. Homo_sapiens.GRCh38.pep.all.fa.gz: https://ftp.ensembl.org/pub/current_fasta/homo_sapiens/pep/
2. Mus_musculus.GRCm39.pep.all.fa.gz: https://ftp.ensembl.org/pub/current_fasta/mus_musculus/pep/
3. Pan_troglodytes.Pan_tro_3.0.pep.all.fa.gz: https://ftp.ensembl.org/pub/current_fasta/pan_troglodytes/pep/:
4. Monodelphis_domestica.ASM229v1.pep.all.fa.gz https://ftp.ensembl.org/pub/current_fasta/monodelphis_domestica/pep/

5. Rattus_norvegicus.mRatBN7.2.pep.all.fa.gz https://ftp.ensembl.org/pub/current_fasta/rattus_norvegicus/pep/
6. Gallus_gallus.bGalGal1.mat.broiler.GRCg7b.pep.all.fa.gz https://ftp.ensembl.org/pub/current_fasta/gallus_gallus/
7. Danio_rerio.GRCz11.pep.all.fa.gz: https://ftp.ensembl.org/pub/current_fasta/danio_rerio/pep/

Create a directory fasta_files and put all these files in there. Then, decompress each file, you can use the following bash command line to decompress all files:

% gunzip **.gz

To make file names more concise and user-friendly, we recommend changing file names to shorten them to species identifiers. The following bash command line can be used:

% mv Homo_sapiens.GRCh38.pep.all.fa Human.fa

Rename all files using the same format as the command line.

Below is the mapping of original file names to their simplified names:

1. Homo_sapiens.GRCh38.pep.all.fa → Human.fa
2. Mus_musculus.GRCm39.pep.all.fa → Mouse.fa
3. Pan_troglodytes.Pan_tro_3.0.pep.all.fa → Chimp.fa
4. Monodelphis_domestica.ASM229v1.pep.all.fa → Mondo.fa
5. Rattus_norvegicus.mRatBN7.2.pep.all.fa → Rat.fa
6. Gallus_gallus.bGalGal1.mat.broiler.GRCg7b.pep.all.fa → Chick.fa
7. Danio_rerio.GRCz11.pep.all.fa → Zebrafish.fa

(b) *Species Tree*: A Newick extended file (species_tree.nhx) containing a species tree of the species under study. This file includes the following tree:

((((Rat,Mouse),(Human,Chimp)),Mondo),Chick),Zebrafish);

Run REvolutionH-tl using the following bash command:

% python -m revolutionhtl -steps 1 2 -F fasta_files

This command will carry out the following:

1. *Alignment Hits Computation*: Compute alignment hits between all proteins in the proteomes.
2. *Best Hits Computation and Orthogroup Detection*: Compute best hits in a species for each protein in another species, as well as detect sets of genes that originated from an ancestral

gene, called orthogroups (OG), referred as well as gene families.

Outputs:

(a) tl_project.best_hits.tsv: Best alignment hits.

(b) tl_project.orthogroups.tsv: Orthogroups.

(c) tl_project.singletones.tsv: List of species-specific genes.

3. *Gene Tree Reconstruction*: To execute this step of the analysis, it is necessary to filter the tl_project.orthogroups.tsv and tl_project.best_hits.tsv using the file the Human-IGmips.csv file. For each file, the following commands retain only the rows corresponding to the orthogroups of interest, along with the header. The filtered data is saved into new files: filtered.best_hits.tsv and filtered.orthogroups.tsv.

Filter orthogroups that contain human IG-miPs Create a file filtered_orthogroups containing the lines that contain IDs extracted from Human-IGmips.csv:

```
awk 'NR==FNR {gsub(/"/, "", $0); if (FNR > 1) ids[$0]; next} FNR==1 {print; next} {for (id in ids) if ($0 ~ id) {print; break}}' Human-IGmips.csv tl_project.orthogroups.tsv > filtered_orthogroups.tsv
```

Output The resulting file, filtered.orthogroups.tsv, will contain only orthogroups (rows) where IG-miP IDs are identified. You can review the file using:

```
% head filtered_orthogroups.tsv
```

Secondly, to filter the file tl_project.best_hits.tsv to obtain only hits of pairs of proteins that belong to an orthogroup of interest, run the following bash command line:

```
cut -f1 filtered_orthogroups.tsv | sed 's/^[[:space:]]//;s/[[:space:]]$//' | awk 'NR==FNR {ids[$1]; next} $1 in ids' - tl_project.best_hits.tsv > filtered_best_hits.tsv
```

Output: After execution, the filtered file will be saved as filtered_best_hits.tsv.

After obtaining the input filtered files, compute step 3:

```
% python -m revolutionhtl -steps 3    best_h filtered_best_hits.tsv
```

Outputs:

(a) tl_project.gene_trees.tsv: Reconstructed gene trees.

(b) tl_project.orthologs.tsv: Orthologous relationships.

(c) tl_project.best_matches.tsv: Best matches between sequences.

4. *Tree Reconciliation*: Reconcile gene and species trees to depict the evolutionary relationships that genes went through along the evolution of species, using the following command:

```
% python -m revolutionhtl -steps 5 -T tl_project.gene_trees.tsv
    -S species_tree.nhx
```

Explanation of Parameters:

- **Step 5**: Specifies that only Step 5 (tree reconciliation) will be executed.
- **T**: Input file containing the gene trees (tl_project.gene_trees.tsv).
- **S**: Input file containing the species tree in extended Newick format (species_tree.nhx).

Outputs:

(a) tl_project.reconciliation.tsv: Contains the reconciled gene and species trees, showing speciation, duplication, and loss events.

(b) tl_project.corrected_trees.tsv: Gene trees corrected based on reconciliation with the species tree.

(c) tl_project.labeled_species_tree.nhx: Annotated species tree with evolutionary events, in Newick extended format.

Important: The file tl_project.corrected_trees.tsv contains all the gene trees under study. However, some gene trees may be inconsistent with the species tree. To address these inconsistencies, the tool edits the gene trees by removing the genes causing the conflict.

In the last column of the file, a value of **0** indicates that the gene tree is consistent with the species tree and was not modified. A value greater than **0** represents the number of genes that were removed to make the gene tree consistent with the species tree.

Visualization. Use the following commands in bash to generate visual summaries of the results:

Summarize Inferred Evolutionary Histories*:*

% python -m revolutionhtl.plot_summary

Example is shown in Fig. 3. Evolutionary reconstruction of IG-Encoded miPs in the Human Genome Across Selected Species.

Visualize the Reconciliation of a Specific Orthogroup*:*

% python -m revolutionhtl.plot_reconciliation <orthogroup_ID>

Important: Make sure to run this command in the directory where files tl_project.reconciliation.tsv and tl_project.

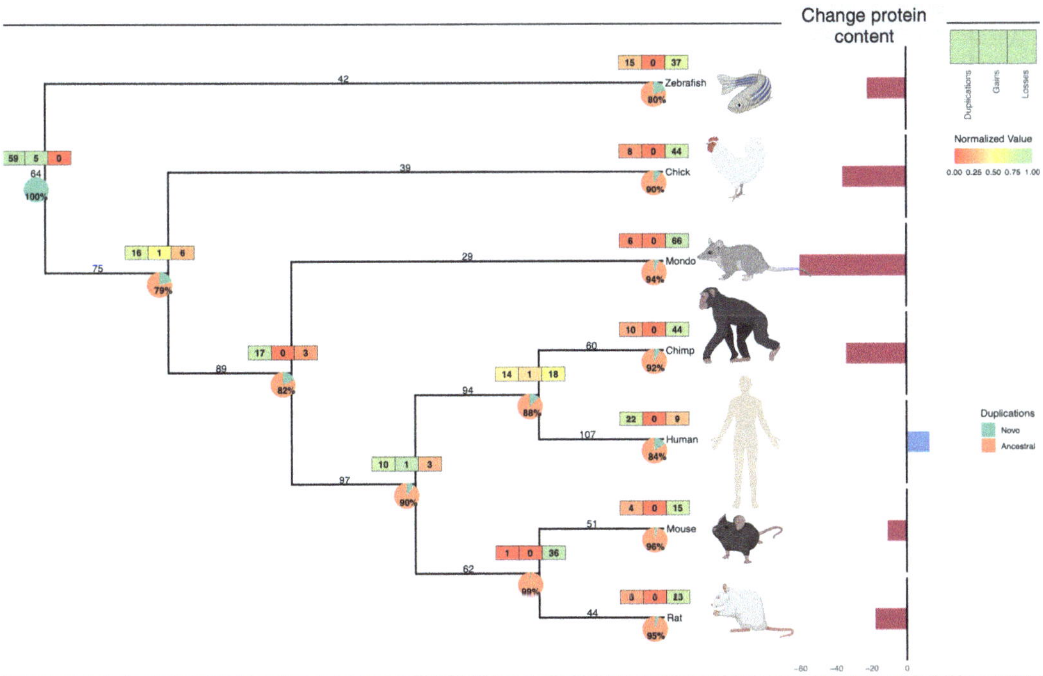

Fig. 3 Evolutionary reconstruction of IG-encoded miPs in the human genome across selected species. This figure highlights the capability of the tool to analyze and visualize the evolutionary reconstruction of gene content across species. The phylogenetic tree tracks gene duplications (green), gains (yellow), and losses (red) along each branch, with nodes representing ancestral states and duplications categorized as *'Novo'* (green) or *'Ancestral'* (orange). The bar plots on the right display normalized changes in gene content, revealing significant variations in clades relevant to human IG-encoded miPs. Species names and number of IG-miPs in each branch enhance clarity and emphasize how the REvolutionH-tl provides insights into the structural, functional, and evolutionary trajectories of these proteins, which could be helpful to advancing research in evolutionary, functional genomics and disease mechanisms. The figure is an output of REvolutionh-tl, adapted to include drawings to illustrate species

labeled_species_tree.nhx are placed, since this command will need them.

Final Expected Outputs. Graphical visualizations summarizing:

(a) *Phylogenetic and orthologous group analysis across species. Example, Fig. 4.*

(b) *Reconciliation of a gene tree and the species tree showing evolutionary events: speciation, duplications and losses. Example, Fig. 5.*

The analyses performed by REvolutionH-tl enable the study of the evolutionary history of gene or protein sets. Specifically, for IGs-miPs, the detection of orthologs is highly valuable for functional genomics, as it helps identify proteins in other vertebrates that might also function as miPs. Additionally, identifying

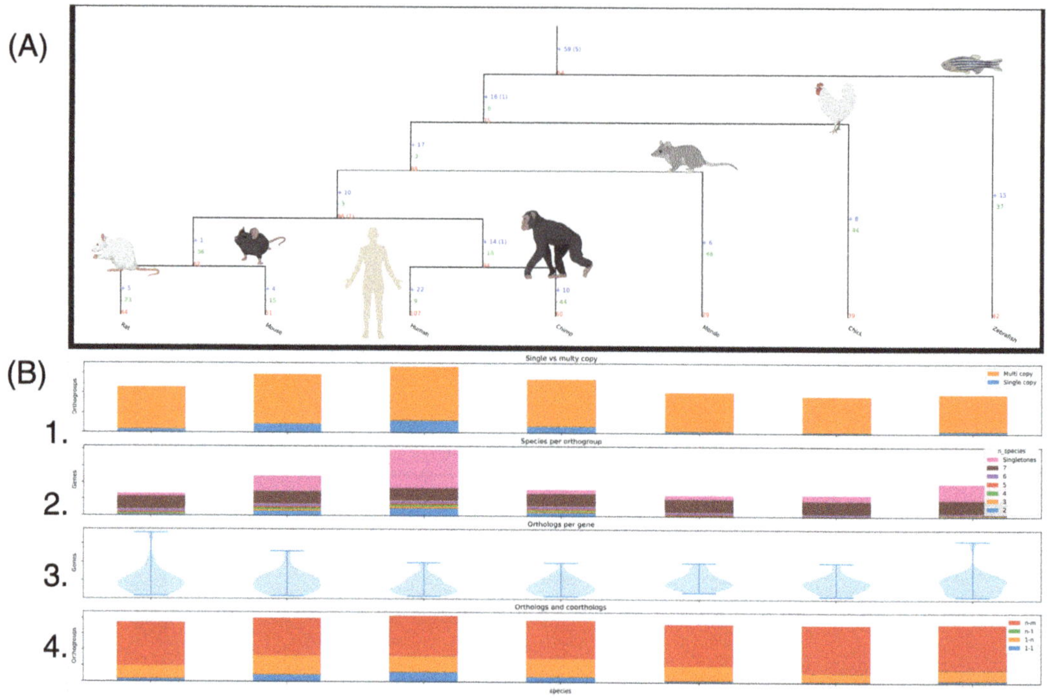

Fig. 4 Phylogenetic and orthologous group analysis across species. (**a**) Phylogenetic tree representing the evolutionary relationships between the analyzed species, including human, chimpanzee, mouse, rat, marsupial, chicken, and zebrafish. Numbers in branch represent evolutionary events: blue for the number of duplication and gain events (between brackets), green the number of losses, and red the number of proteins in the species. (**b**) Summary of orthologous group statistics: (**b**1) Single-copy vs. multi-copy genes in each species. (**b**2) Number of genes of single copy vs. number of genes found in two species, three species, and so on. (**b**3) Distribution of orthologs per gene, represented as violin plots, to illustrate variation across species. (**b**4) Percentage of 1-1, m-1, 1-n, and m-n orthologs for genes in each species, suggesting co-orthologs that may have arisen via gene duplications

orthogroups highlights miPs with multiple orthologs, those that have undergone duplications, and those that were lost in specific species.

The gene trees inferred by REvolutionH-tl are event-labeled, with internal nodes tagged as duplication or speciation events. This labeling provides valuable insights into the origins of specific IG-miPs. Moreover, reconciliation with the species tree facilitates the identification of lost proteins of this type and maps duplication events onto the branches of the species tree. This allows researchers to determine whether certain miPs originated recently or have a more ancestral lineage.

NOTCH-regulated ankyrin repeat protein

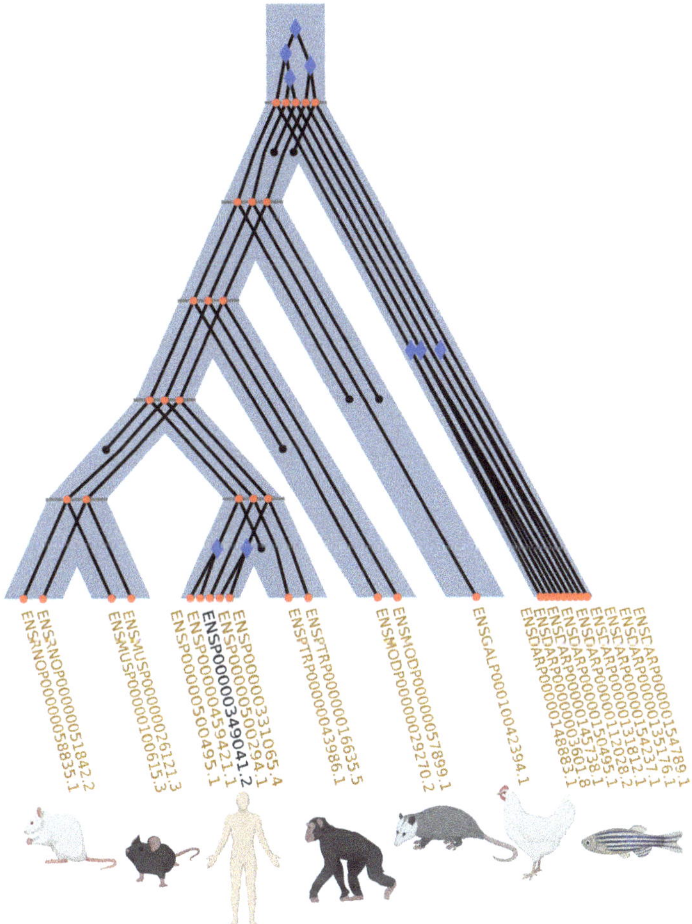

Fig. 5 Reconciled gene tree for an orthogroup: Evolutionary reconstruction of NOTCH-regulated ankyrin repeat protein (NRARP). This figure shows the reconciliation of gene trees with the species tree for a gene family (orthogroup), emphasizing the evolutionary trajectory of the NRARP family (ENSP00000349041.2). Blue nodes represent inferred gene duplication events, black nodes represent gene losses, while red internal nodes indicate ancestral divergence points. Leaves correspond to orthologs from different species. This analysis highlights interspecies relationships, revealing evolutionary adaptations and the critical role of NRARP in regulating Notch-Wnt signaling balance, angiogenesis, and tissue regeneration, particularly in zebrafish, where its duplications suggest robust regeneration and axis formation. Additionally, its conservation across vertebrates underlines its importance in lineage-specific developmental processes mediated by miPs. In humans, NRARP's role as a regulator of Notch signaling links it to cancer progression, specifically tumor angiogenesis and tissue homeostasis

4 Notes

1. Ensure input sequences for miPFinder2 are in a UniProt fasta format with accurate annotations (https://github.com/ku-mip/mipfinder2/tree/master/include).
2. Set detection thresholds for miPFinder2 (e.g., $\leq$*140* aa, conservation scores).
3. Verify miPFinder2 output files include microprotein IDs and functional annotations (https://github.com/ku-mip/mipfinder2/tree/master/include).
4. Validate IGFinder output files include genomic locations and functional annotations. (https://github.com/GEmilioHO/IGFinder/tree/main)
5. Replace [organism_name] with the desired organism's name (default: gallus gallus, e.g., homo sapiens) and specify the path to intron_UTR_db.txt, a required input file (https://github.com/GEmilioHO/IGFinder/tree/main).
6. Ensure all required inputs for REvolutionH-tl are correctly formatted and placed in their respective directory. Forbidden characters: avoid using the following characters in species names, gene names, or other parameters within the tool; and (https://pypi.org/project/revolutionhtl/1.3.0/)
7. For troubleshooting, refer to the REvolutionH-tl documentation. (https://pypi.org/project/revolutionhtl/1.3.0/)
8. Verify all dependencies (e.g., NumPy and Pandas) are properly installed.
9. Follow the recommended procedure for running REvolutionH-tl [19]: (https://pypi.org/project/revolutionhtl/1.3.0/)
   (i) It is preferable to use complete proteomes for each species in the fasta input files, although, the tool can be run for gene families or sets of proteins.
   (ii) Select orthogroups of interest after obtaining best hits. You must also filter your best hits file to work only with genes that are in the orthogroups of interest. Replace the original files tl_project.orthogroups.tsv and tl_project.best_hits.tsv with the filtered files.
   (iii) Avoid working with genomes/proteomes/gene families fasta files of different sizes for steps 3 and 4, as this may result in underrepresented species and an unreliable phylogenetic tree, for example, using a fasta file of *1000* genes and another one of *50* might cause species tree inferences with missing species.

(iv) Outputs from the REvolutionH-tl are in the directory_alignment_all_vs_all/ that contains all computed alignment hits.

(v) If plotting issues occur, confirm output files from previous steps are complete and correctly formatted.

References

1. Graeff M, Wenkel S (2012) Regulation of protein function by interfering protein species. Biomol Concepts 3(1):71–78. https://doi.org/10.1515/bmc.2011.053
2. Staudt AC, Wenkel S (2011) Regulation of protein function by 'microProteins'. EMBO Rep 12(1):35–42. https://doi.org/10.1038/embor.2010.196
3. Couso JP, Patraquim P (2017) Classification and function of small open reading frames. Nat Rev Mol Cell Biol 18(9):575–589. https://doi.org/10.1038/nrm.2017.58
4. Wu Q, Zhong S, Shi H (2022) MicroProteins: dynamic and accurate regulation of protein activity. J Integr Plant Biol 64(4):812–820. https://doi.org/10.1111/jipb.13229
5. Ruiz-Orera J, Albà MM (2019) Translation of small open reading frames: roles in regulation and evolutionary innovation. Trends in genetics: TIG 35(3):186–198. https://doi.org/10.1016/j.tig.2018.12.003
6. Valdivia-Francia F, Sendoel A (2024) No country for old methods: new tools for studying microproteins. iScience 27(2):108972. https://doi.org/10.1016/j.isci.2024.108972
7. Rathore A, Martinez TF, Chu Q, Saghatelian A (2018) Small, but mighty? Searching for human microproteins and their potential for understanding health and disease. Expert Rev Proteomics 15(12):963–965. https://doi.org/10.1080/14789450.2018.1547194
8. Rothnagel J, Gowda H, Menschaert G (2022) Editorial: sORF encoded peptides in health and disease. Front Genet 13:878014. https://doi.org/10.3389/fgene.2022.878014
9. Daisy Precilla S, Biswas I, Anitha TS, Agleshkumar B (2024) Microproteins unveiling new dimensions in cancer. Funct Integr Genomics 24(5):152. https://doi.org/10.1007/s10142-024-01426-8
10. Hofman DA, Prensner JR, van Heesch S (2024) Microproteins in cancer: identification, biological functions, and clinical implications. Trends Genetics S0168-9525(24):00211–00217. Advance online publication. https://doi.org/10.1016/j.tig.2024.09.002
11. Rodrigues P, Bangali H, Ali E, Sharma MK, Abdullaev B, Alkhafaji AT, Deorari MM, Zabibah RS, Haslany A (2024) Microproteins/micropeptides dysregulation contributes to cancer progression and development: a mechanistic review. Cell Biol Int 48(10):1395–1405
12. Jorquera R, González C, Clausen P, Petersen B, Holmes DS (2018) Improved ontology for eukaryotic single-exon coding sequences in biological databases. Database 2018:1–6. https://doi.org/10.1093/database/bay089
13. Aviña-Padilla K, Ramírez-Rafael JA, Herrera-Oropeza GE, Muley VY, Valdivia DI, Díaz-Valenzuela E, García-García A, Varela-Echavarría A, Hernández-Rosales M (2021) Evolutionary perspective and expression analysis of Intronless genes highlight the conservation of their regulatory role. Front Genet 12:654256. https://doi.org/10.3389/fgene.2021.654256
14. Bramblett DE, Copeland NG, Jenkins NA, Tsai MJ (2002) BHLHB4 is a BHLH transcriptional regulator in pancreas and brain that marks the dimesencephalic boundary. Genomics 79:402–412. https://doi.org/10.1006/geno.2002.6708
15. Kataoka K, Matsushima T, Ito Y, Sato T, Yokoyama S, Asahara H (2018) Bhlha9 regulates apical ectodermal ridge formation during limb development. J Bone Miner Metab 36:64–72. https://doi.org/10.1007/s00774-017-0820-0
16. Izzo A, Kamieniarz K, Schneider R (2008) The histone H1 family: specific members, specific functions? Biol Chem 389:333–343. https://doi.org/10.1515/bc.2008.037
17. Wang Z, Shu W, Lu MM, Morrisey EE (2005) Wnt7b activates canonical signaling in epithelial and vascular smooth muscle cells through interactions with Fzd1, Fzd10, and LRP5. Mol Cell Biol 25:5022–5030. https://doi.org/10.1128/mcb.25.12.5022-5030.2005

18. Bhati KK, Kruusvee V, Straub D, Chandran AKN, Jung KH, Wenkel S (2020) Global analysis of cereal microProteins suggests diverse roles in crop development and environmental adaptation. G3 (Bethesda, Md.) 10(10): 3709–3717. https://doi.org/10.1534/g3.120.400794

19. Ramírez-Rafael JA, Korchmaros A, Aviña-Padilla K, López Sánchez A, España-Tinajero AA, Hellmuth M, Stadler PF, Hernández-Rosales M (2024) REvolutionH-tl: reconstruction of evolutionary histories tool. In: Scornavacca C, Hernández-Rosales M (eds) Comparative genomics. Springer Nature Switzerland, pp 89–109

Chapter 12

Bioinformatic Tools for Non-bioinformaticians: A Roadmap to Study Unknown Microprotein Sequences

Patricia Gimeno Le Paih and Iker Valle Aramburu

Abstract

The identification of unknown microprotein sequences is often presented as a FASTA file with hundreds or thousands of unknown RNA or protein sequences. Then, the researcher needs to gain an overall view of all the sequences and identify which ones may be relevant to address their biological question of interest. In the absence of a bioinformatician or certain computational knowledge, this task becomes increasingly challenging proportionally to the number of sequences identified. In this chapter, we provide a guideline of different publicly available online resources that can be used without any computational background to analyze your sequences' structural and physicochemical properties and identify functional domains and motifs. Moreover, we have developed MicrOracle, a publicly available Shiny app that will integrate the different output files from the online prediction tools and generate interactive plots that can contribute to easily visualizing, understanding, and segmenting your data based on structural and functional criteria.

Key words Microproteins, Disordered proteins, Protein bioinformatics, Domain, Motif, Shiny app

1 Introduction

The study of newly identified microproteins or short open reading frames (sORFs) with coding potential often starts with a large FASTA file of multiple protein/RNA sequences from which no information is available. The sequence of amino acids forming a protein can be used to study the physicochemical properties of the polypeptide chain. Parameters such as molecular weight, net charge, hydrophobicity, or isoelectric point can be extracted from the amino acid sequence and can be helpful for the purification of recombinantly expressed microproteins. The protein sequence, to some degree, will also determine the structure of a protein. The proteome is a structural continuum between folded and intrinsically disordered proteins, being the mixed protein group of folded domains and disordered regions the most abundant in the human proteome [1]. Microproteins have been shown to be enriched in

Stephan Wenkel (ed.), *Microproteins: Methods and Protocols*, Methods in Molecular Biology, vol. 2992,
https://doi.org/10.1007/978-1-0716-5013-4_12,

protein disorder [2]. Frequently, folded domains are linked with catalysis and high-specificity binding to substrates. Disordered regions usually form transient interactions, can bind to multiple partners, and play regulatory roles in diverse cellular pathways [3, 4]. Thus, the structural predictions of unknown microprotein sequences can reveal relevant features of the identified microproteins and guide downstream analysis.

When analyzing a pool of hundreds or thousands of unknown microprotein sequences, in addition to the structural features, it is key to identify the functional elements within those sequences. The identification of domains and functional motifs can contribute to increasing the confidence of the newly identified protein sequences. It can also contribute to segmenting your sample to further perform experimental or computational experiments with a smaller sample size, reducing, for instance, the sample size of functional screenings. The broad identification of functional domains and sequence features can be performed with the functional analysis tool InterPro, a service maintained by the European Bioinformatics Institute (EMBL-EBI) [5]. InterProScan, integrates multiple member databases and classifies the sequences of interest into families. It also predicts domains and significant sites, such as transmembrane domains, signal peptides, and DNA-binding domains, among others. These annotated features are linked to other resources, such as gene ontology (GO), providing a valuable context for the identification of potential biological roles. In addition to domains, proteins often contain smaller functional elements known as short linear motifs (SLiMs). These small sequences, typically between 3 and 15 amino acids in length, are often found in intrinsically disordered regions [6, 7]. SLiMs include protein binding motifs, enzyme substrates such as kinase substrates, localization motifs such as nucleocytoplasmic trafficking sequences, and protein stability motifs like degradation signals, among others. The Eukaryotic Linear Motif database contains a catalog of more than 4000 annotated and experimentally validated motifs [8].

The identification of these conserved motifs not only increases the confidence in the authenticity of newly identified microprotein sequences but also offers insights into their potential cellular function through associated GO terms. Additionally, the protein sequence can also be used to directly predict specific functions, such as the antimicrobial probability [9, 10].

The combination of structural, physicochemical, domain analysis, motif identification, and specialized functional predictions, such as antimicrobial potential, provides the researcher with a comprehensive understanding of the newly identified protein sequences. This computational approach not only helps validate the putative newly identified proteins but also provides valuable insights for segmenting samples and directing further experimental investigations.

In this book chapter, we provide instructions on how to perform a large-scale computational analysis of protein sequences, extract structural and physicochemical features, and identify relevant domains and motifs. In the material section, we describe how to obtain different output files from diverse publicly available online tools. In the method section, we describe how to upload these output files into MicrOracle, an R-based Shiny app we have developed to visualize, integrate, and help us interpret the information obtained from the online prediction tools.

2 Materials

You will need an input protein FASTA file containing one or multiple protein sequences. A sequence in FASTA format contains:

1. A line starting with a ">" followed by the name or code to identify the sequence. This line may also contain a description of the sequence.
2. One or multiple lines of the amino acid sequence.

2.1 Prediction of Structural and Physicochemical Properties

1. *Metapredict* [11] is a high-performance, deep learning-based disorder predictor tool that generates per-residue scores from amino acid sequences reflecting the disorder likelihood. For analyzing multiple sequences, we use the Colab link below, but the software can also be downloaded. Load the different steps by clicking on the play button on the left of each step, and load your FASTA file. Follow the instructions on the Colab website (Link), documentation can be found in the following link.
   (a) Input: FASTA file.
   (b) Output: Downloadable CSV file with disorder scores for each amino acid.
   (c) Download the resulting CSV file.
2. *ALBATROSS* is a deep learning model for predicting ensemble dimensions of intrinsically disordered regions (IDRs) [12]. These include the radius of gyration, end-to-end distance, polymer-scaling exponent, and ensemble asphericity, directly from sequences at a proteome-wide scale. Load the different steps by clicking on the play button on the left of each step, and load your FASTA file. Follow the Colab website (Link) instructions and download your CSV output file.
   (a) Input: FASTA file with your protein sequences of interest.
   (b) Output: Downloadable CSV file with sequence length, radius of gyration, end-to-end distance, scaling exponent, asphericity, and prefactor.

3. *MicrOracle* utilizing the R package Peptides [13] calculates the molecular weight, net charge, hydrophobicity, and isoelectric point of each protein sequence in your input FASTA file.

2.2 Functional Prediction

1. InterProScan provides functional analysis of protein sequences, predicting domains and important sites [5]. InterPro integrates signatures from 13 different databases, and it is maintained by the European Bioinformatic Institute (EMBL-EBI). The online tool has a limit of 100 sequences per submitted job (Link). If your input dataset only contains a few hundred sequences, we recommend using the splitting tool in the Shiny app (*see* **Note 1**). If your input FASTA file contains thousands of sequences, the best option is to download Inter-ProScan and run it locally. This can be done following the detailed documentation (Link). This needs to be done on a Linux system.
   - (a) Input: FASTA file with a maximum of 100 sequences when using the web-available InterProScan (*see* **Notes 1** and **2**).
   - (b) Use the search by sequence window. You can upload a file or copy the protein sequence. In the advanced options, all the member databases and additional sequence feature tools are selected by default.
   - (c) Output: Downloadable TSV file containing domain predictions.
2. Wregex is a fast web-based tool for the prediction of SLiMs in protein sequences [14]. Wregex allows the search of predefined Wregex motifs, motifs present in the Eukaryotic Linear Motif database (ELM), and to search for custom motifs. The online tool has a limit of 100 sequences per submitted job when searching all the motifs (Link). If your input dataset only contains a few hundred sequences, we recommend using the splitting tool in the shiny app (*see* **Note 1**). If your input FASTA file contains thousands of sequences, the best option is to download it and run it locally.
   - (a) Input: FASTA file with up to 100 sequences in the "All motifs" search (*see* **Notes 1** and **2**).
   - (b) Go to the search section. Under Motif, select "All motifs". Under Target, select FASTA entry and copy the micro-protein sequence of interest or upload the input FASA file. Leave the default Options unless specific settings are needed for your specific search. Click search.
   - (c) Output: Downloadable CSV file with motif predictions.
3. Antimicrobial Peptide Scanner [15] (Link) is a tool that uses a deep neural network to predict if a protein sequence may be an

antimicrobial peptide (AMP) active against Gram-positive and/or Gram-negative bacteria. Sequences are considered AMP when the prediction probability is >0.5.

(a) Input: FASTA file. The file should contain between 1 and 50.000 protein sequences. Sequences should be between 10 and 200 amino acids in length. For longer sequences, the last 200 amino acids will be considered. The FASTA file should not be bigger than 50 MB. For splitting the input FASTA file based on size, *see* **Note 1**.

(b) Click on "Choose file" and upload your input FASTA file. Then click on start prediction.

(c) Output: Zip folder containing a FASTA file of the analyzed sequences and a CSV file with the AMP prediction probability.

3 Methods

In this section, we provide instructions on how to incorporate the different output files and navigate through MicrOracle to visualize and interact with your data. All the displayed plots are downloadable as PDF files, and the tables as CSV files.

3.1 Starting MicrOracle

1. Open the Shiny App MicrOracle.
2. In the Main tab, on the left side of the window, upload the following files: Input FASTA file, InterProScan TSV file, Wregex CSV file, ALBATROSS CSV file, MetaPredict CSV file, and the Antimicrobial peptide prediction CSV file (Fig. 1a). On the right side, one can find a downloadable updated manual and different tabs displaying your data.

3.2 Protein Structure Tab

1. This tab plots the structural information of your microprotein sequences. It represents the data obtained from Metapredict, from MobiDBLite (a disorder database incorporated in InterProScan), Coils (a database containing known two-stranded coils incorporated in InterProScan) and TMHMM (transmembrane predictor incorporated in InterProScan).
2. The left panel contains the controls where the disorder threshold (0,5 by default) and the length of the protein sequences used can be modified.
3. All the proteins are grouped into three groups: folded, disordered, or mixed based on the % of disordered residues per sequence length (for the Metapredict output, MobiDBLite will only display disordered and not disordered). Proteins are also grouped depending on whether they have a

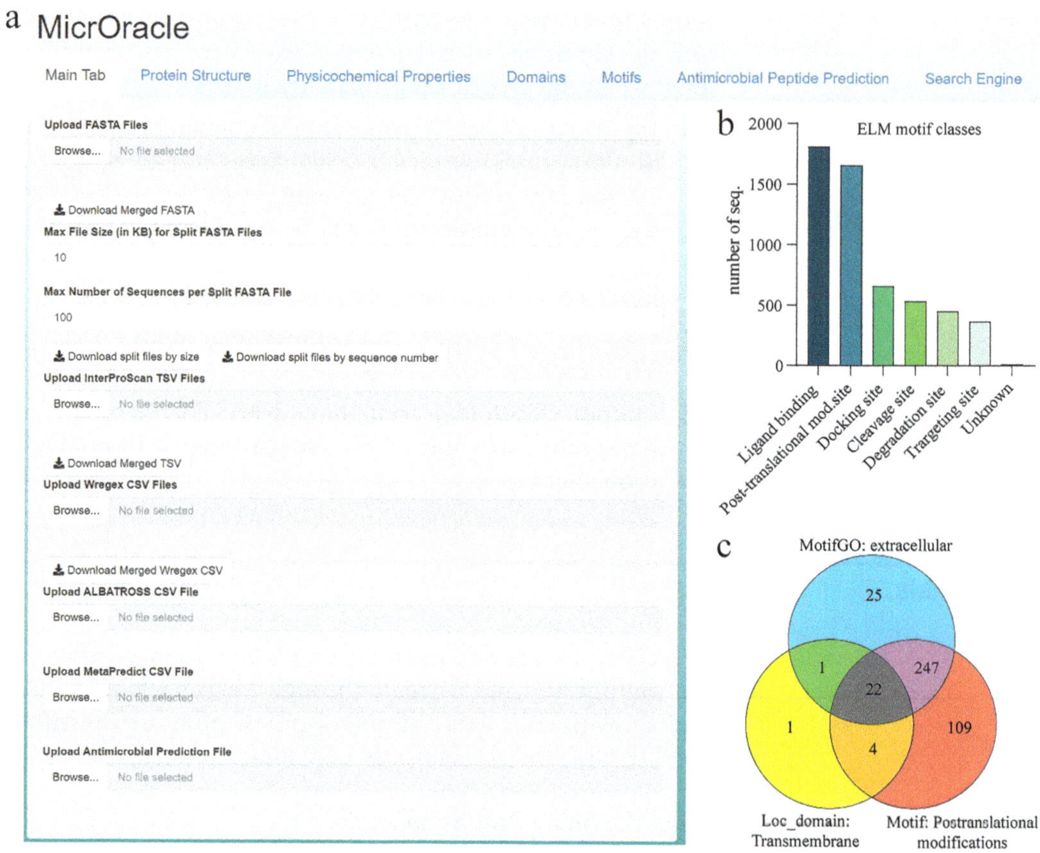

Fig. 1 (**a**) Overview of the initial window of MicrOracle: The main panel contains the windows to upload the input FASTA file, split this file based on size or number of protein sequences. Upload the InterProScan TSV file (s), which could be merged in case multiple files are used. Upload the Wregex CSV file(s), which could also be merged in case multiple files are used. Upload ALBATROSS CSV output file. Upload Metapredict CSV output file and the Antimicrobial Prediction CSV file. On the right side, the different tabs are displayed. From left to right: protein structure, physicochemical properties, domains, motifs, antimicrobial peptide prediction, and search engine. (**b**) Barplot generated from the motif analysis from Wregex, the different identified motifs can be grouped by the motif classes on the eukaryotic linear motif database. (**c**) Example of a Venn diagram that can be created via the search engine tab

transmembrane domain helix (predicted by TMHMM) or a coil (obtained from Coils).

4. FASTA and CSV files of individual groups can be downloaded.
5. The output of MetaPredict and MobiDBLite representing the percentage of disordered and folded proteins is shown on downloadable pie charts. For the Coils and TMHMM outputs, the pie chart will display the percentage of proteins with found transmembrane helices and/and coils.
6. The percentage of disordered, mixed, and folded proteins versus sequence length is plotted as a bar plot.

7. The percentage of disordered, mixed, and folded proteins versus the type of structure, i.e., coils, transmembrane helix or none, is also plotted as a bar plot.

3.3 Physicochemical Properties Tab

1. This tab displays a scatter plot with the mean hydrophobicity versus Net charge from which a selection can be drawn with the cursor. Selected proteins will be displayed in the table below, which is downloadable. The CSV file will contain additional physicochemical and disordered descriptors.
2. The isoelectric point of the input protein sequences is displayed as a histogram.

3.4 Domains Tab

1. This tab visualizes the output from InterProScan.
2. The "Select GO Terms to see in table format" window displays by default the 5 most abundant GO terms identified in your dataset. Additional GO terms can be added to the select window and will be plotted on the bar plot on the right.
3. GO terms can be added to the "Select categories to see in table format" window, which will be displayed as a downloadable table.
4. The pie chart shows the percentage and number of sequences from the input file that have been identified with a domain or sequence feature by InterProScan.
5. Bar plot showing the top 5 most abundant GO domain titles; GO domain titles can be added or removed.
6. The GO terms linked to InterPro entries can be split into Molecular function, Biological Processes and Cellular Component. Bar plots are displayed with GO terms found in these three categories. On the side tab, has three select windows allowing to add of more GO terms to be plotted.
7. InterProScan uses databases that are not integrated into InterPro and do not contain an InterPro entry and a linked GO term. Sequences identified by these search engines will identify signal peptides, whether a protein sequence is cytoplasmic, transmembrane or non-cytoplasmic. A bar plot representing identified domains without an InterPro entry is also displayed.
8. By using the selection window "Select Domains by Localization to see in table format" the user can select GO terms to be displayed in a downloadable table.
9. On the side tab, there is also the possibility of downloading a csv file containing the combined structural, domain and motif data.

3.5 Motifs Tab

1. This tab visualizes the output from Wregex.
2. The top 10 most frequently identified motif-linked GO terms are displayed on a bar plot and on the "Select GO Terms to Visualize".
3. In the window "Select GO Terms to Visualize" additional GO terms can be added or removed from the plot. They appear in order from the most to the least frequent.
4. There is a bar chart displaying motifs classified by which type of site they belong to.
5. In the "Select GO Terms to see in table format" the user can select the GO terms to be displayed at a downloadable table.
6. In the "Select Motif Categories to see in table format" the user can select the motif categories to be displayed in a downloadable table (Fig. 1b).
7. By clicking on the box "Show Sequence Data Table", a downloadable table containing all the motifs for each sequence will be displayed.

3.6 Antimicrobial Peptide Prediction Tab

1. This tab displays the output of the antimicrobial peptide prediction tool. On the left panel, the sequences can be filtered by the probability of AMP. Sequences with a value >0.5 are considered AMP. The selected probability range will be displayed on the downloadable CSV table.
2. The pie chart represents the percentage of AMP and non-AMP from the input file.
3. Histogram representing the distribution of sequences and antimicrobial probability. The red dotted line represents the selected probability range displayed in the tables below.
4. The upper table contains only the sequence ID, the identification as AMP or non-AMP, the prediction probability and the protein sequence. The lower table contains all the integrated data from the other prediction tools.

3.7 Search Engine Tool Tab

1. This tab allows to perform a multi-search in which combinations of domains and motifs can be made to identify protein sequences with specific features.
2. Select a domain based on GO terms and/or based on Localization and select a motif based on GO terms or/and based on Category.
3. A Venn diagram will display the selected domains and motifs, and a table will show the sequence part of the selected categories.

4. When only selecting two categories, by selecting the "Show sequences containing both", the sequences present in both will be displayed in a downloadable table.
5. As an example (Fig. 1c), selecting in a dataset the Domain-based in localization "Transmembrane" and the motifs of interest "Posttranslational modification" and "extracellular", we identify 22 new sequences that fulfill these criteria, which we hypothesize may belong to transmembrane proteins localized in the plasma membrane containing an extracellular side and a signaling motif.

MicroOracle aims to enable the quick visualization of key features of large, unknown microprotein datasets in a stepwise and easy-to-use manner. By introducing interactive plots and the search engine tab, we aim to facilitate data segmentation and contribute to the generation of hypotheses.

4 Notes

1. Certain online tools have a limit to how many protein sequences a FASTA file should contain or to how big the input file should be. To overcome this limitation, in the Main tab of MicrOracle, after uploading the input FASTA file, the user can split the input file. File splitting can be done by size of the FASTA file or by number of sequences. Indicate the maximum size or the maximum number of protein sequences per FASTA file and click "Download split files". A zip folder containing the different sub-FASTA sequences will be generated.
2. When split files are used, you can upload the individual outputs simultaneously and merge them and download them by clicking on the Download Merged FASTA/TSV or Download Merged Wregex CSV bottom on the Main tab.

Acknowledgments

This work was supported by Vetenskapsrådet.

References

1. Tsang B et al (2020) Phase separation as a missing mechanism for interpretation of disease mutations. Cell 183(7):1742–1756. https://doi.org/10.1016/j.cell.2020.11.050
2. Erady C et al (2021) Pan-cancer analysis of transcripts encoding novel open-reading frames (nORFs) and their potential biological functions. NPJ Genom Med 6(1):4. https://doi.org/10.1038/s41525-020-00167-4
3. Holehouse AS, Kragelund BB (2024) The molecular basis for cellular function of intrinsically disordered protein regions. Nat Rev Mol Cell Biol 25(3):187–211. https://doi.org/10.1038/s41580-023-00673-0

4. van der Lee R et al (2014) Classification of intrinsically disordered regions and proteins. Chem Rev 114(13):6589–6631. https://doi.org/10.1021/cr400525m
5. Blum M et al (2024) InterPro: the protein sequence classification resource in 2025. Nucleic Acids Res. https://doi.org/10.1093/nar/gkae1082
6. Van Roey K et al (2014) Short linear motifs: ubiquitous and functionally diverse protein interaction modules directing cell regulation. Chem Rev 114(13):6733–6778. https://doi.org/10.1021/cr400585q
7. Zanzoni A, Ribeiro DM, Brun C (2019) Understanding protein multifunctionality: from short linear motifs to cellular functions. Cell Mol Life Sci 76(22):4407–4412. https://doi.org/10.1007/s00018-019-03273-4
8. Kumar M et al (2024) ELM-the eukaryotic linear motif resource-2024 update. Nucleic Acids Res 52(D1):D442–D455. https://doi.org/10.1093/nar/gkad1058
9. Torres MDT et al (2024) Mining human microbiomes reveals an untapped source of peptide antibiotics. Cell 187(19):5453–5467 e5415. https://doi.org/10.1016/j.cell.2024.07.027
10. Torres MDT, Cesaro A, de la Fuente-Nunez C (2024) Peptides from non-immune proteins target infections through antimicrobial and immunomodulatory properties. Trends Biotechnol. https://doi.org/10.1016/j.tibtech.2024.09.008
11. Emenecker RJ, Griffith D, Holehouse AS (2021) Metapredict: a fast, accurate, and easy-to-use predictor of consensus disorder and structure. Biophys J 120(20):4312–4319. https://doi.org/10.1016/j.bpj.2021.08.039
12. Lotthammer JM et al (2024) Direct prediction of intrinsically disordered protein conformational properties from sequence. Nat Methods 21(3):465–476. https://doi.org/10.1038/s41592-023-02159-5
13. Osorio D, Rondón-Villarreal P, Torres R (2015) Peptides: a package for data mining of antimicrobial peptides. The R Journal 7(1):4–14. https://doi.org/10.32614/RJ-2015-001
14. Prieto G, Rodriguez JA, Fullaondo A (2024) Enhancing prediction of short linear protein motifs with Wregex 3.0. Comput Struct Biotechnol J 23:2978–2984. https://doi.org/10.1016/j.csbj.2024.07.013
15. Veltri D, Kamath U, Shehu A (2018) Deep learning improves antimicrobial peptide recognition. Bioinformatics 34(16):2740–2747. https://doi.org/10.1093/bioinformatics/bty179

Part IV

Function, Localization, and Relevance to Health and Disease

Chapter 13

Functional Testing of Microproteins in a Vertebrate Model of Development

Anthony J. Treichel, Ariel A. Bazzini, and Valerie A. Tornini

Abstract

While thousands of putative microproteins have been identified through ribosome profiling, reporter assays, and mass spectrometry-based methods, their functional testing has remained challenging. Advances in genome sequencing and CRISPR/Cas technologies enable the prioritization and testing of candidate microprotein functions for roles in development, for example, in the maternal-to-zygotic transition or in neurodevelopment. Here, we describe the functional testing of microproteins in vivo using a vertebrate model of early development, *Danio rerio* (zebrafish).

Key words Vertebrate, Microproteins, CRISPR/Cas9, CRISPR/Cas13d, Maternal-zygotic, Development, Behavior

1 Introduction

Sequencing, mass spectrometry-based, and bioinformatics approaches have identified the putative translation of thousands of short open reading frames (sORFs) across species [1–4]. However, understanding the functional roles, if any, that these microproteins may play requires functional testing and translational validation. In multicellular eukaryotic species, it remains challenging to prioritize and test the functions of microproteins due to low-throughput experimental methods and cell-, stage-, or stimulus-specific microprotein roles. The evolutionary conservation, or lack thereof, of candidate vertebrate microproteins lends their study to be of potential biomedical use, including for human disorders or diseases and/or therapeutics.

Danio rerio, or zebrafish, is a valuable vertebrate experimental system to functionally assess the roles of microproteins, particularly in early development. Zebrafish serve as a fast developing model with up to hundreds of externally fertilized embryos per mating, providing statistical power in background-matched samples [5–

Stephan Wenkel (ed.), *Microproteins: Methods and Protocols*, Methods in Molecular Biology, vol. 2992,
https://doi.org/10.1007/978-1-0716-5013-4_13,

7]. For example, within 6 h, one may identify early phenotypes related to the maternal-to-zygotic transition [8–10]; within 6 days, organismal phenotypes, including behavioral phenotypes, may be elucidated in these fully formed and free-living organisms [11–13]. Supported by a fully sequenced genome and genomic tools, embryos may be injected with DNA, RNA, proteins, or drugs for functional testing [14, 15].

Here, we provide methods for functionally assessing the roles of candidate microproteins for two key vertebrate developmental phenotypes: early zygotic genome activation (ZGA) [16] and organismal behavior.

2 Materials

Prepare molecular reagents in clean, sanitized, and RNase-free environments. Prepare all solutions using ultrapure water and analytical grade reagents. Follow the manufacturer's usage and storage instructions. Follow all waste disposal regulations when disposing of waste materials. All animal use and protocols must be approved by your Institutional Animal Care and Use Committee (IACUC) or equivalent before starting any work.

2.1 Microinjection Molds and Needles

1. In an uncapped 500 mL borosilicate glass bottle, microwave 1 g of agarose with 100 mL of 0.5X E2 Media (7.5 mM NaCl, 250 μM KCl, 500 μM $MgSO_4$, 75 μM KH_2PO_4, 25 μM Na_2HPO_4, 500 μM $CaCl_2$, 350 μM $NaHCO_3$); ensure agarose fully dissolved; using a silicone hand protector, pour this liquid halfway up a plastic Petri dish (you may make multiple molds with this volume, adjust as necessary); lay a plastic injection mold (such as mold TU-1, 6 row 45-degree angle from Adaptive Science Tools USA); let cool and solidify; and gently remove the injection mold once cooled. Microinjection molds may be stored by hydrating with a thin layer of 0.5X E2 media at 4 °C and warmed up to room temperature before use. Microinjection plates may be reused multiple times if using chorionated embryos, or only once if using dechorionated embryos.
2. For working with dechorionated embryos: coat the bottom of all plastic Petri dishes with a very thin layer of 1% agarose in 0.5X E2 media. Alternatively, glass Petri dishes may be used.
3. A laser-based micropipette needle puller and thin-wall glass needles/capillaries (1.0 mm) with filament. Microinjection needles should be pulled using custom parameters that depend on individual needle pullers, and conditions optimized to get a fine-tipped needle that can easily pierce through zebrafish chorions or yolks/cells.

2.2 Microinjection Mixes

1. Microinjection setup: A standard light dissecting stereomicroscope with a microinjection setup (pressure microinjector with a back pressure unit, micropipette holder, and foot switch), connected to house air or a compressed air tank.
2. For CRISPR/Cas9 ribonucleoprotein mixes: commercial Cas9 protein, 1 M KCl, and custom-ordered synthetic guide RNAs; 37 °C thermoblock or incubator.
3. For CRISPR/Cas13d microinjections: Cas13d protein at 2–3 ng/1 nL purified from Rosetta(DE3) Competent *Escherichia coli* transformed with pET-28b-RfxCas13d-His (Addgene #141322) [17], using affinity (ÄKTA start, Cytiva #29022094: HisTrap FF, Cytiva 17525501) followed by size exclusion (ÄKTA go, Cytiva #29383015: HiLoad 16/600 Superdex 200 pg SEC column, Cytiva #28989335) chromatography. For Cas13d mRNA: 150–250 pg/1 nL, pT3TS-RfxCas13d-HA (Addgene #141320), XbaI restriction enzyme, a T3 transcription kit (such as T3 mMessage mMachine) to generate capped mRNA, DNA digestion (e.g., Turbo Dnase) & RNA column cleanup kit (e.g., Qiagen RNeasy Miniprep). For homemade Cas13d gRNAs at 400–700 pg/nL, Cas13d universal oligo (5′- TAATACGACTCACTATAGGTACCCC TACCAACTGGTCGGGGTTTGAAAC -3′), Cas13d specific oligo (5′- NNNNNNNNNNNNNNNNNNNNNNNGTTT CAAACCCCGACCAGTT -3′) where "5′-NNN…NNN-3′" is 22–30 nt (recommend 23 nt) of the sense strand of the target gene of interest, high yield T7 (non-capped) transcription kit (such as Ampliscribe T7 Flash Kit, Biosearch Technologies #ASF3507), 3 M Sodium Acetate (pH 5.2), 100% Ethanol.
4. To dechorionate embryos, add 4–5 mL of 0.5X E2 media to a 1 mL aliquot of 5 mg/mL Pronase (made in 0.5X E2 Media) (for example, Millipore Sigma #10165921001) and warm to room temperature before exposing to embryos; use 35 mm x10 mm Petri dishes (Greiner Bio-One # 627102), 250 mL glass beaker, and glass Pasteur transfer pipettes with at least 0.11 cm tip opening.

2.3 Zebrafish Rearing

1. A heated incubator to be maintained at 28.5 °C. For phenotypes past 24 h postfertilization (hpf), a headed incubator with circadian lighting (such as a DigiTherm 38 liter Heating/Cooling Incubator from Tritech) should be used and set to a 14:10 light:dark cycle (for example, lights on at 9:00 AM, lights off at 11:00 PM).
2. Embryo media should be changed daily, and debris removed using plastic and/or glass transfer pipettes in plastic or glass Petri dishes. 0.5X E2 media can be used until 24 (hpf); blue water (fish system water with 0.5 mg/L of methylene blue) may also be used after 24 hpf.

2.4 Genotyping for CRISPR/Cas-Induced DNA Cuts

1. 100 mM NaOH; 1 M Tris-HCl (pH 7.5); PCR strip tubes or plates.
2. A standard thermocycler with a heated lid.
3. PCR Mix: Using the ThermoPol system from New England Biolabs. Note that other DNA polymerases and mixes may be used. Use 0.2 mL plastic PCR strip tubes with lids or plates with sealing lids. Per well, mix: 2.5 μL of 10X ThermoPol Reaction Buffer; 0.5 μL 10 mM dNTPS; 0.5 μL each of 10 μM forward and reverse primers; 2 μL template DNA (from Subheading 3.6 **step 2**); 0.125 μL *Taq* DNA Polymerase; 18.9 μL nuclease-free water.
4. Agarose Gel: Adjust the percentage of gel for the expected change in band sizes. We suggest starting with around 2% agarose gel. In an uncapped 500 mL borosilicate glass bottle, microwave 2 g of agarose powder in 100 mL 1X TAE buffer (50X TAE commercially available or: 242 g Tris base in double-distilled water; 57.1 mL glacial acetic acid; 100 mL 0.5 M EDTA solution (pH 8.0), volume adjusted to 1 L with water. Dilute 1:50 using ultrapure water for 1X TAE) for about 1: 30 min (timing will vary by microwave); ensure agarose fully dissolved; using a silicone hand protector, remove from microwave, add 1uL of 10 mg/mL ethidium bromide (alternatively, SYBR Safe), swirl, and pour this liquid into an enclosed gel mold for DNA gels (size will vary, fill slightly above halfway up); add combs rigs and allow to solidify; once ready, pull out combs straight up, and place gel tray in an electrophoresis rig; add 6X loading dye to your respective samples, then pipette individual samples into wells; connect the electrophoresis rig to a power pack and run the gel for approximately 30 min at 100 V; image on a UV nucleic acid imaging system.

2.5 RT-qPCR

1. RNase-free water; a thermocycler; a quantitative PCR (qPCR) machine; liquid nitrogen and an appropriate dewar; TRIzol reagent, chloroform, and isopropanol OR TRIzol with Zymo Direct-zol Microprep Kit with DNase I; RNase-free pestles and pellet pestle motor; 100% ethanol; vortexer; room temperature centrifuge with rotor for 1.5–2 mL tubes (up to at least 16,000× *g*); ice and ice buckets; SuperScript IV Reverse Transcription Kit (Invitrogen) or similar for cDNA synthesis (should include 50 μM oligo dTs and/or 50 ng/μL random hexamers, 10 mM dNTP mix (10 mM each), concentrated buffer, 100 mM DTT, RNase inhibitor, and reverse transcriptase enzyme); PerfeCTa SYBR Green FastMix (Quantabio); appropriate consumables such as PCR tubes, filter tips, and 1.5 mL tubes appropriate for flash-freezing (for example, safe-lock tubes).

2.6 Larval Behavioral Profiling

1. A ZebraBox (from Viewpoint Behavior Technology) equipped with a high-quality camera (at least 30 frames per second), ZebraLab software, and a high-memory desktop computer, or a comparable system. Please consult with your IACUC to meet animal testing and housing requirements for this hardware and experimental setup.
2. Whatman UNIPLATE 96-well clear microplates; plastic transfer pipettes; a standard dissection stereomicroscope; blue water; tricaine or other approved euthanasia method.

3 Methods

3.1 Identifying CRISPR Cut Sites for Candidate Microprotein Sequences

1. After obtaining your candidate microprotein(s), identify their location, DNA sequence, and (if applicable) annotations in a publicly available genome browser such as ensembl.org or ucsc.genome.edu. At the time of publication, the most recent zebrafish genome assembly is GRCz11/danRer11. Your sequence of interest may or may not be annotated in these browsers. Depending on your sequence, you may also be able to identify all possible open reading frames from any given transcript [11] (*see* **Note 1**).
2. After identifying the genomic sequence of your candidate microprotein, input that sequence into crisprscan.org [18] and select your Cas of choice. For targeting genomic sequences for zygotic phenotypes, CRISPR/Cas9 may be ideal [11, 19]. For targeting RNA and DNA for maternal-zygotic phenotypes, CRISPR/Cas13d may be best [9]. The results output in crisprscan.org will provide the top recommended target sequences. This output includes the ranking/CRISPRscan score, genomic locus (and orientation), DNA target sequence, off-target prediction scores (including for seed or whole sequence). You should examine your target site so that your guides are ideally nonoverlapping, have no or minimal predicted off-targets, and high efficiency scores within the putative coding sequence (*see* **Note 2**).
3. Select three top nonoverlapping CRISPR/Cas9 target sites, such that targeting these sequences would interrupt the putative coding sequence. If there are not three optimal sequences, aim for at least two target sites.
4. Using the "oligo" or "gRNA" sequences provided next to each selected guide in crisprscan.org, order custom chemically modified synthetic guides (*see* **Note 3**).

3.2 Designing Primers for Testing

1. Once you have selected your target sites, identify ideal primer sites to test the efficiency of CRISPR/Cas targeting. We suggest using https://primer3.ut.ee/, which is designed to provide optimized primer sequences flanking your target sites of interest. You may indicate the start and end of your CRISPR target sites using brackets ([and]). Ideally, your wild type and targeted amplicon sizes will be amenable for screening by electrophoresis and/or Sanger sequencing.
2. Order these primers and save for after CRISPR/Cas injections (Subheadings 3.6 or 3.7).

3.3 Zebrafish Husbandry

1. All use of animals for research purposes must first be submitted and approved by your Institutional Animal Care and Use Committee or equivalent, all users must be approved for working with live vertebrate animals (zebrafish) for the methods described here, and users must comply with all relevant ethical regulations. Adults should be kept at about 28.5 °C in 14:10 light:dark circadian lighting.
2. The afternoon before injections, set up sexually mature adult mating pairs of zebrafish in a mating tank with a removable divider. Set up only 1 male and 1 female per cross, with a divider between them, and the mating tank filled about halfway with fish system water (*see* **Note 4**).
3. On the morning of injections, do not disturb the fish until the circadian lights turn on. Once the lights are on, remove the divider and allow the fish to mate. It may help to raise and tilt the inner tank to mimic shallow water and induce mating.
4. As soon as you observe eggs being released into the water, set a timer for 10 min, and proceed immediately to Subheadings 3.4.1 or 3.4.2.

3.4 CRISPR/Cas Microinjections

We provide two microinjection protocols for targeting with either Cas9 or Cas13d. For assessing effects of targeting the DNA that codes for your putative microprotein (for example, for later zygotic phenotypes), opt for CRISPR/Cas9 microinjections as described in Subheading 3.4.1. This step may be done with chorionated embryos. To directly target microprotein messenger RNA (mRNA), the CRISPR/Cas13d system is becoming a potent tool in zebrafish, especially in the interrogation of maternal and maternal/zygotic gene function (*see* **Notes 5–8**) [17, 20, 21]. We present this protocol in 3.4.2, preferably using dechorionated embryos.

3.4.1 CRISPR/Cas9 Microinjections

1. Form ribonucleoproteins (RNPs) by making separate mixtures of 3 μM Cas9 protein, 300 mM KCl, and 10 mM of each synthetic sgRNA targeting one gene *or* scramble sgRNAs, incubating at 37 °C for 10 min, and cooling to room temperature for 5 min.

2. Mix each of your RNPs equally into one injection mixture, to be kept at room temperature to avoid precipitation (the unused single sgRNA mixes may be stored at −20 °C for up to three freeze–thaw cycles, with decreasing efficiency each cycle.)
3. Use a loading pipette tip to load a glass microinjection needle (with filament, pulled using a needle puller) with each of these mixes.
4. Once 10 min have passed from Subheading 3.3 **step 4**, collect embryos from mating tanks into a plastic or glass Petri dish with system water.
5. Observe embryos under a standard dissecting microscope. Healthy embryos should appear with a chorion, a large yolk, and a cell on one side that inflates after fertilization.
6. Using a plastic transfer pipette, load embryos onto an injection mold. Remove excess water from the plate before microinjecting.
7. Microinject one-cell stage zebrafish embryos with 100 pl of each respective mix early after fertilization into the yolk. Injections should take place before the first cell division, usually around 30 min postfertilization.
8. After injections, place embryos into a clean Petri dish with fresh 0.5X E2 medium (described in Subheading 3.4.2 #4) into a 28.5 °C incubator.

3.4.2 CRISPR/Cas13d Microinjections

1. Cas13d can be injected as either protein (2–3 ng) or mRNA (150–250 pg) in a 1 nL injection (*see* **Note 8**). We recommend co-injecting 400–700 pg of gRNA against your microprotein of interest, also in a 1 nL injection. For the most homogenous knockdown, aim to hit the yolk/cell interface or the cell prior to the first cell division.
2. For Cas13d protein injections, inject all embryos first with Cas13d protein to prevent diluting the Cas13d protein. Then, gRNAs (or water) are injected into individual lanes. Cas13d protein is best kept on wet ice during injections.
3. For Cas13d mRNA injections, make mixes of Cas13d mRNA ± gRNA to inject, which lessens the speed requirement to inject a given number of experimental conditions when compared to injecting Cas13d as protein. Keep RNA injection mixes on dry ice, and speed thaw for each injection.
4. For the most consistent and highest throughput Cas13d injections, we recommend injecting enzymatically dechorionated embryos. However, special care is needed to prevent the dechorionated embryos from contacting either plastic or air. If working with dechorionated embryos, use glass transfer pipettes, agarose-coated plastic Petri dishes or glass Petri dishes.

Prepare three agarose-lined dishes per condition if keeping past 8 h postfertilization (hpf), and two per condition if stopping analyses at 8 hpf. For dechorionated embryo husbandry, we also recommend using 0.5X E2 media supplemented with 0.5 mg/L of methylene blue or blue water.

5. Add 4–5 mL of 0.5X E2 media to a 1 mL aliquot of 5 mg/mL Pronase and warm to room temperature before exposing to embryos.
6. Pull dividers, let zebrafish breed for 7–9 min, then collect embryos (this will ensure all embryos are synchronous).
7. To dechorionate embryos:
   - (a) Put embryos in a 35 mm × 10 mm petri dish (Greiner Bio-One # 627102) and incubate for 4 min in 3–4 mL of 800–1000 μg/mL Pronase in 0.5X E2 media.
     - (i) Make stocks of 5 mg/mL Pronase and freeze in 1 mL aliquots. Then add 4–5 mL of 0.5X E2 media to each aliquot and warm to room temperature before exposing to embryos.
     - (ii) Depending on embryo number, you can split a single batch into two dishes to maximize Pronase effectiveness.
   - (b) Gently submerge the 35 mm dish in a 250 mL beaker full of 0.5X E2 media and dump embryos.
   - (c) Mechanically dislodge chorions by either: (i) pipetting the media up and down in the beaker, (ii) stirring the media with a glass pipette, or (iii) decanting most media, then replacing with a steady stream (avoid bubble incorporation).
8. Load embryos into an agarose injection plate with a glass pipette. We generally inject one lane (~60 embryos) per condition per replicate and transfer one lane into an agarose-coated 100 mm Petri dish for care postinjection.
9. Transfer all embryos to new, warm (28.5 °C) agarose-lined dishes with 0.5X E2 media at ~2 hpf to remove damaged/unfertilized embryos. Ensure to use glass transfer pipettes and that the embryos do not come in contact with air.
10. Transfer all embryos to new, warm (28.5 °C) agarose-lined dishes again between 6 and 8 hpf.
11. After 24 hpf, dechorionated embryos can be cared for as usual.

3.5 Embryo Development

1. The same number of experimental and control sibling embryos should be grown per dish, to ensure identical growing conditions.

2. For experiments using later embryonic/larval stages, zebrafish embryos should be raised at 28.5 °C in Petri dishes at densities of 70 embryos/dish on a 14 h:10 h light:dark cycle in an incubator with circadian lighting.
3. Clean out the dishes of embryos once in the afternoon after microinjections, and then once per day with new embryo water. CRISPR/Cas-targeted embryos and control embryos should be raised in identical conditions.
4. Assess development for any obvious phenotypes, compared to Cas9/scrambled guides sibling controls, noting that these phenotypes may be relevant to your targeted microprotein function.
5. For behavioral phenotyping, any larvae exhibiting abnormal developmental features, including curved trunks or no inflated swim bladder, should not be used and should be euthanized with Tricaine (if over 2 dpf).

3.6 Testing Efficiency of CRISPR/Cas-Induced DNA Cuts

1. Collect up to 8 embryos at 24 hpf per experimental or control condition and place each embryo into one well of a PCR strip tube or plate.
2. Incubate each embryo in 50 μL of 100 mM NaOH at 95 °C for 20 min. Then, add 25 μL of 1 M Tris-HCl (pH 7.5) to neutralize this mix.
3. Use two μL of these crude DNA extracts for genotyping with the corresponding forward and reverse primers (from Subheading 3.2), using a standard PCR mix. For example, for a 25 μL reaction per sample:
   (a) 10X ThermoPol reaction buffer—2.5 μL.
   (b) 10 mM dNTPS—0.5 μL.
   (c) 10 μM forward and reverse primers—0.5 μL each.
   (d) Template DNA—2 μL (from Subheading 3.6 **step 2**).
   (e) *Taq* DNA polymerase—0.125 μL.
   (f) Nuclease-free water—18.9 μL.
4. Adjust thermocycling conditions to your selected primer properties. For example:
   (a) Initial denaturation: 95 °C for 3 min.
   (b) 30–35 cycles of: (i) 95 °C for 20 s, (ii) 50–65 °C for 30 s, (iii) 68 °C for 1 min/kb.
   (c) Final extension: 68 °C for 5 min.
   (d) Hold at 4–12 °C.
5. PCR can be stored at −20 °C, or samples can be combined with 6X loading dye and run on a 2% agarose gel immediately.

6. Once gel is done running, assess for differences in band sizes between control and experimental samples. Fuzzy or distinct multiple bands in experimental, but not control, samples indicate that the DNA at that locus has been successfully targeted.
7. (optional) To assess the types and extent of mutations induced, PCR products, or experimental and some control samples may be submitted for Sanger sequencing and analyzed for mutation efficiency, for example, using Synthego Performance Analysis (ICE Analysis. 2019. v3.0. Synthego).
8. Proceed with phenotyping only for batches of injected embryos with efficient DNA targeting.

3.7 Testing Knockdown Efficiency with RNA-Targeting CRISPR/Cas

1. Using the cDNA sequence from your microprotein of interest, generate 3–5 oligo pairs that give an amplicon ranging from 80 to 200 bp with a primer design software (e.g., Primer3, IDT PrimerQuest) or by hand.
   (a) Ideally, design oligos so that the forward and the reverse sit on opposite sides of a splice junction to mitigate any trace gDNA contamination leftover after RNA extraction.
2. Validate RT-qPCR assay using cDNA from wildtype zebrafish embryos at the stage you wish to measure the knockdown to ensure a primer efficiency of 90–110% (*see* **Note 9**) [22–24].
   (a) Start with a 1:5–1:10 dilution (to prevent too much carryover of RT reagents) of the wildtype cDNA and serially dilute it 1:5 7 times. To generate a standard curve, plot the log10 concentration of cDNA (x-axis) against the cycle number (y-axis) and fit a simple (first-degree) linear regression line. It is advisable to exclude any cycle numbers above 30–32 from this regression line.
   (b) Standard curves should have slopes ranging from −3.103 to −3.587. If their slope is greater than −3.103, the oligos are amplifying multiple species. If the slope is less than −3.587, the oligos are not amplifying enough copies at each cycle (*see* **Note 10** on calculating primer efficiencies).
   (c) Most RT-qPCR equipment can also be set up to run melting curves. Melting curves should give a single, sharp peak without any shoulders to indicate a single, specific amplicon. Agarose electrophoresis of amplicons can serve as an alternate/supplement to melting curves.
3. Perform RNA-extraction from 10–20 zebrafish embryos for each condition/replicate.
   (a) Regardless of the extraction method chosen, snap-freezing embryos in liquid nitrogen gives the highest quality RNA samples. Further, freezing embryos directly

in TRIzol (or comparable lysis buffer) and storage at − 80 °C preserves the highest quality RNA.

(b) TRIzol/Chloroform total RNA extraction (1 mL of TRIzol/ 200 μL of chloroform) followed by isopropanol precipitation is effective, but it is tedious and prone to variability with moderate to high sample numbers.

(c) The Zymo Direct-zol Microprep kit is a column-based, TRIzol extraction that is a highly effective and reproducible method for RNA-extraction from 0–24 hpf zebrafish embryos as follows (*see* **Note 11** for increasing yield):

(i) Fill 1.5 mL microfuge tubes with 350 μL of ice-cold TRIzol.

(ii) Collect 10–20 zebrafish embryos per condition in each tube and snap-freeze using liquid nitrogen. Store embryos at -80 °C until ready for RNA extraction.

(iii) When ready for RNA extraction, thaw on wet ice until slushy.

(iv) Homogenize each sample with a pestle and pellet pestle motor.

(v) Add 350 μL of 100% ethanol to each sample.

(vi) Vortex for 30 s.

(vii) Add 700 μL of each sample to a labeled *Zymo IC Spin Column* in a collection tube and process according to manufacturer's protocol.

(viii) Elute RNA into 1.5 mL tubes with 6–20 μL of RNase-free water and centrifugation for 30–60 s @ 16,000× *g* at room temperature. In general, it is useful to elute in 15 μL, moving 5 7 μL to a strip-tube or 96-well plate for easier RNA quantification and cDNA synthesis.

4. Setup a 10 μL reverse transcription reaction with 1 μg of total RNA with SuperScript IV Reverse Transcription Kit to generate cDNA from total RNA. Make sure to include at least one sample without reverse transcriptase (RT-minus) per batch of RNA extraction to control for genomic DNA contamination.

(a) Thaw RNA and set up all reactions/mastermixes on ice.

(b) Assemble an "annealing mastermix" for each reaction as containing the following:

(i) 0.5 μL of 50 μM oligo dT AND/OR 0.5 μL of 50 ng/μL random hexamers.

(ii) 0.5 μL of 10 mM dNTP mix (10 mM each).

   (iii) Nuclease free water to 6.5 μL (calculated based upon volume of RNA).

(c) Add appropriate volume of "annealing mastermix" to each sample tube.

(d) Add 1 μg of total RNA to each sample tube.

(e) Flick to mix, spin down, and incubate at 65 °C for 5 min.

(f) Incubate on ice for at least 2 min.

(g) While annealing oligos, assemble RT positive and RT negative mastermixes for each reaction containing the following:

   (i) 2 μL 5X SSIV buffer.

   (ii) 0.5 μL 100 mM DTT.

   (iii) 0.5 μL RNase inhibitor.

   (iv) 0.5 uL SSIV reverse transcriptase enzyme (RT positive) *OR* 0.5 uL nuclease free water (RT negative, critical control for gDNA carryover).

(h) Add 3.5 μL of RT positive OR RT negative mastermix to each sample.

(i) Flick to mix, spin down, and incubate at 50 °C for 10 min for reverse transcription, then 80 °C for 10 min to inactivate SSIV RT.

   (i) If using random hexamers, add an initial incubation step of 23 °C for 10 min to the beginning of this RT protocol.

5. For RT-qPCR, dilute experimental cDNA at least 1:5. For targeted genes, choose a dilution where control samples give cycle values of 19–23 to enable a broad dynamic range of knockdown measurements. Aim to have housekeeping genes within 2–4 cycles of targets to minimize error based on differences in expression level.

6. Set up RT-qPCR with PerfeCTa SYBR Green FastMix (Quantabio) with recommended volumes and cycling parameters.

7. Set thresholds in the exponential amplification phase where curves for all samples are parallel to call cycle numbers (most software does this automatically).

8. Calculate relative RNA levels using at least one housekeeping gene and the ΔΔCT method.

3.8 Assessing Early Maternal or Maternal–Zygotic Phenotypes

1. For the first 2.5 h, zebrafish embryos undergo rapid, synchronous cell divisions to generate two distinct populations of blastomeres, the enveloping layer (external) cells and the deep (internal) cells. From 2.5 to 4 hpf, cell divisions begin to slow down and become asynchronous. Cells at the margin between

the yolk and the blastomeres cease cytokinesis and begin to form what is known as the yolk-syncytial layer. Around 3.25 hpf, the blastomeres start to compact until 4 hpf where they complete a roughly spherical shape (with a flat margin) in coordination with the yolk cell. Any phenotype arising during this period is most readily observed through a decrease in cell number, anomalies in the cell cycle, or abnormalities in cellular structure/adhesion. During the first 2.5 hpf, quantifying cell number based on transmitted light is possible. However, nuclear counterstains such as Hoechst are especially helpful to quantify cell cycle time and synchrony. Co-expression of membrane-tethered reporters with nuclear ones illuminates yolk-syncytial layer formation.

2. Around 4.25 hpf, zebrafish epiboly begins. The yolk cell domes, while the blastomeres and yolk-syncytial layer spread out and begin to encircle the yolk cell. They reach the equator of the yolk cell by 5.25 hpf, where they pause while gastrulation initiates with the formation of germ ring and the shield (the zebrafish version of the organizer). Blastomeres complete their encasement at 10 hpf. Delays in epiboly are associated with defects in gastrulation and/or the cytoskeletal machinery in the yolk/yolk-syncytial layer that drives epiboly movements. With transmitted light, stages are easy to distinguish between 30%, 50%, and 75% epiboly, but require careful observation to determine if embryos are in 50% epiboly, germ-ring, or shield stages [7]. Further, in situ hybridization for germ-layer markers can determine if gastrulation is affected.
3. Between 10 and 24 hpf, the embryo begins to differentiate tissues, organs, and elongate its body axis through somitogenesis. The optic cup and the otic vesicle become visible in the anterior of the embryo, while the tail bud forms at the posterior. Here, anterior/posterior patterning defects become apparent through missing structures. More subtle defects can be observed through examining the hindbrain structure, anterior/posterior axis length, and yolk shape. There are many transgenic lines that are available to look at cell-type-specific defects during and after this window.

3.9 Assessing Later Zygotic Developmental Phenotypes

1. By 24 hpf, organogenesis in the zebrafish embryo is well underway. Under a dissecting microscope, compare the broad developmental morphology between control and experimental embryos.
2. Depending on your microprotein candidate expression or predicted function, you may focus on specifically analyzing that tissue, cell, or organ type. Some phenotypes may appear later in development. As appropriate, antibodies, in situ hybridization, or transgenic lines may be used for specific phenotyping.

3.10 Assessing Later Zygotic Behavioral Phenotypes

Baseline behavioral phenotyping may be an appropriate readout for brain or central nervous system-enriched microprotein candidates (*see* **Note 12**).

1. Between 1 and 4 days postfertilization (dpf), control and experimental embryos should be grown in identical densities and conditions in the 14:10 circadian lighting incubator, up to 70 embryos per Petri dish. Clean embryos from debris and replace fresh "blue water".
2. At 4 dpf, transfer single larvae from experimental or control plates into individual wells of a clear 96-square well flat plate (Whatman) filled with 650 μL of blue water warmed to 28.5 ° C. Ensure to note which wells contain experimental or control larvae, and note that no larvae with obvious developmental defects should be used. Developmental defects may include no inflated swim bladder, curved spines, or cardiac edemas, among other defects.
3. Gently place plates in a Zebrabox (ViewPoint Life Sciences) in an approved room, set to 28.5 °C. Set up the software tracking protocol to the same 14:10 circadian lighting schedule. Track each well using ZebraLab software (Viewpoint Life Sciences) in "quantization" mode for up to 3 nights and 3 days, up to 7 dpf. Alternative tracking hardware and software may be appropriate at this step as well.
4. Replace evaporated blue water every morning (shortly after lights on) with a plastic transfer pipette, ensuring to not overfill the wells such that they bubble up.
5. Once the run is finished, remove the plates with larvae and euthanize with an approved method according to your animal protocol (for example, tricaine overdose). Individual larvae may be collected for DNA analysis.
6. Transfer results files and analyze them with custom software and instructions provided open source [25, 26]. After inputting the genotypes for control or experimental larvae, analyze behavioral data for any statistical significance in measured parameters (for example, how active fish are during the day or night, how long they sleep, how long it takes them to wake up, etc.) using one-way ANOVA followed by Tukey's post hoc test ($\alpha = 0.05$) [26].
7. Prioritize the most statistically significant F0 mutants as top microprotein candidates for functional testing in stable mutants.

3.11 Establishing Stable Mutants for Zygotic or Maternal–Zygotic Phenotyping

1. To assess a full knockout phenotype, grow up injected zebrafish ("F0" fish) to adulthood. You may also choose to identify new target sequences to generate stable functional lines targeting the start codon, full transcript, or the coding sequence (*see* **Notes 13** and **14**).
2. Outcross adult F0 fish to wild-type fish, using the breeding protocol in Subheading 3.3.
3. Collect pools of embryos, and adapt Subheading 3.6 to identify possible germline mutants with putative loss-of-function (LOF) microprotein mutations.
4. Grow up to adulthood clutches of embryos that contained putative LOF sequences. Sanger sequencing of individual fish (by fin clipping or embryo genotyping, as permitted by your animal protocol) is required to identify successful LOF mutants.
5. Phenotyping as described in Subheadings 3.8, 3.9, and 3.10 should be performed using prioritized stable mutants (*see* **Notes 13–16**).

4 Notes

1. We suggest prioritizing microprotein candidates via F0 mutants by targeting the putative coding sequence with three CRISPR guides, before proceeding to generating stable functional lines that delete the full transcript (to assess for regulatory function), the start codon (to distinguish between non-coding and coding functions), or the coding sequence (LOF protein function).
2. Other Cas systems may be more amenable to your sequences. CRISPRscan.org provides target sequences to use with Cas12a (Cpf1), SpG or SpY. Another reliable resource for identifying CRISPR/Cas target sites in zebrafish is available at https://chopchop.cbu.uib.no/ [27].
3. For the highest and most reliable targeting, we suggest synthetically modified guides (for example, from IDT or Synthego) [11, 19] with some nuances for the CRISPR-RfxCas13d system (described in [20] and in **Notes 6–8** below). If you prefer to test using in vitro transcribed guides, crisprscan.org provides oligo sequences to order; please refer to [28] and **Note 6** for more information. If going the in vitro transcribed route, we highly recommend PAGE purification for the universal oligo because we have observed higher efficiency with both fill-in-PCRs and in vitro transcription using a PAGE-purified universal.

4. When choosing adult zebrafish mating pairs, it is important to remove any maternal/background effects per individual MZ experiment or behavior run, so it is best not to set up multiple fish per cross (for example, do not set up two females and two males per mating). It is preferable to use clean wild-type background lines (for example, no transgenics in background), unless a mutant or transgenic line may aid in your phenotyping. If you are screening for effects on specific cell types, you may inject into transgenic reporter embryos.
5. RfxCas13d represents the pioneer CRISPR/Cas13d system adapted to zebrafish [17, 20] and exhibits rapid, potent depletion of target mRNA levels. To address shortcomings with the RfxCas13d system (*see* **Note 6**), alternate CRISPR/Cas-based RNA knockdown systems have been adapted in zebrafish [21] (*see* **Note 7**). Still, each RNA knockdown tool has trade-offs, and their employment is context-dependent (*see* **Note 8**).
6. RfxCas13d can cleave non-target RNA (denoted "collateral activity") when employed against highly expressed reporters and/or endogenous genes [21]. Cleavage of the 28S rRNA is a hallmark of this RfxCas13d collateral activity and is associated with embryonic lethality that begins as early as 2 hpf [21]. Second, although in vitro transcription of RfxCas13d gRNAs is efficient and cost-effective, we have observed that a subset of in vitro transcribed RfxCas13d gRNAs (but not their synthetic counterparts) induces 28S rRNA cleavage and embryonic lethality identical to that observed in collateral activity [21]. Therefore, RfxCas13d requires costly, synthetic gRNAs or meticulously screened, in vitro transcribed gRNAs.
7. Alternate CRISPR/Cas-based RNA knockdown systems adapted to zebrafish include Cas7–11 and DjCas13d. Based upon initial work, DjCas13d exhibits reduced collateral activity compared to RfxCas13d, but without the 28S rRNA cleavage observed in RfxCas13d. Cas7–11 appears to lack collateral activity, but achieves weaker mRNA depletion [27].
8. The choice of Cas protein and gRNA synthesis method for mRNA knockdown in zebrafish remains highly case-dependent. If targeting highly expressed microproteins, it is advisable to employ Cas7–11 or DjCas13d to minimize collateral activity. When targeting a few, moderately expressed microproteins, RfxCas13d will maximize the knockdown efficiency. Finally, in vitro transcribed gRNAs minimize experimental cost but require screening when employed with RfxCas13d to mitigate risk of 28S rRNA cleavage.
9. In general, RT-qPCR is a robust method to determine knockdown when targeting RNA with CRISPR/Cas systems. In zebrafish, it can be tricky to design oligo pairs that give a single,

quantifiable amplicon. Therefore, as the cost of RNA-seq continues to drop, it is possible that RT-qPCR will fall out of favor. Nonetheless, we presented above how to design a good RT-qPCR assay.

10. Primer efficiency (1 = 100% efficiency, perfect exponential amplification) can be calculated based on the slope of the regression line obtained from the standard curve as follows: $\textit{Primer efficiency} = -1 + 10^{(-1/\text{Slope of Standard Curve})}$. Similarly, for any primer efficiency, the expected slope can be calculated as follows: $\textit{Slope of Standard Curve} = -1 * \log(10)/\log\,(\textit{Primer efficiency} + 1)$.
11. For the Zymo Direct-zol Microprep kit, you can moderately increase the yield of the extraction by doubling the volume of TRIzol (and ethanol) added to the embryos. In this case, still homogenize in 350 μL of TRIzol and then add 350 μL of TRIzol to the homogenate (this prevents spilling TRIzol) and vortex briefly. Add 700 μL of ethanol and process as you otherwise would, using two spins to bind all of the RNA to the column (max loading volume is 700 μL). Further, we have found that the Zymo Direct-zol Miniprep kit is effective if larger quantities of RNA are needed.
12. We provide a general protocol for baseline behavioral profiling of larval zebrafish. Alternative stimulus-driven or other behavioral profiling protocols of larval zebrafish may be more appropriate for your microprotein candidates of interest.
13. Stable mutant lines are still considered the standard for testing zygotic and some maternal–zygotic phenotypes. In some cases, zygotic phenotypes may be lethal; this is a particularly challenging aspect of working with maternal-zygotic mutants. Recent efforts to screen and prioritize using F0 approaches have yielded promising results, often supported by similar stable mutant line phenotypes (for example, [9, 11, 19]. LOF mutants for microproteins that are required for germ cell development or early development may not be viable, such that F0 DNA—or RNA-targeting may be most appropriate.
14. Once stable mutants are generated, they may be used for numerous other functional tests, including into adulthood, with additional stressors or small molecules, in other transgenic or mutant backgrounds, or other experimental tests.
15. Beyond the scope of this chapter, the standard for elucidating the specific and sufficient translation and function of a microprotein is through reporter assays or antibody detection (*see* **Note 13**) and rescue experiments (using a transgenic line expressing the microprotein sORF coding sequence). For rescue experiments, the exact nuances of the tagged or polycistronic construct will depend on the properties of your

candidate microprotein. This may include N- or C-terminal tagging using small tags (including FLAG, HA, or newer epitopes like ALFA), and/or using a T2A/P2A polycistronic system for construct expression but not translational detection (as demonstrated in [11]). Larger tagged fluorescent proteins, such as GFP may be too large relative to the microprotein size, such that they may interfere with microprotein function.

16. Detecting translated microproteins remains challenging, as custom antibodies can be cost-prohibitive, may not be specific, or may not work in zebrafish. Newer alternatives, such as endogenous tagging with a small epitope, may be a reasonable alternative (for example, [29]), with the same caveats presented in **Note 15**.

Acknowledgments

This work was supported by the Stowers Institute for Medical Research (A.J.T. & A.A.B.), the US National Institute of Health F31HD110268 (A.J.T), R01GM136849 (A.A.B.), and NIH R21OD034161 (A.A.B.), and by a *Eunice Kennedy Shriver* National Institute of Child Health and Human Development grant 4R00HD105001 and funds from the Division of Life Sciences at UCLA (V.A.T.) This work was completed as part of thesis research for A.J.T. for the Graduate School of the Stowers Institute for Medical Research.

References

1. Bazzini AA, Johnstone TG, Christiano R et al (2014) Identification of small ORFs in vertebrates using ribosome footprinting and evolutionary conservation. EMBO J 33:981–993. https://doi.org/10.1002/embj.201488411
2. Duffy EE, Finander B, Choi G et al (2022) Developmental dynamics of RNA translation in the human brain. Nat Neurosci 25:1353–1365. https://doi.org/10.1038/s41593-022-01164-9
3. Ingolia NT, Ghaemmaghami S, Newman JRS, Weissman JS (2009) Genome-wide analysis in vivo of translation with nucleotide resolution using ribosome profiling. Science 324:218–223. https://doi.org/10.1126/science.1168978
4. Slavoff SA, Mitchell AJ, Schwaid AG et al (2013) Peptidomic discovery of short open reading frame–encoded peptides in human cells. Nat Chem Biol 9:59–64. https://doi.org/10.1038/nchembio.1120
5. Choi T-Y, Choi T-I, Lee Y-R et al (2021) Zebrafish as an animal model for biomedical research. Exp Mol Med 53:310–317. https://doi.org/10.1038/s12276-021-00571-5
6. Kimmel CB (1989) Genetics and early development of zebrafish. Trends Genet TIG 5:283–288. https://doi.org/10.1016/0168-9525(89)90103-0
7. Kimmel CB, Ballard WW, Kimmel SR et al (1995) Stages of embryonic development of the zebrafish. Dev Dyn 203:253–310. https://doi.org/10.1002/aja.1002030302
8. Lee MT, Bonneau AR, Takacs CM et al (2013) Nanog, Pou5f1 and SoxB1 activate zygotic gene expression during the maternal-to-zygotic transition. Nature 503:360–364. https://doi.org/10.1038/nature12632
9. Hernandez-Huertas L, Moreno-Sanchez I, Crespo-Cuadrado J, et al (2024) CRISPR-RfxCas13d screening uncovers Bckdk as a post-translational regulator of the maternal-

to-zygotic transition in teleosts. 2024.05.22.595167

10. Miao L, Tang Y, Bonneau AR et al (2022) The landscape of pioneer factor activity reveals the mechanisms of chromatin reprogramming and genome activation. Mol Cell 82:986–1002.e9. https://doi.org/10.1016/j.molcel.2022.01.024
11. Tornini VA, Miao L, Lee H-J et al (2023) linc-mipep and linc-wrb encode micropeptides that regulate chromatin accessibility in vertebrate-specific neural cells. eLife 12:e82249. https://doi.org/10.7554/eLife.82249
12. Thyme SB, Pieper LM, Li EH et al (2019) Phenotypic landscape of schizophrenia-associated genes defines candidates and their shared functions. Cell 177:478–491.e20. https://doi.org/10.1016/j.cell.2019.01.048
13. Mendes HW, Neelakantan U, Liu Y et al (2023) High-throughput functional analysis of autism genes in zebrafish identifies convergence in dopaminergic and neuroimmune pathways. Cell Rep 42. https://doi.org/10.1016/j.celrep.2023.112243
14. Sakai C, Ijaz S, Hoffman EJ (2018) Zebrafish models of neurodevelopmental disorders: past, present, and future. Front Mol Neurosci 11. https://doi.org/10.3389/fnmol.2018.00294
15. Veldman MB, Lin S (2008) Zebrafish as a developmental model organism for pediatric research. Pediatr Res 64:470–476. https://doi.org/10.1203/PDR.0b013e318186e609
16. Treichel AJ, Bazzini AA (2022) Casting CRISPR-Cas13d to fish for microprotein functions in animal development. iScience 25:105547. https://doi.org/10.1016/j.isci.2022.105547
17. Kushawah G, Hernandez-Huertas L, Abugattas-Nuñez Del Prado J et al (2020) CRISPR-Cas13d induces efficient mRNA knockdown in animal embryos. Dev Cell 54:805–817.e7. https://doi.org/10.1016/j.devcel.2020.07.013
18. Moreno-Mateos MA, Vejnar CE, Beaudoin J-D et al (2015) CRISPRscan: designing highly efficient sgRNAs for CRISPR-Cas9 targeting in vivo. Nat Methods 12:982–988. https://doi.org/10.1038/nmeth.3543
19. Kroll F, Powell GT, Ghosh M et al (2021) A simple and effective F0 knockout method for rapid screening of behaviour and other complex phenotypes. eLife 10:e59683. https://doi.org/10.7554/eLife.59683
20. Hernandez-Huertas L, Kushawah G, Diaz-Moscoso A et al (2022) Optimized CRISPR-RfxCas13d system for RNA targeting in zebrafish embryos. STAR Protoc 3:101058. https://doi.org/10.1016/j.xpro.2021.101058
21. Moreno-Sánchez I, Hernández-Huertas L, Nahón-Cano D et al (2025) Enhanced RNA-targeting CRISPR-Cas technology in zebrafish. Nat Commun 16:2591. https://doi.org/10.1038/s41467-025-57792-9
22. Bustin SA, Benes V, Garson JA et al (2009) The MIQE guidelines: minimum information for publication of quantitative real-time PCR experiments. Clin Chem 55:611–622. https://doi.org/10.1373/clinchem.2008.112797
23. Taylor SC, Laperriere G, Germain H (2017) Droplet Digital PCR versus qPCR for gene expression analysis with low abundant targets: from variable nonsense to publication quality data. Sci Rep 7:2409. https://doi.org/10.1038/s41598-017-02217-x
24. Taylor SC, Nadeau K, Abbasi M et al (2019) The ultimate qPCR experiment: producing publication quality, reproducible data the first time. Trends Biotechnol 37:761–774. https://doi.org/10.1016/j.tibtech.2018.12.002
25. Rihel J, Prober DA, Arvanites A et al (2010) Zebrafish behavioral profiling links drugs to biological targets and rest/wake regulation. Science 327:348–351. https://doi.org/10.1126/science.1183090
26. Rihel J (2023) JRihel/Sleep-Analysis: Sleep_Code_v1
27. Labun K, Montague TG, Krause M et al (2019) CHOPCHOP v3: expanding the CRISPR web toolbox beyond genome editing. Nucleic Acids Res 47:W171–W174. https://doi.org/10.1093/nar/gkz365
28. Vejnar CE, Moreno-Mateos MA, Cifuentes D, et al (2016) Optimized CRISPR–Cas9 System for Genome Editing in Zebrafish. Cold Spring Harb Protoc 2016:pdb.prot086850. https://doi.org/10.1101/pdb.prot086850
29. Boswell CW, Hoppe C, Sherrard A, et al (2023) Genetically encoded affinity reagents (GEARs): a toolkit for visualizing and manipulating endogenous protein function in vivo. 2023.11.15.567075

Chapter 14

Functional Validation of Transcriptionally Active Microproteins/sORFs In Planta

Ashish Sharma and Prabodh Kumar Trivedi

Abstract

Advancements in biotechnology, bioinformatics tools, and high-throughput sequencing techniques have greatly enhanced our ability to identify small but significant molecules in plants that were previously thought to be junk. Small molecules, including microproteins (miPs) and micoRNA-encoded peptides (miPEPs), play a crucial role in various aspects of plant biology, such as growth, development, stress responses, hormonal signaling, and primary and secondary metabolism. Despite their importance, many of these molecules still need to be characterized due to the challenges associated with validating them in plants. Traditional validation methods, like protein sequencing and western blotting using specific antibodies, are often expensive and labor-intensive, which limits their use in large-scale studies. To address these challenges, alternative methods like promoter-reporter lines linked to a reporter gene such as β-glucuronidase (GUS) can offer a more efficient and cost-effective solution to validate transcriptionally active small peptides/ORFs *in planta*. The GUS assay allows for the direct visualization of gene activity in plant tissues, simplifying the validation of candidate molecules (small peptides/ORFs) and alternative transcripts. This approach speeds up the functional characterization of novel genes and molecules, ultimately advancing plant research and biotechnology applications focused on their functional characterization.

Key words Alternative transcript, GUS (β-glucuronidase), Hormonal signalling, Microproteins, miPEP, ORF, Stress responses, Small peptides

1 Introduction

With the rapid advancement of bioinformatics tools and the application of high-throughput techniques in molecular biology, significant progress in identifying and characterizing a new class of small yet crucial molecules (small peptides) in plant systems has been made [1–5]. These molecules, often known as small regulatory molecules or metabolites, were previously overlooked due to their subtle yet essential roles in plants. They are involved in various biological processes, such as modulating plant growth and development, stress responses, and hormone signalling [6–14]. These

Stephan Wenkel (ed.), *Microproteins: Methods and Protocols*, Methods in Molecular Biology, vol. 2992, https://doi.org/10.1007/978-1-0716-5013-4_14,

molecules have a vast potential to serve as new targets for crop improvement and plant biotechnology [15–20].

Despite their importance, these molecules represent only the tip of the iceberg. Most small peptides in plants, especially those with novel functions, still need to be characterized or discovered [1, 2, 14]. One of the primary challenges in advancing this research lies in the validation of these newly discovered molecules. Traditional validation methods, such as protein sequencing or Western blot analysis, are commonly used to confirm the identity and functional relevance of proteins or metabolites. However, these approaches come with several limitations. First, they can be quite expensive and labor-intensive. For example, protein sequencing requires high-quality samples and advanced equipment, which can be expensive, especially when working with a large number of candidate molecules. Secondly, western blotting relies on antibodies to detect specific proteins, but these antibodies can bind to off-target proteins, leading to false positives or inaccurate results. As a result, there is a need for more efficient, cost-effective, and accurate validation strategies.

To overcome these challenges, alternative methods like the use of promoter-reporter lines can be used for candidate molecule validation. This strategy involves incorporating specific initiation codons from various open reading frames (ORFs) into a reporter construct. By linking these ORFs to a reporter gene, such as the β-glucuronidase (GUS) gene, we can create a system where the reporter gene's activity serves as an indicator of functional gene expression or molecular activity [21–23]. The GUS assay is a powerful tool for detecting functional ORF because the GUS enzyme produces a blue color for direct visualization of gene activity in plant tissues through histochemical staining [24, 25].

This promoter-reporter approach offers several benefits:

- It enables the direct detection of functional ORFs in a living organism, eliminating the need for time-consuming and expensive protein extraction or sequencing procedures.
- The GUS assay provides a simple and reliable way to monitor gene activity, making it easier to determine whether a given ORF is functional.

The method can be modified to explore the potential of alternative transcripts, which are becoming increasingly important in plant genomics. Many plant genes produce multiple transcript isoforms through alternative splicing or other mechanisms, and some of these isoforms may encode proteins with distinct functions [26, 27]. By employing the GUS assay at the early stages of transcript screening, researchers can assess whether an alternative transcript can encode a functional protein without using complex techniques like mass spectrometry or western blotting.

In summary, while traditional methods of validation have their place, the use of promoter-reporter lines and the GUS histochemical assay offers a cost-effective and accurate alternative for validating new peptides and genes in plant research. This strategy will enable researchers to accelerate the discovery and functional characterization of small molecules, alternative transcripts, and novel genes, ultimately leading to a deeper understanding of plant biology and the development of new tools for crop improvement.

2 Materials

Prepare all buffers/reagents/solutions using Milli-Q water (The Milli-Q water is deionized water with a conductivity of 18.2 MΩ, at room temperature) and analytical/molecular grade chemicals. Ensure all reagents are prepared and stored at room temperature unless otherwise specified.

2.1 Development of Construct

1. Genomic DNA: High-quality genomic DNA (gDNA) extracted from selected plant species using a specific extraction method [for example, CTAB or by genomic DNA isolation kit]; having an A260/A280 ratio of approximately 1.8.
2. Primer: Forward & Reverse primers (20–24 bp):
   - GC content 40–60%.
   - Tm 55–60%.
   - Forming no hairpin loop, self-dimerization, or cross-dimerization.
3. DNA Polymerase: Q5® High-Fidelity DNA Polymerase (Cat No. M0491).
4. T4 DNA Ligase & Ligase Buffer Thermo Scientific (Cat No. EL0011).
5. Restriction enzymes (NEB).
6. Cloning vector: pTZ 57 R/T or pJET Thermo Scientific (Cat No. PGK121 or K1232).
7. Expression vector having reporter (GUS) gene—pBI121.
8. Competent cell (*E. coli* DH5α and *Agrobacterium* GV3101/LBA4404).

2.2 Transient Expression Assay

1. Luria broth (2% w/v): For 100 mL, 2 g of Luria broth solution is prepared in Milli-Q water, autoclaved before use.
2. 0.1 M MES (pH 5.6).
3. 0.1 M $MgCl_2$
4. 200 μM Acetosyringone.

5. *Arabidopsis thaliana* or *Nicotiana benthamiana* plants (3 weeks old for Arabidopsis and 4–6 weeks old Nicotiana) properly grown in soil in glass house or growth chamber under controlled conditions [28].
6. Blunt syringe (2 mL or 5 mL).

2.3 Histochemical GUS Assay

1. 100 mM sodium phosphate buffer (pH = 7.0): Mix 72 mL of $Na_2HPO_4.7H_2O$ solution with 28 mL of $NaH_2PO_4.H_2O$ solution. Adjust the pH to 7.0 by using one of them.
   - 100 mM sodium dihydrogen phosphate ($NaH_2PO4.H_2O$): Mol. Wt 137.99; For 100 mL, dissolve 1.38 g in 100 mL Milli-Q water.
   - 100 mM Disodium hydrogen phosphate (-$Na_2HPO_4.7H_2O$): Mol. Wt 177.99; For 100 mL, dissolve 1.78 g in 100 mL Milli-Q water.
2. 0.5 M EDTA.
3. 0.1% Triton X-100.
4. 0.1 M potassium ferricyanide.
5. 0.1 M potassium ferrocyanide.
6. 1 mM X-Gluc (in DMF).

3 Methods

3.1 Development of Construct

1. Designed a forward primer from approx. 1.8 Kb away from the translational initiation codon and reverse primer having a translational initiation codon (ATG) in frame with the GUS gene. Amplification of promoter along with translational initiation codon using forward & reverse primers and DNA polymerase (Fig. 1).
2. Clone the amplified product into a cloning vector using T4 ligase.
3. Confirm the cloned fragment through sequencing.
4. Digest the cloned vector having amplified product and Promoter: reporter vector with the same restriction enzymes and ligate them using T4 ligase [30] (Fig. 1).
5. Transform the cloned Promoter: reporter vector into GV for transient expression in plants (*see* **Notes 1–8**).

3.2 Transient Expression Assay

1. Perform primary inoculation by incubating a positive colony of *Agrobacterium* (GV or LBA) having promoter: reporter construct/plasmid in 5/10 mL LB with selectable antibiotics at 28 °C with shaking (150–180 rpm) overnight.

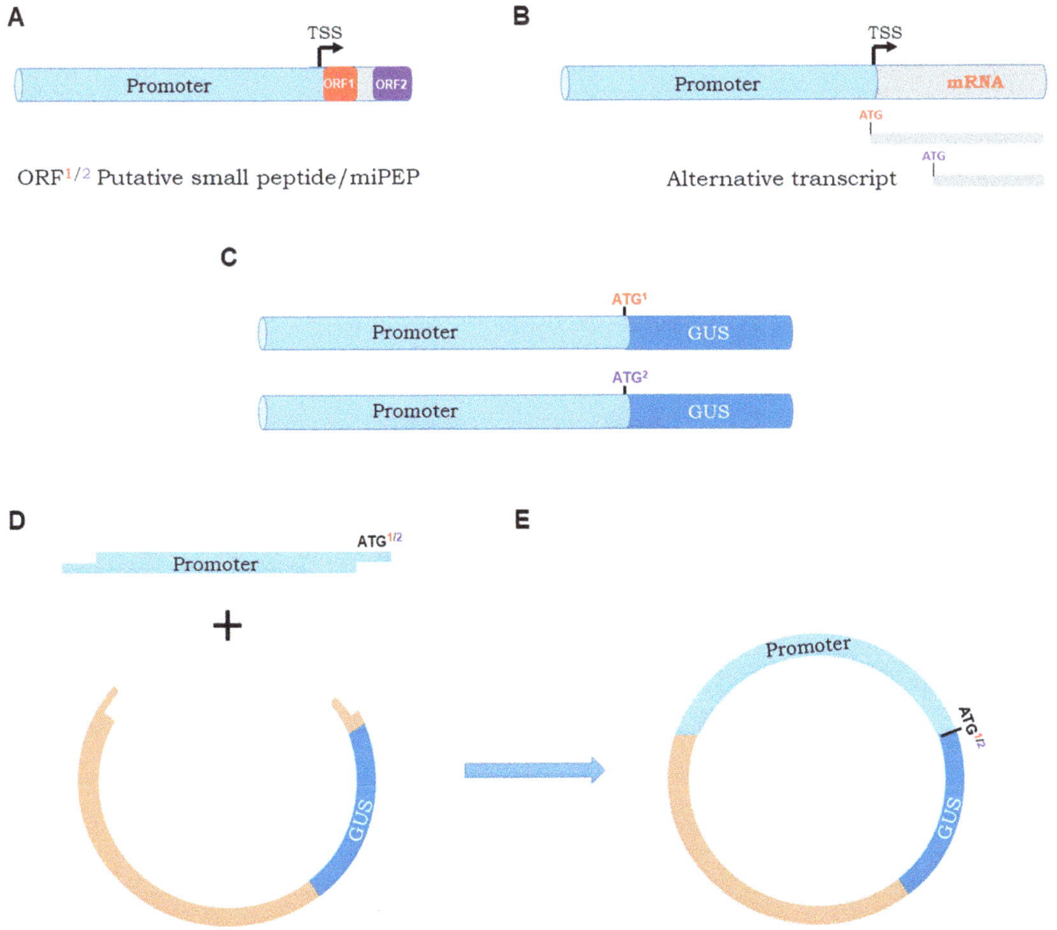

Fig. 1 (**a**) Schematic representation of promoter along with putative ORFs. (**b**) Schematic representation of a promoter with alternative transcripts. (**c**) Schematic representation of different promoter: reporter fusion constructs. (**d**) Ligation of restriction-digested promoter (along with the translation initiation codon) amplicon to restriction digested expression vector. (**e**) Schematic representation of a complete cloned promoter: reporter construct in an expression vector

2. Conduct secondary inoculation by transferring the primary culture to fresh 50/100 mL LB medium (49 mL LB+ 1 mL 0.1 M MES as mentioned above) and allowing it to grow until the optical density reaches 0.8–1.0. (It generally requires overnight growth to reach the required optical density, and also depends on the amount of primary culture taken for inoculation.
3. Prepare 100 mL infiltration buffer: (10 mM $MgCl_2$, 10 mM MES pH 5.6, 150 μM acetosyringone make up volume by autoclaved Milli-Q water).
4. Centrifuge the *Agrobacterium* culture at 900 g for 5 min to harvest bacteria.

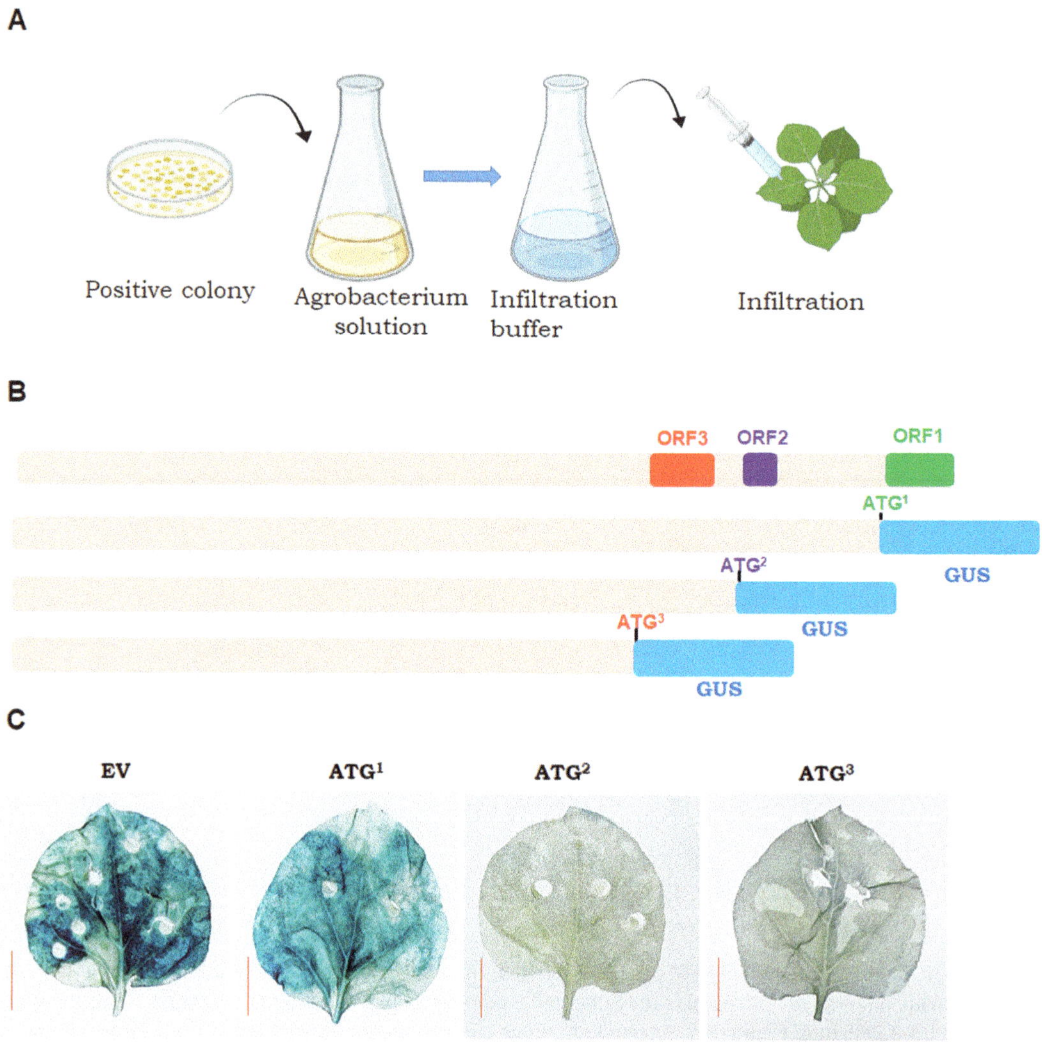

Fig. 2 (**a**) Schematic representation of different processes of agroinfiltration. (**b**) Schematic representation of different constructs used for the GUS assay. (**c**) Representative images of histochemical staining showing GUS activity in *Nicotiana benthamiana* leaves transiently expressing EV and different promoters: reporter construct

5. Dissolve the pellet in a small volume of infiltration buffer (as mentioned above) to achieve an OD_{600} of around 0.5 to 0.8. (~0.6 will be ideal for agroinfiltration).
6. Incubate the suspension solution at 28 °C with shaking (70–80 rpm) for 2 h (it helps to enhance transformation efficiency).
7. Take suspension solution in a blunt syringe (2 mL) without a needle for infiltration.
8. Carefully insert the syringe tip into the abaxial side (lower side) of the leaf, applying gentle pressure while it is positioned on the

underside. You will observe the solution spreading through the leaf tissue [29] (Fig. 2a).

9. Infiltrate three to four leaves from each plant, also to ensure that the solution covers a large portion of each infiltrated leaf.
10. For the GUS histochemical assay, an infiltrated leaf can be used after 2–3 days of infiltration (Fig. 2b, c) (*see* **Notes 9–16**).

3.3 Histochemical GUS Assay

1. Prepare 100 mM sodium phosphate buffer (pH − 7.0) by mixing 72 mL $Na_2HPO_4.7H_2O$ solution with 28 mL $NaH_2PO_4.H_2O$ solution. Adjust pH by using one of them.
2. For 100 mL GUSGUS solution, add the following and make up the volume to 100 mL using Milli-Q:
   - 50 mL of 100 mM sodium phosphate buffer (pH = 7.0)
   - 2 mL of 0.1 M potassium ferricyanide
   - 2 mL of 0.1 M potassium ferrocyanide
   - 100 μL 0.1% Triton X-100
   - 250 μL 0.5 M EDTA
   - 1 mL of 1 mM X-Gluc (as mentioned in materials)
   - Store in a cool, dark place.
3. Firstly, dip the agroinfiltrated leaves in GUS solution in a tube/ Falcon.
4. Cover tube/Falcon with aluminum foil and incubate at 37 °C for 24-48 h.
5. Sometimes, visualizing the blue color against the dark green leaves can be challenging. To enhance visibility, remove the plants from the GUS assay solution and place them in 70% ethanol.
6. To prepare 70% ethanol, mix 30 mL of distilled water with 70 mL of 100% ethanol. This will extract the chlorophyll into the ethanol, making it easier to observe the blue staining of the plant tissue.
7. Transfer the infiltrated leaves to fresh 70% ethanol multiple times, as needed, until the plant tissues are light green or clear.
8. Observe the bleached leaves under the microscope to see the GUS staining (*see* **Notes 17–21**).

4 Notes

1. During the cloning process, keep enzymes like DNA polymerase and T4 ligase on ice or in a cool container immediately after use to maintain their functionality.

2. Make fusion constructs (promoter along with translation initiation codon and reporter), ensuring that the translation initiation codon is in frame with the GUS gene.
3. Always confirm that the PCR product, plasmid, or genomic DNA is of high purity and free from contamination, as this could affect the success of cloning.
4. Ensure the use of appropriate antibiotics in the selection media to prevent contamination or excessive growth.
5. Always confirm that the plasmid containing the fusion construct is successfully transformed into bacteria (*Escherichia coli*) or *Agrobacterium*, verified by colony PCR or sequencing.
6. Ensure that all DNA samples, PCR products, plasmids, constructs, and plates are labelled clearly and accurately to avoid any mix-ups.
7. The media used for *Agrobacterium* inoculation should be freshly prepared to avoid any contamination.
8. Confirm that *Agrobacterium* cultures are healthy, actively growing, and well-prepared before use. A fresh overnight culture is often the most suitable.
9. During the preparation, the concentration of *Agrobacterium* must be optimized carefully. A concentration that is too high can damage plant tissue, whereas a concentration that is too low may lead to poor transformation efficiency.
10. Always use healthy, actively growing plants for infiltration. Young leaves that are fully expanded but not yet mature, as they lack a robust vascular system, are considered the best for agroinfiltration.
11. Avoid placing stress on plants before or during infiltration. Stress factors, such as extreme temperatures, drought, or poor lighting, can lower the success rate of transient expression.
12. The pressure should be carefully regulated when using a syringe to prevent damage to the leaves. Excessive pressure can cause punctures or breakage of leaf cells.
13. To ensure successful infiltration, it is important that the *Agrobacterium* suspension penetrates adequately into the leaf tissue without causing any damage. Typically, an appropriate volume of 1–2 mL of the suspension should be enough per leaf. Infiltrating too much solution can lead to wilting or damage to the tissue.
14. To prevent cross-contamination and ensure proper infiltration of each construct into the plant leaf, change the syringe for each construct during the agroinfiltration process.

15. Always use an empty vector as a positive control during agroinfiltration to ensure that the infiltration procedure was successful and did not produce nonspecific effects in the plant tissue.
16. After agroinfiltration, it is essential to keep the plants in suitable conditions (proper light, temperature, humidity, and water).
17. Some of the chemicals used in the preparation of GUS solution, like DMSO, EDTA and detergent (TritonX100), require careful handling. Use gloves, goggles, and a lab coat to protect exposure.
18. Proper disposal of the used GUS solution is required, mainly if it is mixed with plant material, according to your institution's biosafety protocols.
19. Keep the GUS solution in dark conditions and in the refrigerator/cold condition.
20. Take fresh leaf/tissue for histochemical GUS assay.
21. Always use freshly prepared GUS solution.

Acknowledgments

Research in our laboratory was supported by the Council of Scientific and Industrial Research (CSIR), New Delhi, India in the form of NCP project no. MLP006. P.K.T. also acknowledges the Science and Engineering Research Board, New Delhi, for the JC Bose National Fellowship (JCB/2021/000036). A.S. acknowledges DST/INSPIRE faculty fellowship (GAP509).

References

1. Domínguez-Rosas E, Hernández-Oñate MÁ, Fernandez-Valverde SL, Tiznado-Hernández ME (2023) Plant long non-coding RNAs: identification and analysis to unveil their physiological functions. Front Plant Sci 14:1275399
2. Mattick JS, Amaral PP, Carninci P, Carpenter S, Chang HY, Chen LL, Chen R, Dean C, Dinger ME, Fitzgerald KA, Gingeras TR (2023) Long non-coding RNAs: definitions, functions, challenges and recommendations. Nat Rev Mol Cell Biol 24:430–447
3. Anyatama A, Datta T, Dwivedi S, Trivedi PK (2024) Transcriptional junk: waste or a key regulator in diverse biological processes? Curr Opin Plant Biol 82:102639
4. Datta T, Kumar RS, Sinha H, Trivedi PK (2024) Small but mighty: peptides regulating abiotic stress responses in plants. Plant Cell Environ 47:1207–1223
5. Budnick A, Franklin MJ, Utley D, Edwards B, Charles M, Hornstein ED, Sederoff H (2024) Long-and short-read sequencing methods discover distinct circular RNA pools in Lotus japonicus. P Genome 17:20429
6. Posner BI, Laporte SA (2010) Cellular signalling: peptide hormones and growth factors. Prog Brain Res 181:1–16
7. Staudt AC, Wenkel S (2011) Regulation of protein function by 'microProteins'. EMBO Rep 12:35–42
8. Matsubayashi Y (2014) Posttranslationally modified small-peptide signals in plants. Annu Rev Plant Biol 65:385–413
9. Eguen T, Straub D, Graeff M, Wenkel S (2015) MicroProteins: small size–big impact. Trends Plant Sci 20:477–482
10. Zhang P, Wu W, Chen Q, Chen M (2019) Non-coding RNAs and their integrated networks. J Integr Bioinform 16:20190027

11. Kim JS, Jeon BW, Kim J (2021) Signaling peptides regulating abiotic stress responses in plants. Front Plant Sci 12:704490
12. Badola PK, Sharma A, Gautam H, Trivedi PK (2022) MicroRNA858a, its encoded peptide, and phytosulfokine regulate Arabidopsis growth and development. Plant Physiol 1189: 1397–1415
13. Gautam H, Sharma A, Trivedi PK (2023) Plant microProteins and miPEPs: small molecules with much bigger roles. Plant Sci 326:111519
14. Petri L, Van Humbeeck A, Niu H, Ter Waarbeek C, Edwards A, Chiurazzi MJ, Vittozzi Y, Wenkel S (2024) Exploring the world of small proteins in plant biology and bioengineering. Trends Genet. https://doi.org/10.1016/j.tig.2024.09.004
15. Sanchita, Trivedi PK, Asif MH (2020) Updates on plant long non-coding RNAs (lncRNAs): the regulatory components. PCTOC 140: 259–269
16. Bhati KK, Dolde U, Wenkel S (2021) Micro-Proteins: expanding functions and novel modes of regulation. Mol Plant 14:705–507
17. Ormancey M, Guillotin B, Merret R, Camborde L, Duboé C, Fabre B, Pouzet C, Impens F, Van Haver D, Carpentier MC, Clemente HS (2023) Complementary peptides represent a credible alternative to agrochemicals by activating translation of targeted proteins. Nat Commun 14:254
18. Kumar RS, Sinha H, Datta T, Asif MH, Trivedi PK (2023) microRNA408 and its encoded peptide regulate sulfur assimilation and arsenic stress response in Arabidopsis. Plant Physiol 192:837–856
19. Sharma A, Anyatama A, Gautam H, Gaddam SR, Singh D, Sinha H, Trivedi PK (2024) Enhancing nutritional quality in plants using complementary peptide for sustainable agriculture. Plant Physiol 196:711–715
20. Ormancey M, Guillotin B, Ribeyre C, Medina C, Jariais N, San Clemente H, Thuleau P, Plaza S, Beck M, Combier JP (2024) Immune-enhancing miPEPs reduce plant diseases and offer new solutions in agriculture. Plant Biotechnol J 22:13–15
21. Lauressergues D, Couzigou JM, Clemente HS, Martinez Y, Dunand C, Bécard G, Combier JP (2015) Primary transcripts of microRNAs encode regulatory peptides. Nature 520:90–93
22. Sharma A, Badola PK, Bhatia C, Sharma D, Trivedi PK (2020) Primary transcript of miR858 encodes regulatory peptide and controls flavonoid biosynthesis and development in Arabidopsis. Nat Plants 6:1262–1274
23. Lauressergues D, Ormancey M, Guillotin B, San Clemente H, Camborde L, Duboé C, Tourneur S, Charpentier P, Barozet A, Jauneau A, Le Ru A (2022) Characterization of plant microRNA-encoded peptides (miPEPs) reveals molecular mechanisms from the translation to activity and specificity. Cell Rep 38:110339
24. Jefferson RA (1989) The GUS reporter gene system. Nature 342:837–838
25. Liu Z, Wang W, Zhang CG, Zhao JF, Chen YL (2017) GUS staining of guard cells to identify localised guard cell gene expression. Bio Protoc 7:2446
26. Campalans A, Kondorosi A, Crespi M (2004) Enod40, a short open reading frame–containing mRNA, induces cytoplasmic localization of a nuclear RNA binding protein in *Medicago truncatula*. Plant Cell 16:1047–1059
27. Edwards A, Chiurazzi MJ, Blaakmeer A, Vittozzi Y, Sharma A, Matton S, Kruusvee V, Straub D, Sessa G, Carabelli M, Morelli G, Wenkel S (2024) A shade-responsive micro-Protein in the Arabidopsis ATHB2 gene regulates elongation growth and root development. eLife 13:RP96725. https://doi.org/10.7554/eLife.96725.1
28. Sharma A, Badola PK, Gautam H, Gaddam SR, Trivedi PK (2022) HY5 regulates light-dependent expression and accumulation of miR858a-encoded peptide, miPEP858a. Biochem Biophys Res Commun 589:204–208
29. Vergish S, Wolf R, Song WY (2024) Simplified protocol to demonstrate gene expression in *Nicotiana benthamiana* using an agrobacterium-mediated transient assay. Bio Protoc 14:4987
30. Sambrook J, Fritsch EF, Maniatis T (1989) Molecular cloning: a laboratory manual, 2nd edn. Cold Spring Harbor Laboratory Press

Chapter 15

Elucidating the Localization and Interactome of Mitochondrial Microproteins

Jiemin Nah, Pooja Sridharan, and Lena Ho

Abstract

A large number of novel microproteins discovered to date are nuclear encoded, mitochondrial proteins, pointing to their widespread roles in metabolic regulation. In this chapter, we provide a workflow of how to verify if a candidate microprotein is localized to the mitochondria, its submitochondrial localization (i.e., outer, inner membrane, or matrix) and how to determine its interactome in order to elucidate its molecular function.

Key words Mitochondria, OXPHOS, Microproteins

1 Introduction

Microproteins, typically defined as polypeptides comprising fewer than 100 amino acids, have garnered increasing interest in the field of molecular biology due to their potential roles in cellular regulation and function. Despite their small size, microproteins have been implicated in diverse biological processes throughout evolution [1–3], including transcription regulation, signal transduction, and cellular stress responses [4, 5]. Among these, microproteins appear to play a prominent role in metabolic regulation, as many microproteins have been found to be localized to the mitochondria, where they interact with metabolic enzymes and transporters to regulate respiration, energy production, reactive oxygen species (ROS) production, apoptosis, and cellular signaling [6]. Given the mitochondria's role as a hub of energy production and metabolic regulation, disruptions in microprotein functions can have far-reaching implications for cell viability and organismal health, potentially contributing to disease states such as neurodegeneration, metabolic disorders, and cancer.

For microprotein biologists interested in defining small proteins that contribute to mitochondrial function, the first hurdle to

Stephan Wenkel (ed.), *Microproteins: Methods and Protocols*, Methods in Molecular Biology, vol. 2992,
https://doi.org/10.1007/978-1-0716-5013-4_15,

overcome is assessing the potential for mitochondrial localization. Furthermore, because the mitochondrion is compartmentalized into several distinct regions, including the outer membrane (OMM), intermembrane space (IMS), inner boundary membrane (IMM), cristae membrane, intercristal space, and matrix—each with specialized functions [7]—determining submitochondrial locational information is critical for subsequent functional elucidation. Hence, methods to localize a microprotein to the mitochondria must be followed by ascertaining which mitochondrial subdomains—such as the OMM, IMM, or matrix—a microprotein candidate of interest operates in. Thereafter, to define the molecular function of an unknown protein, a very helpful approach is to define its interacting partners, since intracellular microproteins rarely work as free-standing entities. If interacting proteins have known functions, one can immediately hypothesize that the candidate microprotein is in the same pathway.

This chapter aims to provide a comprehensive protocol detailing the methods for investigating the mitochondrial localization of a microprotein of unknown function. The first piece of evidence of mitochondrial localization is typically an immunofluorescence experiment, where an antibody against the endogenous protein or against a tag (if using tagged fusion proteins) is used to determine colocalization with established mitochondrial markers such as TOMM20 or Citrate Synthase (CS). As these are basic techniques, they will not be covered here. The protocol starts with the assumption that the microprotein of interest is known to be localized to the mitochondria. We will outline strategies for identifying whether the protein resides in the OMM, IMM, or matrix compartment using proximity labelling with the APEX system, as well as a proteinase K protection assay. Furthermore, the chapter will detail the application of native immunoprecipitation coupled with mass spectrometry (IP/MS) to uncover potential interacting partners of the microprotein, offering insights into its functional network within the mitochondrial environment. By systematically characterizing the localization and interaction landscape of mitochondrial microproteins, we hope to bridge the current knowledge gap in mitochondrial biology and contribute to the broader understanding of microprotein functions and their roles in cellular homeostasis.

2 Materials

All chemicals used are of analytical grade, and buffers and dilutions are made with Milli-Q water (18 Mohm/cm resistance).

2.1 Mitochondrial Isolation for Cells

- Tightly fitting glass/Teflon Potter Elvehjem homogenizer (Sigma–Aldrich, P7734).
- Motor to drive homogenizer, either a hand-drill (e.g., BOSCH GSR120-LI) or a benchtop mixer (e.g., JOANLAB overhead stirrer mixer) that can go up to 2000 rpm.
- Sucrose (Sigma–Aldrich, 84100). Make 1M solution.
- HEPES (Sigma-Aldrich, H4034). Make 1M solution. Adjust pH to 7.4 with 1M KOH.
- 0.5M Ethylenediaminetetraacetic acid (EDTA) solution (Invitrogen, 15575020).
- Hydrochloric acid for pH adjustment.
- Mitochondrial isolation buffer (MIB): Combine 1 mL of 1 M HEPES-KOH, 0.2 mL of 0.5M EDTA and 25 mL of 1 M sucrose. Bring the volume to 100 mL with distilled water. Adjust pH to 7.4 with 1 M HCl. For tissues, please *see* **Note 1**.
- Resuspension buffer (RB): Combine 3.2 mL of 1 M sucrose and 0.1 mL of 1 M HEPES-KOH, 0.02 mL of 0.5M EDTA, bring the volume up to 10 mL with distilled water. Adjust pH to 7.4.
- Pierce BCA assay kit (ThermoFisher, A55864).
- cOmplete, EDTA-free protease inhibitor cocktail (Roche, COEDTAF-RO).

2.2 Proteinase K Digestion

- Proteinase K (20 mg/mL) RNA Grade (Invitrogen, 25530049).
- Digitonin, high purity (Millipore, 300410), make up to 5% in water, incubate at 95C for 5 min to fully dissolve.
- Triton X-100 (Sigma-Aldrich X100).
- Phenylmethanesulfonyl fluoride (PMSF) (Roche, PMSF-RO), dissolved in isopropanol.
- NuPAGE™ Bis-Tris Mini Protein Gels, 4–12%, 1.5 mm and an electrophoresis apparatus.
- Polyvinylidene fluoride (PVDF) transfer membranes and semi-dry transfer apparatus.

2.3 APEX Proximity Labeling

- MitoV5-APEX2 (Addgene #72840) for marking the mitochondrial matrix.
- IMS-APEX2 (Addgene #79058) for marking the IMS.
- APEX2-OMM (Addgene #79056) for marking the cytosol-facing OMM.
- 20% (5.88 M) hydrogen peroxide.
- Sodium ascorbate (Supelco, PHR1279).
- Sodium azide (Sigma–Aldrich, 71289).

- Trolox/(±)-6-hydroxy-2,5,7,8-tetramethylchromane-2-carboxylic acid (Sigma-Aldrich, 238813) (dissolve in ethanol or DMSO).
- Quenching buffer: PBS + 5 mM Trolox, 10 mM sodium ascorbate, 10 mM sodium azide.
- Quenching additives: 5mM Trolox, 10 mM sodium ascorbate, and 10 mM sodium azide.
- RIPA buffer: 50 mM Tris HCl, 150 mM NaCl, 1.0% (v/v) IGEPAL® CA-630, 0.5% (w/v) sodium deoxycholate, 1.0 mM EDTA, 0.1% (w/v) SDS and 0.01% (w/v) sodium azide at a pH of 7.4. or Sigma R0278.
- Pierce strepavidin magnetic beads (ThermoFisher, 88816).
- Streptavidin–HRP (Invitrogen, S911).
- SDS–PAGE electrophoresis and transfer setup.
- Biotin (Sigma B4501).
- Pierce™ 660 nm Protein Assay Reagent (ThermoFisher, 22660).

2.4 Native Immunoprecipitation (IP)

- IP buffer: 20 mM Tris–HCl, 100 mM NaCl, 0.5 mM EDTA, 10% glycerol, adjust to pH 7.4. Add 1X cOmplete protease inhibitor cocktail fresh.
- Digitonin, high purity (Millipore 300410), make up to 10% in water (100 mg/ml), incubate at 95C for 5 minutes to fully dissolve.
- Eppendorf® Protein LoBind tubes (Merck, EP0030108116).
- Anti-FLAG® M2 magnetic beads (Millipore, M8823).
- PureProteome™ magnetic stand (Millipore) or equivalent.
- FLAG peptide (Sigma-Aldrich, F3290).

3 Methods

A typical workflow starts with isolating mitochondria from cells or tissues of interest, either with knockout or overexpression of the microprotein-of-interest, or following APEX labeling. The following section offers an end-to-end protocol for deciphering the submitochondrial topology and interactome of a mitochondrial microprotein.

3.1 Mitochondrial Isolation

This protocol describes how to obtain functional, purified and intact mitochondria from cultured cells and was adapted from Frezza and Scorrano [8]. The following description is for isolation from cultured cells such as fibroblasts or HEK293T. For isolation from animal tissues, please refer to modifications in the Notes section. Grow cells to a high confluence (80–90% for a higher yield of mitochondria). A confluent 10-cm dish of HEK293T cells typically yields 0.2–0.5 mg of purified mitochondria, while an adult heart yields 2–3 mg. We recommend collecting a minimum of 0.5 mg of mitochondria for one proteinase K experiment and 1 mg for native immunoprecipitation.

1. Prepare mitochondrial isolation buffer (MIB) and resuspension buffer (RB) as detailed. For PK digestion, we do not supplement MIB or RB with protease inhibitors. Instead, we work fast and keep the sample on ice at all times during the isolation process.
2. Harvest cells by trypsinization, followed by FBS-mediated neutralization. Pellet cells by centrifugation and wash once with PBS.
3. Resuspend cells in 1 mL of ice-cold MIB for up to 20 million cells.
4. Homogenize the cell suspension in a glass potter fitted with 20–25 strokes of the Teflon pestle, driven by a drill or stirrer operating at 1600 rpm. Keep the potter on an ice bath during the homogenization (*see* **Note 2**).
5. Transfer the homogenate to an Eppendorf tube and pellet the nuclear and membrane debris at 1000 g (600 g for tissues) for 5 min at 4 °C.
6. Transfer the supernatant containing mitochondria to a fresh tube, careful not to pick up any debris from the loose pellet.
7. Pellet the mitochondria at 12,000 g (7000 g for tissues) for 5 minutes at 4 °C. Discard supernatant.
8. Wash the mitochondria pellet with 1 mL of ice-cold MIB.
9. Remove all supernatant and resuspend the mitochondrial pellet in a small volume (50–100 μL) Resuspension Buffer (RB) using a p200 pipette tip with the tip cut off to increase the size of the orifice. This induces less stress on the mitochondria and preserves their integrity. Pipette up and down at least 20 times to achieve a homogenous suspension.
10. Measure the concentration of the mitochondria using a BCA protein assay according to the manufacturer's instructions. Adjust the concentration of the mitochondria to 1 mg/ml using RB. Do not add protease inhibitors for 3.2, but to all other applications, supplement RB with 1X cOmplete protease inhibitors.
11. To assess the quality of the mitochondrial prep, *see* **Note 3**.

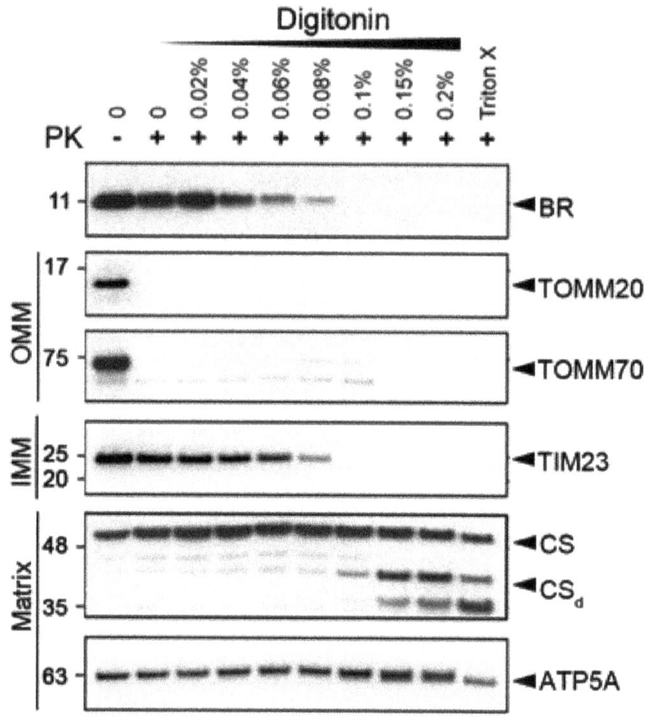

Fig. 1 PK protection assay demonstrating that BRAWNIN (UQCC6) is located in the IMM (from Zhang et al. 2020 [9]). BR is a 73aa microprotein that mediates early assembly of the electron transport chain (ETC) Complex III [10]. It contains a single transmembrane domain (TMD) which spans the inner membrane. Using an antibody against a C-terminal epitope of BR, we found that its PK susceptibility closely mirrors that of TIM23, another IMM transmembrane protein. OMM proteins like TOMM20/70 show immediate sensitivity to PK, whereas matrix proteins such CS and ATP5A are resistant to digitonin, requiring Triton-X for increased accessibility to enable digestion. Note that smaller degraded fragments, such as those of CS (CS_d) are sometimes visible, indicating partial digestion. (Figure reproduced from [9])

3.2 Proteinase K Protection Assay

The proteinase K protection assay is a technique used to determine the topology of mitochondrial proteins. It relies on the enzyme's ability to selectively degrade proteins based on their accessibility. Intact mitochondria are treated with proteinase K under increasing concentrations of the selective nonionic detergent digitonin to permeabilize the outer and inner membranes. Proteins that are susceptible to degradation in intact mitochondria (i.e., no digitonin) are located on the outer membrane (OMM) exposed to the cytosol, whereas proteins located in the intermembrane space (IMS), inner membrane (IMM) require increasing amounts of digitonin to become accessible and susceptible to PK digestion (Fig. 1). Accessing matrix proteins requires complete solubilization by nonselective nonionic detergents like Triton-X100 to

Table 1
Reagent formulation for the digitonin series used in PK digestion

Reaction #	Digitonin stock (%)	Digitonin volume (μL)	Final digitonin (%)	Triton X-100 volume (μL)	Final triton X-100 (%)
1	0	0	0	0	0
2	0	0	0	0	0
3	1	1	0.02	0	0
4	1	2	0.04	0	0
5	5	0.6	0.06	0	0
6	5	0.8	0.08	0	0
7	5	1	0.1	0	0
8	5	1.5	0.15	0	0
9	5	2	0.2	0	0
10	0	0	0	0.5	1

completely disrupt both the IMM lipid bilayer. Proteins with well-defined submitochondrial localization are used as reference points to indicate the transition from the OMM to IMS to IMM to matrix as the digitonin gradient increases. For this assay to work well, it is important to have a specific antibody against the microprotein of interest. Ideally, the epitope of the antibody is known, which can give further information about the topology of the microprotein, especially if it is membrane spanning (*see* **Note 4**).

1. Dilute the 5% digitonin stock to the appropriate concentrations listed in column 2 of Table 1.
2. Aliquot equal amounts of mitochondria (50ug in 50 μL RB) into ten 1.5 ml Eppendorf tubes.
3. Solubilize mitochondria with increasing concentrations of digitonin (0–0.2%) or Triton X-100, as detailed in Table 1.
4. Incubate on ice for 15 min to allow solubilization, agitating midway to prevent mitochondria from settling to the bottom.
5. Add proteinase K or Triton X100 to reactions 2–10 to a concentration of 100 μg/ml.
6. Incubate on ice for 30 min to enable complete digestion of accessible proteins. Agitate midway.
7. Prepare fresh PMSF, a pan-serine protease inhibitor, in isopro panol and add to all reactions to a final concentration of 10 mM to stop the reaction. Mix thoroughly.
8. Add 1X protein loading dye to all reactions and heat at 95 degrees for 10 min. Spin down at top speed to remove all

insoluble material. Load 40 μL (~30 μg) into an SDS–PAGE gel. We use Invitrogen's NuPAGE™ Bis–Tris Mini Protein Gels, 4–12% with a MES buffer system, but this protocol is compatible with any SDS–PAGE system of your choice.

9. Transfer SDS–PAGE to a PVDF or nitrocellulose membrane for western blotting.
10. We recommend using the following antibodies to demarcate the submitochondrial domains:
    (a) OMM: TOM20 (Proteintech 11,802–1-AP).
    (b) IMM: TIM23 (Proteintech 11123-1-AP).
    (c) Matrix: Citrate Synthase (Santa Cruz SC-390693).

3.3 APEX Proximity Labeling for Determining Submitochondrial Localization

Ascorbate peroxidase (APEX) proximity labeling is a powerful technique for determining the submitochondrial localization of uncharacterized proteins [11, 12]. In this method, APEX is fused to known mitochondrial markers targeting the OMM, IMS and the matrix and expressed in live cells. Upon addition of biotin-phenol and hydrogen peroxide, APEX generates reactive biotin-phenoxyl radicals that label nearby proteins, including the microprotein of interest, which can be endogenous or expressed with a tag. These biotinylated proteins are then isolated and identified by western blotting or mass spectrometry. By comparing the labelling patterns of the microprotein with those of known submitochondrial markers, one can infer the localization. This approach offers high spatial resolution and the ability to capture both stable and transient protein interactions, making it valuable for creating detailed submitochondrial proteome maps. APEX proximity labeling complements and enhances the proteinase K protection assay for determining submitochondrial protein localization. It offers higher spatial resolution, enabling nanometer-scale mapping within mitochondrial compartments. One important advantage over the PK methods is that APEX labelling occurs in living cells instead of lysates, preserving native conditions and capturing transient interactions that might be missed by other methods. It allows for comprehensive proteome mapping, identifying multiple proteins simultaneously, and is versatile enough to work with both soluble and membrane-bound proteins. While the proteinase K assay provides some initial topology information, APEX labeling can offer a more accurate picture of mitochondrial protein organization (*see* **Note 6**), making it particularly useful for studying complex spatial arrangements and protein import pathways in mitochondria.

1. Seed 300,000 HEK293T cells in a 6 cm dish overnight.
2. Co-transfect each well with 2 μg Microprotein expression vector (if using) and 2 μg of either mitoV5-APEX2, IMS-APEX2, and APEX2-OMM in separate wells. Change the media after

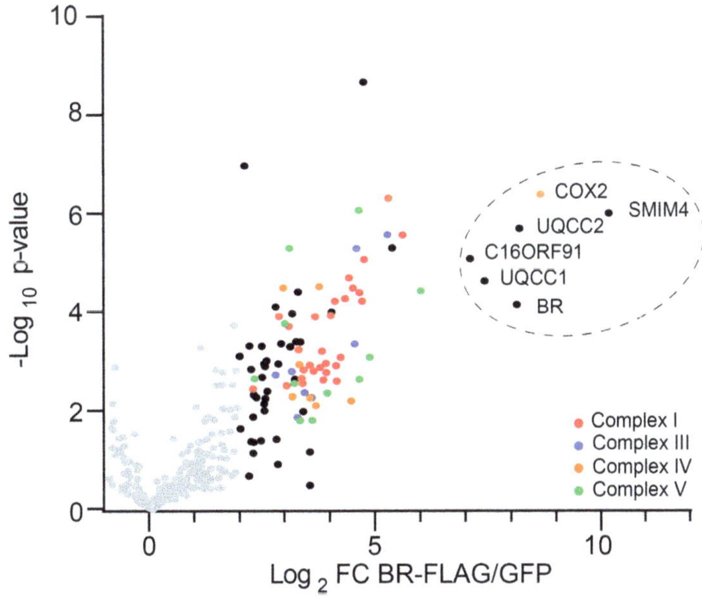

Fig. 2 APEX labeling demonstrates that the microprotein CRISTA is accessible both from the matrix and from the IMS. Previous PK studies determined that CRISTA is bound to the IMM, but its topology was unclear. To resolve this, we exposed cells expressing CRISTA-FLAG to APEX labeling. mitoV5 labels matrix proteins such as CS, IMS-APEX2 labels IMS proteins such as MICU2, and APEX2-OMM labels OMM proteins such as TOM20. CRISTA is labeled by both mitoV5 and IMS-APEX2, indicating that the single-transmembrane containing microprotein has domains that project into both the matrix and the IMS. Strep-HRP blot confirms successful APEX labeling, although the efficiency of labelling is highest in the OMM and lowest in the IMS. (Figure reproduced from [13])

6 h and allow 24 h for protein expression to peak and for cells to recover from the transfection.

3. Add Biotin-phenol to each plate to a final concentration of 0.5 mM. Mix well. Incubate for 30 min at 37 °C.
4. Start the biotinylation reaction by adding H_2O_2 to a final concentration of 1 mM for exactly 1 min. *See* **Note 5**.
5. Stop the reaction by adding 2 mL of quenching buffer to each well, and wash one more time with 2 mL quenching buffer.
6. Trypsinize the cells and quench the trypsin reaction with 1 mL of media with quenching additives.
7. Pellet the cells and wash the cell pellet once with quenching buffer. Optional: freeze cell pellet in −80 °C.
8. Isolate mitochondria in MIB supplemented with quenching additives.
9. Quantify the mitochondria in RB with quenching additives by measuring the absorbance at 660 nm using a Pierce™ 660 nm

Protein Assay Reagent. The BCA assay is not compatible with quenching additives.

10. Aliquot 15 μg of mitochondria per immunoprecipitation reaction. Spin down and remove supernatant.
11. Resuspend 15 μg of mitochondria in 100 μL of RIPA with quenching reagents, 0.5% digitonin, and protease inhibitors.
12. Add 100 μL of 2% SDS in RIPA (final concentration of 1% SDS).
13. Spin at top speed to clear lysate. Remove 2 μL of clarified lysate as input control.
14. Add lysate to 25 μL of Pierce streptavidin magnetic beads (pre-washed 2x with 500 μL RIPA).
15. Incubate overnight with lysate in a cold room on rotation.
16. Let beads stand on the magnet for 1–2 mins, remove supernatant (flowthrough).
17. Wash beads with 300–500 μL of 1% SDS in RIPA for 5 min on a rotator. Repeat wash four times.
18. After the last wash, remove all wash buffer from the beads. To elute bound proteins, resuspend beads in 30 μL of 25 mM biotin in RIPA, and incubate at 95 °C for 15 min.
19. Let eluate stand on the magnet for 1–2 min and transfer eluate into a new tube. Repeat to let stand on the magnet for 1–2 min, and transfer eluate again to new tube to remove excess magnetic beads.
20. Add protein loading dye to 1x into eluted sample and incubate at 95C for another 5–10 min.
21. Load onto an SDS–PAGE gel for electrophoresis.
22. Transfer proteins from the membrane to a PVDF membrane.
23. Probe membrane with Streptavidin-HRP (1:1000 dilution) to check for successful APEX labeling and pull-down of biotinylated proteins. A successful experiment should have a robust and specific signal with Streptavidin–HRP (Fig. 2).
24. Proceed to blot for the microprotein of interest. We recommend using the following antibodies to demarcate the submitochondrial domains:
    (a) OMM: TOM20 (Proteintech 11802-1-AP).
    (b) IMM: MICU (Abcam ab101465).
    (c) Matrix: Citrate Synthase (Santa Cruz SC-390693).

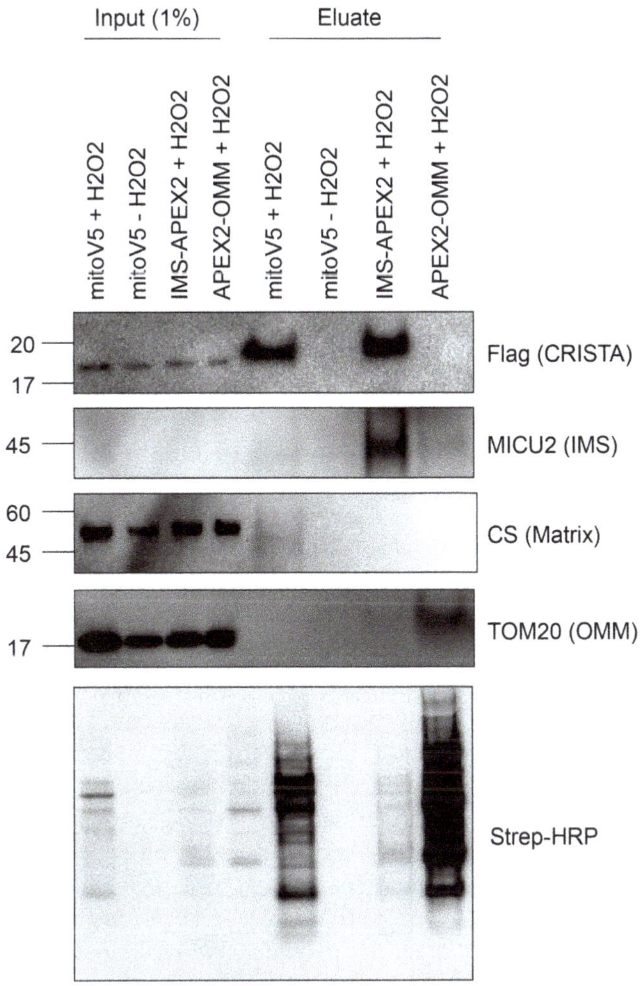

Fig. 3 Native interactome of BR determined by a native FLAG co-IP in cardiac mitochondria expression BR-FLAG. Using native Co-IP, we determined the interactome of BR in cardiac mitochondria expressing a BR-FLAG fusion protein. GFP-expressing tissues were used as a negative control. The experiment was performed in triplicate, and the LC-MS results were analyzed in MaxQuant [14]. Label-free quantitation (LFQ) intensities were Log_2 transformed. Missing values were imputed at the limit of detection based on normal distribution (width = 0.2, down shift = 1.8) and the other group was filtered to have at least two valid values before a two-sample t-test, which used p-value = 0.05 for truncation. Results were visualized as a volcano plot with significance set $p < 0.05$ and the indicated log_2 fold-change. Only proteins quantified in at least two of the three IP samples from the BR-FLAG cells (with no restrictions on control identifications in GFP cells) were included. This led to the identification that BR is located in the COMB complex consisting of SMIM4, UQCC1, UQCC2, and C16ORF91, which is responsible for the assembly of Complex III. (Figure reproduced from [10])

3.4 Native Immunoprecipitation to Decipher Microprotein Interactome

Native co-immunoprecipitation (Co-IP) is a crucial technique for studying protein–protein interactions of microproteins, particularly those localized to membranous organelles such as the mitochondria. This method preserves the natural state of protein complexes by using non-denaturing conditions throughout the process. In native Co-IP, a specific antibody is used to isolate the microprotein of interest (or a tagged version thereof, *see* **Note 7**), along with its interacting partners, from cells or tissues that are solubilized with the mild nonionic detergent digitonin. Digitonin selectively interacts with cholesterols in membranes and effectively solubilizes membrane proteins while preserving their native structure and function (*see* **Note 8**). The technique involves sample preparation, antibody binding, immunoprecipitation, washing, and elution steps. Native Co-IP can reveal valuable information about protein–protein interactions, complex composition, functional associations, post-translational modifications, and interaction dynamics. It is especially useful for mitochondrial microprotein studies as it maintains the integrity of mitochondrial membranes and protein complexes, allows detection of weak interactions, and can be combined with mass spectrometry for unbiased identification of interacting partners. This protocol describes the use of a FLAG-tagged version of the microprotein as bait, with the eluted sample used for unbiased identification by mass spectrometry. However, it can be adapted to other tags or endogenous proteins based on the user's preference (*see* **Note 7**), and the eluted proteins and input can be subjected to Western blotting to interrogate candidate interacting proteins. We do not provide the protocol for mass spectrometry protein sample preparation, as this may vary depending on your service provider. We recommend performing at least four replicates per sample, in order to robustly detect enrichment of interactors in the FLAG-tagged mitochondria over wild-type control.

1. Prepare mitochondrial pellets from cells stably expressing a FLAG-tagged version of the microprotein of interest using protocol 2.1. You will need 1 mg of purified mitochondria per IP reaction. Similarly, prepare mitochondrial pellets from control cells (no FLAG-tag or GFP-tagged versions of microprotein) as a control.
2. Solubilize the mitochondria pellet in IP buffer such that the final digitonin/protein ratio is 8 g/g and the final concentration of protein is 1 mg/mL (i.e., for 1 mg of mitochondria, resuspend in 1 ml of IP buffer with 8 mg of digitonin (80 μL of 10% digitonin stock)).
3. Incubate the lysate at 4 °C for 1 h on a nutator for complete solubilization.
4. After 1-hour incubation, clear the insoluble fraction by centrifugation at 20,000 g for 20 min at 4 °C.

5. Remove the soluble fraction (supernatant) to a LoBind 1.5 mL Eppendorf tube and keep 5% of the supernatant for input analysis.
6. Wash 50 μL (per sample) of Anti-FLAG® M2 Magnetic Beads with 1 mL IP buffer supplemented with 0.1% digitonin, twice. Add buffer to beads, invert the tube a couple of times and remove the wash buffer with the magnetic stand.
7. Add the supernatant from step 5 to the M2 magnetic beads incubate overnight in a cold room on a nutator.
8. Remove the supernatant (flowthrough) from the bound M2 magnetic beads with the magnetic stand. Wash beads with 1 mL of IP buffer supplemented with 0.1% digitonin by inverting the tube 7–8 times or nutating for three minutes. Repeat four times.
9. Wash beads with IP buffer without glycerol and protease inhibitors, but still containing 0.1% digitonin (check with your mass spec facility about the compatibility of their workflow with digitonin).
10. Remove all the buffer after the last wash, until the beads look dry.
11. Elute the bound proteins using 30 μL of elution buffer per sample for a final concentration of 250 ng/ μL of FLAG peptide in IP buffer without glycerol or protease inhibitors.
12. Incubate at 4 °C for 30 min, with gentle agitation.
13. Process the eluate and input for data-dependent acquisition (DDA)-based liquid chromatography mass spectrometry (LC–MS), calculating the enrichment of the IP from the FLAG-containing sample over that of the no-FLAG control for each detected protein. For LCMS sample preparation, instrumentation details and data analysis, we refer the user to Liang et al. 2021 [10] for a detailed description. Expected results are excerpted from this publication and shown in Fig. 3.

4 Notes

1. To adapt this protocol for isolation from muscle or heart, adjust the MIB to the following concentration: Add 6.7 mL of 1 M sucrose, 5 mL of 1 M Tris/HCl, 5 mL of 1 M KCl, 1 mL of 1 M EDTA and 2 mL of 10% BSA and bring up to a final volume of 100 mL with Milli Q water. Adjust pH to 7.4.
2. When performing mitochondrial isolation for the first time, or for a new cell type, check that plasma membranes have been effectively broken. This is done by mixing 10 μL of lysate with 5 μL of Trypan Blue, and checking under a coverslip with a

phase contrast microscope that almost 100% of nuclei are blue, indicating a compromised plasma membrane that allows mitochondria to be released from the cells. At 40X magnification, you will also see very small dark and mobile specks, which are the released mitochondria.

3. If you are performing this technique for the first time in your lab, we recommend that you perform some quality control steps to ensure that the mitochondrial preparation is of acceptable purity. Run the same cell-equivalent amount of the whole cell homogenate, nuclear/debris pellet, purified mitochondria, and cytosolic fraction (supernatant from Subheading 3.1, **step** 7). Probe for typical markers (TOMM20—mitochondria, beta-Actin or Tubulin - cytosol, and Histone H3—nucleus). A successful isolation should not have cytosolic contamination in the mitochondria, and vice versa. It is, however, not uncommon to see some nuclear or ER membrane contamination in the mitochondrial fraction due to the co-sedimentation of membrane fragments.
4. Proteinase K digestion assays can give information about the topology of proteins that span the OMM or the IMM. Knowing the topology of membrane-spanning microproteins can be very helpful in determining if and how they bind to interacting proteins. Deciphering topology is most effectively done if 2 different antibodies, one with a C-terminal epitope and one with an N-terminal epitope, are available. For instance, if the N terminus is in the cytosol and the C terminus is in the IMS, PK digestion might show complete sensitivity without PK with the N-terminal antibody, while the C-terminal antibody shows partial sensitivity without PK (but migrating at a smaller size with the N-terminus cleaved off) and requires low amounts of PK for complete digestion. In the absence of such bespoke antibodies, one can arrive at similar conclusions using transfected copies of the microprotein tagged separately at the N- and C-terminus, and probing using an antibody to the tag. Proteins with N- and C-termini in different submitochondrial compartments sometimes show a different digestion pattern, with the terminus that is more protected being in an inner compartment. However, due to the small size of microproteins, this method might have limited resolution. For more accurate topology determination, we suggest using the APEX assay instead (*see* **Note 6**).
5. To control the timing of APEX labelling, work in batches and stagger the quenching reactions across wells to maintain exactly 1 min of labelling per well, accounting for the quenching and washing steps.

6. APEX labelling can be used to determine the topology of microproteins by tagging the microprotein at the N- and the C-terminus with APEX. The respective eluates are analysed using mass spectrometry. For a protein that has an N-terminus in the matrix and a C-terminus in the IMS, the N-terminal and C-terminal APEX fusions should enrich for matrix proteins and IMS proteins, respectively.
7. Because of the small size of microproteins, antibody epitopes are often occluded by interacting partners in the native state. As such, using an antibody raised against the microprotein might disrupt interactions with binding partners. We therefore favor using a FLAG-tagged version of the microprotein expressed stably in the cell type of interest, at levels that are comparable to the endogenous protein. FLAG-based immunoprecipitation is clean and can be easily eluted by the addition of free FLAG protein. Note that the placement of the FLAG tag (N or C terminus, with or without linker) should be optimized based on the availability of a phenotypic readout (i.e., for the tagged version to rescue a knockout phenotype).
8. An alternative to digitonin in Co-IP experiments is Triton X-100, a stronger non-ionic detergent that can disrupt membrane protein interactions and multisubunit complexes but generally maintains the secondary structure of solubilized proteins. This method can also be considered "native", but from our experience, does not maintain the integrity of many of the protein complexes in the IMM. However, co-IP in Triton X-100 can be cleaner, yielding stronger interactors, and works particularly well for soluble proteins. Hence, if the candidate microprotein is located in the matrix, a co-IP in Triton X-100 can be a good complementary approach to a co-IP in digitonin.

References

1. Burton AT, Zeinert R, Storz G (2024) Large roles of small proteins. Ann Rev Microbiol 78: 1–22. https://doi.org/10.1146/annurev-micro-112723-083001
2. Chanut-Delalande H, Zanet J (2024) Small ORFs, big insights: drosophila as a model to unraveling microprotein functions. Cells 13: 1645. https://doi.org/10.3390/cells13191645
3. Petri L, Humbeeck AV, Niu H, Waarbeek CT, Edwards A, Chiurazzi MJ, Vittozzi Y, Wenkel S (2024) Exploring the world of small proteins in plant biology and bioengineering. Trends Genet. https://doi.org/10.1016/j.tig.2024.09.004
4. Hofman DA, Ruiz-Orera J, Yannuzzi I, Murugesan R, Brown A, Clauser KR, Condurat AL, van Dinter JT, Engels SAG, Goodale A et al (2024) Translation of non-canonical open reading frames as a cancer cell survival mechanism in childhood medulloblastoma. Mol Cell 84:261–276.e18. https://doi.org/10.1016/j.molcel.2023.12.003
5. Wright BW, Yi Z, Weissman JS, Chen J (2022) The dark proteome: translation from noncanonical open reading frames. Trends Cell Biol 32:243–258. https://doi.org/10.1016/j.tcb.2021.10.010
6. Kamradt ML, Makarewich CA (2023) Mitochondrial microproteins: critical regulators of protein import, energy production, stress response pathways, and programmed cell death. Am J Physiol-Cell Physiol 325:C807–

C816. https://doi.org/10.1152/ajpcell.00189.2023

7. Spinelli JB, Haigis MC (2018) The multifaceted contributions of mitochondria to cellular metabolism. Nat Cell Biol 20:745–754. https://doi.org/10.1038/s41556-018-0124-1
8. Frezza C, Cipolat S, Scorrano L (2007) Organelle isolation: functional mitochondria from mouse liver, muscle and cultured filroblasts. Nat Protoc 2:287–295. https://doi.org/10.1038/nprot.2006.478
9. Zhang S, Reljić B, Liang C, Kerouanton B, Francisco JC, Peh JH, Mary C, Jagannathan NS, Olexiouk V, Tang C et al (2020) Mitochondrial peptide BRAWNIN is essential for vertebrate respiratory complex III assembly. Nat Commun 11:1312. https://doi.org/10.1038/s41467-020-14999-2
10. Liang C, Zhang S, Robinson D, Ploeg MV, Wilson R, Nah J, Taylor D, Beh S, Lim R, Sun L et al (2022) Mitochondrial microproteins link metabolic cues to respiratory chain biogenesis. Cell Rep 40:111204. https://doi.org/10.1016/j.celrep.2022.111204
11. Lee S-Y, Kang M-G, Park J-S, Lee G, Ting AY, Rhee H-W (2016) APEX fingerprinting reveals the subcellular localization of proteins of interest. Cell Rep 15:1837–1847. https://doi.org/10.1016/j.celrep.2016.04.064
12. Hung V, Zou P, Rhee H-W, Udeshi ND, Cracan V, Svinkina T, Carr SA, Mootha VK, Ting AY (2014) Proteomic mapping of the human mitochondrial intermembrane space in live cells via ratiometric APEX tagging. Mol Cell 55:332–341. https://doi.org/10.1016/j.molcel.2014.06.003
13. Nah J, Mahendran S, Kerouanton B, Cui L, Hock D.H., Cabrera-Orefice A, Dunlap K, Robinson D, Tung DWH, Leong SH, Tan K, Chothani SP, Sun J, Dziegowska A, Leszczynska G, Rackham OJL, Wittig I, Dedon P, Ducker GS, Stroud DA, Ho L (2025) Microprotein SMIM26 drives oxidative metabolism via serine-responsive mitochondrial translation. Molecular Cell 85(14) 2759-2775.e12 https://doi.org/10.1016/j.molcel.2025.05.033
14. Cox J, Mann M (2008) MaxQuant enables high peptide identification rates, individualized p.p.b.-range mass accuracies and proteome-wide protein quantification. Nat Biotechnol 26:1367–1372. https://doi.org/10.1038/nbt.1511

Chapter 16

Illuminating Small Proteins Through HiBiT Blotting

Laure Simoens, Veronique Jonckheere, Dominiek Catteeuw, and Petra Van Damme

Abstract

Epitope tag immunoblotting represents a routine method for targeted surveys of protein abundance and expression. Especially for microproteins, defined by an arbitrary cutoff of 100 amino acids in length, blotting-based approaches are indispensable as the small size of microproteins often renders them elusive to mass spectrometry. Nonetheless, the blotting of microproteins introduces a set of technical challenges, leading to microprotein losses, which significantly affect the sensitivity of blotting-based approaches like epitope tag immunoblotting.

We introduce HiBiT blotting, an antibody-free luminescent detection method for HiBiT-tagged proteins, offering an alternative blotting-based protein detection method to improve microprotein analysis. The availability of an anti-HiBiT antibody enabled a comparative analysis of HiBiT versus classical epitope tag immunoblotting, overall demonstrating the superior sensitivity of HiBiT blotting in detecting HiBiT-tagged microproteins. By offering a more direct and sensitive approach for small protein analysis, HiBiT blotting represents a substantial contribution to the field, enabling the effective study of microproteins and addressing the longstanding challenge of their detection.

Key words HiBiT blotting, Immunoblotting, Luminescence, Microproteins, Protein blotting, SDS-PAGE

1 Introduction

Microproteins, vital in the regulation of physiology and (infectious) diseases, have been identified across all domains of life [1–4]. Microproteins, also known as small open reading frame (sORF)-encoded polypeptides, or SEPs, are arbitrarily defined as proteins of translational origin capped at 100 amino acids (AA) in length. Recent reannotation efforts, facilitated by advanced next-generation sequencing (NGS) techniques such as ribosome profiling or Ribo-seq [5], have highlighted sORFs as a significantly under-annotated category of genomic elements in prokaryotic, eukaryotic, and even viral genomes [6–8]. This oversight is largely due to their historical dismissal as mere noise by conventional annotation algorithms [9].

Stephan Wenkel (ed.), *Microproteins: Methods and Protocols*, Methods in Molecular Biology, vol. 2992,
https://doi.org/10.1007/978-1-0716-5013-4_16,

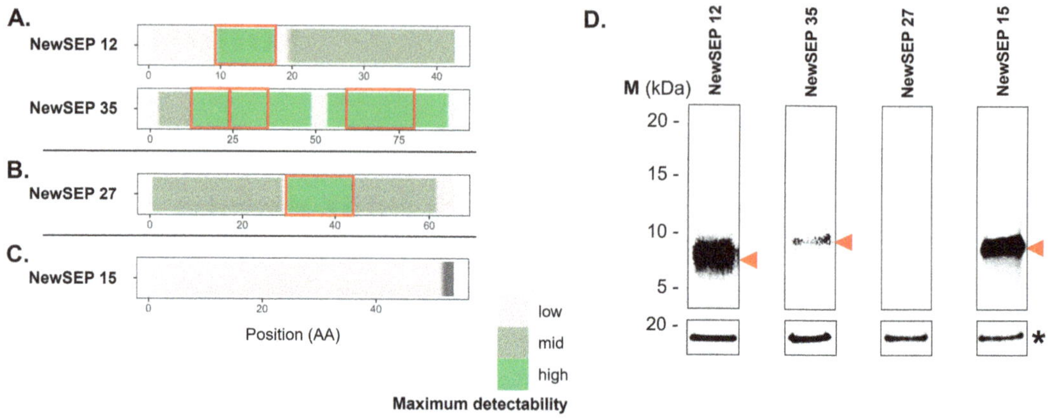

Fig. 1 Analysis of small ORF (sORF)-encoded polypeptide (SEP) expression in *Salmonella* Typhimurium through mass spectrometry (MS) and HiBiT blotting. The theoretical tryptic MS-identifiability of SEPs presented in the left figure panel was calculated using the Advanced Proteotypic Peptide Predictor (AP3) [26]. Identifiability is depicted on a color scale denoting "Maximum detectability" from low to high. Peptides experimentally identified by MS are highlighted in red boxes [10]. (**a**) For a subset of newly discovered SEPs, MS-based peptide evidence was obtained by searching shotgun proteomics data against a six-frame translation database of the SL1344 genome [10]. The expression of these SEPs was further validated by targeted detection using HiBiT blotting (**d**). (**b**) For another subset, only MS-based peptide evidence is currently available. (**c**) Conversely, some SEPs lacked MS-based peptide evidence but were detected through HiBiT blotting (**d**). (**d**) HiBiT detection in protein lysates from endogenously HiBiT-tagged sORF *Salmonella* strains expressing the HiBiT-tagged SEPs NewSEP 12 (6.2 kDa), NewSEP 35 (11.3 kDa), NewSEP 27 (8.4 kDa), and NewSEP 15 (6.8 kDa) [10] through HiBiT blotting. NewSEP 12 was detected in the lysate of an exponential phase culture (~ $OD_{600} = 0.6$), whereas the other two detected SEPs were in lysates from stationary phase cultures (~ $OD_{600} = 3.6$). The detection of HiBiT-tagged SEPs is indicated with red arrowheads. The biotin carboxyl carrier protein (BCCP; 17 kDa), used as a loading control, was probed with a fluorescent streptavidin conjugate (marked with an asterisk). *AA* amino acid, *M* molecular weight marker

Hence, the notable underrepresentation of microproteins in protein databases is primarily a consequence of the under-annotation of the corresponding sORFs. As Ribo-seq data increasingly confirmed the effective translation of previously unannotated sORFs, the focus has shifted towards enhancing small protein detection using conventional protein methods like mass spectrometry (MS) and immunoblotting. These techniques are notably more challenging for small proteins than larger ones, due to their low molecular weights (MWs) and relatively higher hydrophobicity [10, 11]. MS-based microprotein detection is particularly hampered by the limited number—and in some cases, the complete absence—of theoretical MS-identifiable tryptic peptides [10] (Fig. 1), highlighting the need for suitable methods to accurately identify these small yet significant proteins.

Consequently, targeted protein detection through blotting methods retains its relevance in the omics era (Fig. 1). However, small protein blotting encounters numerous technical challenges at various stages of the procedure. Notably, the most significant losses

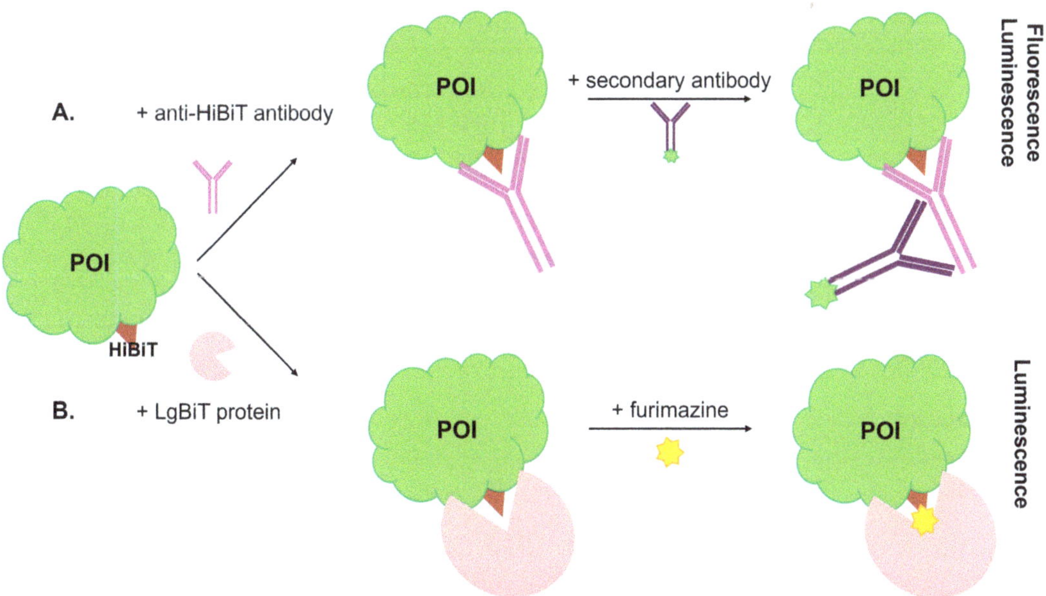

Fig. 2 Comparison of HiBiT blotting with traditional epitope tag immunoblotting. HiBiT is a compact peptide tag comprising 11 amino acids with a molecular weight of only 1.3 kDa. (**a**) The availability of an anti-HiBiT antibody permits the detection of HiBiT-tagged proteins of interest (POI) via immunoblotting. This method involves the recognition of the HiBiT tag by the primary antibody and subsequent detection using a fluorescently labeled or horseradish peroxidase (HRP)-conjugated secondary antibody. (**b**) The high-affinity interaction between HiBiT and the complementing Large BiT (LgBiT; 18 kDa) protein facilitates the reconstitution of NanoLuc luciferase, enabling the detection of a POI through the catalytic conversion of a luciferase substrate (e.g., furimazine), producing a luminescent signal

of small proteins typically occur during the blotting and immunodetection phases, where proteins may completely pass through the membrane—a phenomenon known as membrane blow-through—or detach from it [12, 13], reducing the sensitivity of the method and often resulting in diffuse or missing signals.

To address these challenges, rather than resorting to the traditional use of bulkier epitope tags, which increase the MW of target proteins and potentially alter their properties (e.g., stability, localization, and function), we propose the use of a small luminescent protein tag. More specifically, the sensitivity of small protein detection through blotting can be enhanced by transitioning from antibody-based epitope tag immunoblotting to affinity complementation-based HiBiT blotting [14] (Fig. 2). HiBiT, a compact peptide tag comprising 11 AA (VSGWRLFKKIS) with an MW of only 1.3 kDa, engages in high-affinity interactions with the complementing Large BiT (LgBiT; 18 kDa) protein, facilitating the reconstitution of NanoLuc luciferase [15, 16] (Fig. 2b). This system enables highly sensitive and specific detection of HiBiT-tagged proteins upon the addition and conversion of a dedicated luciferase substrate, such as furimazine. Given the femtogram sensitivity of luminescence-based HiBiT blotting [17] compared to epitope tag blotting methods, the protocol is particularly effective for small protein detection, streamlining the process into just three steps: tag exposure, HiBiT-LgBiT complementation, and substrate addition. This streamlined method circumvents the need for numerous washes and incubations associated with traditional epitope tag immunoblotting methods, thereby offering a more direct and sensitive approach to small protein analysis.

2 Materials

2.1 Standard Material and Equipment

1. Gloves.
2. Tweezers.
3. Spatula.
4. Micropipettes (P10, P20, P200, and P1000) and specialized gel-loading micropipette tips (*see* **Note 1**).
5. Pipette controller and sterile pipettes (5 mL, 10 mL, 25 mL, and 50 mL).
6. Distilled water.
7. Bottles and tubes for buffer preparation.
8. 50 mL centrifuge tubes (sterile).
9. 15 mL centrifuge tubes (sterile).
10. 1.5 mL Eppendorf tubes.
11. Test tube rack.
12. Eppendorf rack.
13. Floating foam Eppendorf rack.
14. Vortex mixer.
15. Thermomixer for 1.5 mL Eppendorf tubes.
16. Centrifuge for centrifuge tubes.
17. (Tabletop) Eppendorf centrifuge.
18. Ice bucket.
19. −80 °C freezer.
20. Growth incubator.
21. Laminar flow cabinet.

2.2 Bacterial Strains and Cultivation

2.2.1 Bacterial Strains

1. The *Salmonella enterica* subspecies *enterica* serovar Typhimurium (*S.* Typhimurium) wild-type (WT) strain SL1344 (genotype: *hisG46*, phenotype: His(−); biotype 26i) was obtained from the *Salmonella* Genetic Stock Center (*Salmonella* Genetic Stock Center (SGSC), Calgary, Canada; cat n° 438) [18].

Table 1
Information for genomic manipulation of new sORFs and expression analysis of their encoded new SEPs [10]

Protein name	Genomic coordinates	Flanking genes	Strand orientation	AA sequence	Length (AA)	MW (kDa)
NewSEP 12	Chromosome:901221-901346	*ybiP/rybA*	> > >	MNEFKRCIRVFSHSPFKVRLMLLSMLCDMINGKPEQDNPSTK	42	5.0
NewSEP 15	Chromosome:4837208-4837366	*osmY/yjjU*	< < <	MFRWGIIFLVIALIAAALGFGGLAGTAAGAAKIVFVVGIVLFLVSLFMGRKRP	53	5.5
NewSEP 27	Chromosome:4629376-4629570	*hflC/purA*	< < <	MNSTIWLALALVLVLEGLGPMLYPGAWKKMVSALAQLPENVLRRFGGGLVVAGVVVYYMLRKTIG	65	7.1
NewSEP 33	Chromosome:2893731-2893988	*CBW18816/CBW18817*	> > >	MIASLLANEYVSAFGYIVGVISGCIAIYQTTQVTKKNNEIKQLNVTIKDLNSQIINITTNRNNINQGERSQYFQDNNGPVNIDNRG	86	9.6
NewSEP 35	Chromosome:2885591-2885863	*CBW18807/*cllc	> < >	MKNNEQQNEALKQLTNVLTEAGNQTRVDVLAHSILLQAIFSVLSEEQKNQIIKILQTATVNQHAATAGVEAEVKMSLAQLLSGFLTPQKLN	91	10.0
NewSEP 37	Chromosome:3417503-3417805	*yqjC/yqjE*	< < <	MSKDNTTEHLRAELKSLTDTLEEVLSSSGEKSKEELSKIRSKAERALKESRYRLGETGDVIAKQTRVAAARADDYVRENPWTGVGIGAAVGLVLGVLLTRR	101	11.1

Strand orientation is denoted with “>” and “<”, representing the clockwise (Watson) and counterclockwise (Crick) strands of the *Salmonella* chromosome, respectively, including for the flanking genes. The protein length is given in amino acids (AA) and the molecular weight (MW) in kDa, relating to the corresponding non-tagged protein form

2. Endogenously HiBiT- [19] and TurboID-tagged strains of *newORF 12*, *newORF 15*, *newORF 33*, *newORF 35*, and *newORF 37*, reported in Fijalkowski et al. [10] and listed in Table 1, and *mrcB*, *fruR*, and *xthA* were constructed by λ Red recombineering using the pKD4 plasmid with the FRT-flanked kanamycin resistance gene [20] (*see* **Note 2**).

2.2.2 Media and Buffers

1. Luria-Bertani (LB) broth according to Miller's formulation: 10 g/L Bacto peptone, 5 g/L Bacto yeast extract, and 10 g/L NaCl.
2. LB agar: 10 g/L Bacto peptone, 5 g/L Bacto yeast extract, 10 g/L NaCl, and 20 g/L agar.
3. LB medium with 50 μg/mL kanamycin sulphate monohydrate.
4. 1X Dulbecco's phosphate-buffered saline (DPBS), without calcium and magnesium (e.g., Gibco™; cat n° 14190094).

2.2.3 Materials and Equipment

1. Sterile Petri dishes.
2. Inoculation needles.
3. Round-bottom snap-cap tubes.
4. (Baffled) Erlenmeyer flasks.
5. Cuvettes.
6. Spectrophotometer.

2.3 Protein Extraction, Protein Quantification, and SDS–PAGE

2.3.1 Reagents and Buffers

1. Lysis buffer: 9 M urea, 50 mM ammonium bicarbonate, pH 7.9 [21] (*see* **Note 3**).
2. 1X Bradford reagent (e.g., Bio-Rad Protein Assay Dye Reagent Concentrate; cat n° 5000006).
3. Bovine Serum Albumin (BSA) protein standards (e.g., Thermo Scientific™ Pierce™ BSA Standards Pre-Diluted Set; cat n° 23208) and BSA standard curve.
4. 4X XT sample buffer (Bio-Rad; cat n° 1610791) (*see* **Note 4**).
5. 20X XT reducing agent (Bio-Rad; cat n° 1610792) (*see* **Note 4**).

2.3.2 Materials and Equipment

1. Cryogenic storage dewar and liquid nitrogen.
2. Safety goggles.
3. Insulated, waterproof cold-protection gloves (compliant with EN 511 standard).
4. Microprobe sonicator (e.g., LABSONIC® M; type BBI 853 5027 with 1 mm diameter probe).
5. Ear protection.

2.4 Expression Analysis Using HiBiT Blotting and Immunoblotting

2.4.1 Reagents and Buffers

1. 1X XT MES running buffer (Bio-Rad; cat n° 1610789) (*see* **Note 4**).
2. Prestained Protein Standards (e.g., Bio-Rad Precision Plus Protein™ Dual Xtra Prestained Protein Standards; cat n° 1610377) (*see* **Note 5**).
3. Transfer buffer: 25 mM Tris, 192 mM glycine, 15% methanol (v/v), pH ~ 8.3 (without pH adjustment) [22] (*see* **Note 6**).
4. Tris-buffered saline (TBS) solution: 50 mM Tris–HCl, pH 7.5, 150 mM NaCl.
5. TBS solution with 0.1% (v/v) Tween® 20 (TBS-T).
6. LgBiT solution: in-house recombinant 6xHis-LgBiT protein expressed from a pAG87 vector backbone in *Escherichia coli* BL21DE3 cells, used at a final concentration of 25 nM in TBS-T (*see* **Note 7**).
7. Furimazine solution: furimazine (TargetMol; cat n° T15359) at a final concentration of 5 nM in TBS-T (*see* **Note 7**).

2.4.2 Materials and Equipment

1. Midi-format vertical electrophoresis tank.
2. Electrophoresis power supply.
3. 12% Criterion XT Bis-Tris polyacrylamide protein gel (Bio-Rad; cat n° 3450117 [12 + 2 wells], cat n° 3450118 [18 wells], or cat n° 3450119 [26 wells]) (*see* **Note 4**).
4. Horizontal tank.
5. Trans-Blot Turbo Transfer device (Bio-Rad).
6. Trans-Blot Turbo Midi 0.2 μm nitrocellulose transfer pack (Bio-Rad; cat n° 1704159).
7. Blot roller.
8. (Midi-format) blot incubation box.
9. Tabletop shaking platform.
10. Imaging device capable of measuring luminescence and fluorescence (e.g., LI-COR Odyssey infrared imaging system).

3 Methods

In this chapter, we outline a detailed, step-by-step protocol for the detection of HiBiT-tagged small proteins at endogenous levels within total protein extracts through HiBiT blotting. The protocol encompasses the collection of bacterial cells, cell lysis and total protein extraction, and the separation of proteins through one-dimensional sodium dodecyl sulfate-polyacrylamide gel electrophoresis (1D SDS–PAGE). It also covers the transfer of proteins to membranes using semi-dry blotting and concludes with the chemiluminescence-based detection of the target (small) proteins.

3.1 Bacterial Protein Extraction

3.1.1 Bacterial Cultivation and Pellet Collection

1. Pick a single colony from a (selective) LB agar plate using an inoculation needle and inoculate it into 2 mL of LB medium (supplemented with 50 μg/mL kanamycin for endogenously tagged strains) in a round-bottom snap-cap tube.
2. Incubate the liquid bacterial cultures overnight (~ 16 h) in an incubator at 37 °C with agitation (180 rpm).
3. Measure the optical density at 600 nm (OD_{600}) of the overnight cultures using a spectrophotometer (*see* **Note 8**).
4. Dilute the overnight cultures (~ OD_{600} 2.5–3.0) to an OD_{600} of 0.02 in LB medium in a baffled Erlenmeyer flask (*see* **Notes 9** and **10**).
5. Continue to incubate the cultures at 37 °C and 180 rpm.
6. Once the desired OD_{600} representative of a specific bacterial growth phase (*see* Table 2) is reached, transfer the required amount of culture to an appropriate centrifuge tube (15 mL or 50 mL, depending on volume). Keep the bacteria on ice to halt growth.
7. Pellet the bacteria by centrifuging at 6000× *g* for 10 min in a cooled (4 °C) centrifuge.
8. Carefully discard the supernatant from the pellet (*see* **Note 11**).

Table 2
Cultivation, harvest, and lysis parameters for conditional protein expression analysis in *Salmonella* Typhimurium (strain SL1344)

Growth phase	OD_{600}	Estimated time to reach phase	Culture volume (ml)	Estimated CFU count	Estimated protein content (mg)
Exponential phase (EP)	0.6	2 h 15–2 h 30	40	7.5×10^9–1.0×10^{10}	1.5–2
Late exponential phase (LEP)	1.0	2 h 45–3 h	20	7.5×10^9–1.0×10^{10}	1.5–2
Early stationary phase (ESP)	2.0	3 h 30–4 h	10	7.5×10^9–1.0×10^{10}	1.5–2
Stationary phase (SP)	3.6	4 h 30–5 h	5	7.5×10^9–1.0×10^{10}	1.5–2

This table details the relationship between bacterial growth phases and culture concentration, expressed as optical density at 600 nm (OD_{600}). An estimation of the growth times in hours (h) required to reach each phase from a starting culture at OD_{600} 0.02 is given, assuming growth in a baffled Erlenmeyer flask at 37 °C and 180 rpm agitation. Required culture volumes are given for preparing total protein extracts at a concentration of about 3 mg/mL when lysing in a volume of 600 μL, along with the bacterial enumeration given in colony-forming units (CFU) and protein content in mg

9. Wash the pellet through resuspension in 1 mL of 1X DPBS and transfer the resuspension to a 1.5 mL Eppendorf tube.
10. Pellet the bacteria again by centrifuging at 6000× *g* for 5 min in a cooled (4 °C) Eppendorf centrifuge.
11. Discard the supernatant and store the bacterial pellets at −80 °C pending lysis and protein extraction.

3.1.2 Protein Extract Sample Preparation

1. Resuspend the frozen bacterial pellets in 600 μL of lysis buffer by vortexing.
2. Perform three freeze–thaw cycles by submerging the samples in liquid nitrogen (*see* **Note 12**) and reheating them in a water bath in a floating foam Eppendorf rack. Vortex the samples after each thawing step to ensure complete lysis.
3. Sonicate the samples for 30 seconds using a microprobe sonicator with 0.5-second pulses at an amplitude of 40 (optimized conditions for the indicated sonicator).
4. Clear the lysates of bacterial debris by centrifuging at 16,100× *g* (the maximum speed of the available tabletop Eppendorf centrifuge) at 20 °C for 10 min (*see* **Note 13**). Carefully transfer the supernatant to a fresh 1.5 mL Eppendorf tube.
5. Determine the protein concentration through a Bradford assay. Dilute 4 μL of the protein extracts (or lysis buffer as a blank) with 2 mL of Bradford reagent in a cuvette. Pipet up and down to mix thoroughly, and measure the absorbance at 595 nm using a spectrophotometer after blanking with lysis buffer.
6. Calculate the protein concentration considering the dilution factor (500×) and the slope of the BSA standard curve.
7. Identify the sample with the lowest protein concentration and dilute the other samples to this concentration using lysis buffer to achieve total sample volumes of 100 μL.
8. Supplement the protein samples with 35 μL 4X XT sample buffer and 7 μL 20X XT reducing agent and boil for 5 min at 95 °C with agitation at 600 rpm to denature the proteins.

3.2 SDS–PAGE

1. Fill a midi-format electrophoresis tank with 1X MES buffer just below the fill line indicated on the tank. To conduct the electrophoresis of two gels simultaneously, a total volume of 1 L of buffer should be sufficient.
2. Remove a precast midi Bis-Tris gel from its packaging and peel off the protective strip from the bottom of the cassette. Insert the gel cassette into the electrophoresis tank.
3. Fill the well reservoir with 1X MES running buffer, and then carefully remove the comb to expose the wells.

4. Vortex the protein samples to ensure they are well mixed before loading them into the wells. The maximum loading volume varies by gel format: up to 45 μL for 12 + 2 well gels, 30 μL for 18 well gels, and 15 μL for 26 well gels (*see* **Note 1**). Allocate one well to load 5 μL of Precision Plus Protein™ Dual Xtra Prestained Protein Standards (*see* **Note 5**).
5. Secure the lid on the electrophoresis cell and connect it to a power supply. Electrophorese at a constant voltage of 150 V for approximately 70 min. Monitor the migration of the dye front, ensuring that the dye does not run off the gel to avoid losing low-MW proteins (*see* **Notes 5** and **14**).
6. Use a spatula to open the cassette carefully. Determine which side the gel is adhering to, and hold the cassette horizontally with this side facing downward.
7. Place the gel, still attached to one side of the cassette, into a horizontal tank filled with transfer buffer to facilitate the release of the gel from the cassette. Submerge and equilibrate the gel in transfer buffer before initiating the blotting-based protein transfer.

3.3 Semi-Dry Blotting Procedure

1. Turn on the Trans-Blot Turbo semi-dry blotter device and unlock a cassette.
2. Open a midi-format transfer pack containing a 0.2 μm nitrocellulose membrane. Use tweezers and the finger tab on the right side of the stack to carefully transfer the entire anode stack, including the membrane, to the center of the cassette base. Ensure there are no air bubbles by gently rolling a blot roller from the center outward in all directions.
3. Wet the fingertips of your gloves with transfer buffer before lifting the gel from the container. Carefully position the gel on the membrane at the top of the anode stack. Use the blot roller to eliminate any air bubbles.
4. Clamp the finger tab of the cathode stack (on the left) with tweezers, and lay this stack on top of the gel. Again, use the blot roller to remove any air bubbles.
5. Secure the cassette with its lid and lock it. Then, install the cassette into the Trans-Blot Turbo semi-dry blotter.
6. Program a new protocol by selecting 'NEW'. Adopt the optimized parameters (Fig. 3): a maximal voltage of 25 V, a constant amperage of 2.5 A, and 10 min of blotting (*see* **Note 15**). Press "RUN" and specify which tray contains the loaded cassette to start the protocol.
7. Once the protocol is finished, remove, and unlock the cassette. Discard the cathode stack and the gel to expose the membrane. Confirm the success of the transfer by evaluating the protein

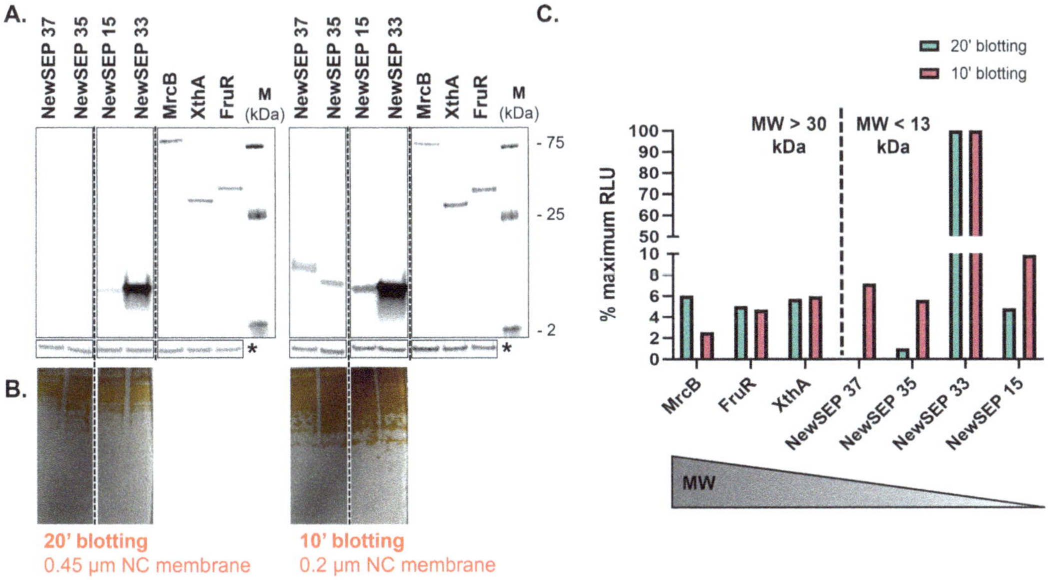

Fig. 3 Quantitative molecular weight (MW)-associated biases introduced by varying blotting parameters. Protein extracts were prepared from endogenously HiBiT-tagged *Salmonella* strains expressing various HiBiT-tagged proteins, including NewSEP 37 (12.4 kDa), NewSEP 35 (11.3 kDa), NewSEP 15 (6.8 kDa), NewSEP 33 (10.9 kDa), MrcB (95 kDa), FruR (39 kDa), and XthA (32 kDa), harvested at stationary growth phase (~ $OD_{600} = 3.6$). A total of 50 µg of protein was loaded for fractionation by 1D SDS-PAGE alongside 3 µL of Precision Plus Protein™ Dual Xtra Protein Standards as molecular weight marker (M). (**a**) This panel compares two semi-dry blotting protocols to maximize the transfer and membrane retention of low-MW proteins. On the left, proteins were transferred onto a 0.45 µm nitrocellulose (NC) membrane for 20 min (20′) at a constant amperage of 2.5 A (and a maximal voltage of 25 V). On the right, proteins were transferred onto a 0.2 µm NC membrane for 10 min (10′) at the same amperage and voltage settings. Proteins were visualized through HiBiT blotting. The biotin carboxyl carrier protein (BCCP; 17 kDa), used as a loading control, was probed with a fluorescent streptavidin conjugate (marked with an asterisk). (**b**) Following blotting, proteins remaining in the gel were visualized through silver staining. (**c**) Differential chemiluminescence signal intensities, indicative of protein abundance, were quantified using the analysis application of the Odyssey LI-COR Fc infrared imaging system and were normalized based on the detected signal for the BCCP control protein. Results are given as percentages of the highest signal detected for NewSEP 33. Relative abundances were compared for the two methods in function of decreasing MW. *RLU* relative luminescence unit

standards. Then, use tweezers to transfer the membrane from the stack to a blot incubation box prefilled with TBS-T solution.

3.4 NanoLuc Reconstitution and Detection

1. Incubate the membrane in TBS-T for 10 min to remove the transfer buffer and render the HiBiT tag accessible for complementation with LgBiT.
2. Discard the TBS-T from the membrane within the incubation box and replace it with the LgBiT solution. Incubate with gentle agitation at room temperature for 1 h (Fig. 4, *see* **Note 16**). This step facilitates the complementation between the

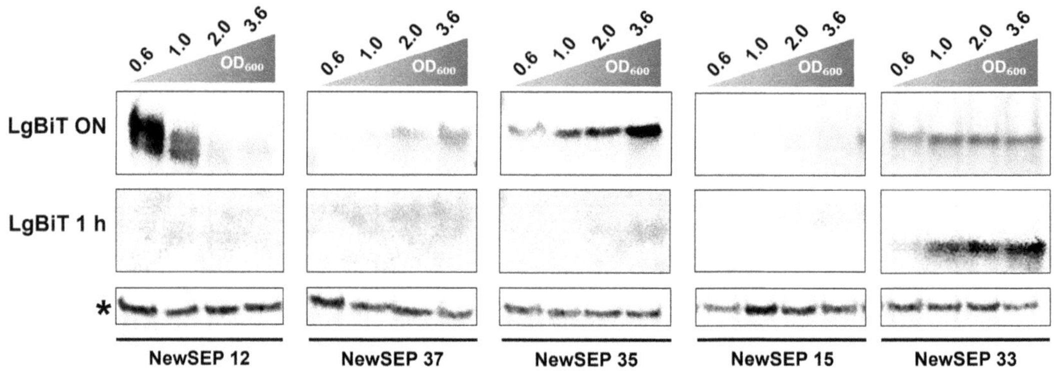

Fig. 4 Effect of Large BiT (LgBiT) incubation time on HiBiT detection. Total protein lysates from endogenously HiBiT-tagged sORF *Salmonella* strains, collected at various optical densities (OD_{600}), were subjected to 1D SDS-PAGE fractionation. Subsequently, HiBiT blotting was performed to assess the expression of HiBiT-tagged SEPs. The luminescence signal intensities from a 1-hour incubation at room temperature (LgBiT 1 h) and an overnight incubation at 4 °C (LgBiT ON) were evaluated. The luminescence signals shown originate from a single protein blot to allow direct comparison. The biotin carboxyl carrier protein (BCCP; 17 kDa), used as a loading control, was probed with a fluorescent streptavidin conjugate (marked with an asterisk)

LgBiT protein and the accessible HiBiT tag, thereby reconstituting the functional NanoLuc luciferase [16].

3. Remove the LgBiT solution from the membrane, replacing it with the furimazine solution. Incubate for 5 min with gentle agitation at room temperature.
4. While keeping the membrane submerged in the furimazine solution, transfer it to an imaging device capable of measuring chemiluminescence (*see* **Note 17**).
5. Capture images of the blot using a luminescence measuring channel for 10 min. Additionally, use fluorescence channels at 600 nm and 700 nm for visualization of the protein standards, with exposure times ranging from 30 s to 2 min (Fig. 5) (*see* **Note 18**).

4 Notes

1. Due to the generally low sensitivity of endogenous small protein blotting, it is recommended to load larger total protein amounts (50–60 μg) for SDS–PAGE. Given the use of a less concentrated sample buffer (i.e., the addition of 4X XT sample buffer to the samples), loading volumes often approach the maximum loading capacity of the protein gel. To manage these volumes, use specialized gel-loading pipette tips for precise well-filling.
2. In the strains described, the HiBiT tag was directly fused to the C-terminus of the SEPs. N-terminal or internal HiBiT-tagging,

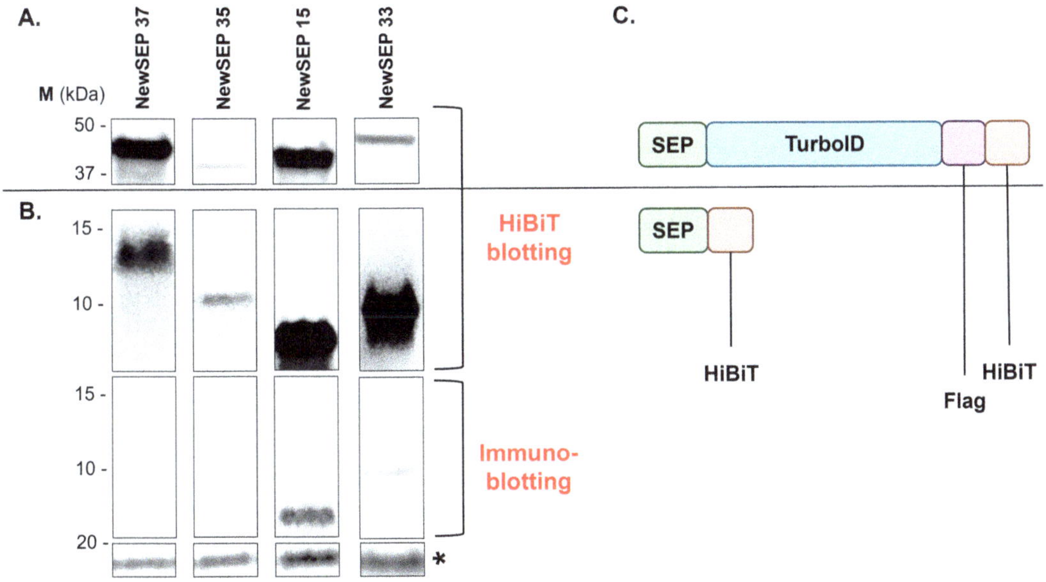

Fig. 5 Evaluation of the HiBiT tag in verifying the expression of newly discovered, putative small ORF (sORF)-encoded polypeptides (SEPs). (**a**) HiBiT blotting analysis of protein lysates from endogenously TurboID-tagged (incorporating Flag and HiBiT tags) (**c**) sORF *Salmonella* strains harvested during the stationary phase (~OD_{600} = 3.6). Analyzed strains encode TurboID-tagged SEPs NewSEP 37, NewSEP 35, NewSEP 15, and NewSEP 33 of 49.2, 48.1, 43.6, and 47.7 kDa, respectively. (**b**) HiBiT detection in protein lysates from endogenously HiBiT-tagged (**c**) sORF *Salmonella* strains harvested in the stationary growth phase (~OD_{600} = 3.6). Analyzed strains encode HiBiT-tagged SEPs NewSEP 37, NewSEP 35, NewSEP 15, and NewSEP 33 of 12.4 kDa, 11.3 kDa, 6.8 kDa, and 10.9 kDa, respectively. Detection methods included immunoblotting using an anti-HiBiT antibody with a fluorescent secondary antibody and HiBiT blotting through the reconstitution of an active NanoLuc luciferase through complementation between the HiBiT tag and LgBiT protein. Immunoblotting and HiBiT blotting signals (**b**) originate from the same blot, allowing direct comparison of method sensitivities. The biotin carboxyl carrier protein (BCCP; 17 kDa), used as a loading control, was probed with a fluorescent streptavidin conjugate (marked with an asterisk). *M* molecular weight marker

which others have reported [23, 24], was not explored. For TurboID-tagged strains, a linker sequence was introduced between the SEP and TurboID biotin ligase coding sequence, as reported by Arora et al. [25]. Additionally, the TurboID tag includes a Flag (DYKDDDDK) and HiBiT tag (Fig. 5).

3. The use of the urea-based lysis buffer is versatile and not limited to protein extraction from bacterial cells, as it has been successfully used in the context of mammalian and plant samples. With minor adaptations, the described protocol can be extended to detect small proteins at endogenous levels in lysates originating from various sample sources.
4. Various SDS–PAGE gel systems, along with (their corresponding) running and sample buffers, were tested and compared in terms of final protein blot quality. The gel systems tested included 16.5% polyacrylamide Tris–Tricine gels and

10–20% polyacrylamide Tris–HCl gels, each used with their recommended buffers. Based on our experiences and observations, we specifically mentioned Bio-Rad's XT 12% polyacrylamide Bis–Tris gel system for the SDS–PAGE step preceding the blotting-based membrane transfer of the proteins.

5. Include protein standards that feature an MW fragment below 10 kDa, or ideally below the MW of the smallest microprotein analyzed. To prevent the loss of low-MW proteins, ensure that the dye front from the 4X XT sample buffer does not run off the gel. Monitor the migration of the low MW protein ladder fragment to determine the optimal SDS–PAGE running time.
6. Blotting transfer buffer, according to Towbin's formulation, is recommended for wet tank blotting and some semidry blotting systems due to its low ionic strength and conductivity.
7. The working concentration of our in-house recombinant LgBiT protein was determined through comparative analysis with the activity of Promega's LgBiT (from the Nano-Glo HiBiT Blotting System; cat n° N2410, part n° N401C) [16] at the recommended 200x dilution. Analogously, the working concentration of the furimazine substrate was optimized by comparing it with the efficiency of Promega's Nano-Glo Luciferase Assay Substrate (from the Nano-Glo HiBiT Blotting System; cat n° N2410, part n° N113A) at the recommended 500× dilution.
8. When measuring OD_{600}, be aware to remain within the linear range of detection of the used spectrophotometer. Dilution of the overnight cultures in LB medium might be required. Typically, a 1:5 dilution (200 μL overnight culture mixed with 800 μL LB medium) is used in a spectrophotometer-compatible cuvette.
9. The amount of culture grown should reflect the desired growth phase for expression analysis. To achieve optimal protein concentrations (approximately 3 mg/mL) and sufficient yield (Table 2), more bacterial culture is required if harvesting is done during the early phases of the bacterial growth curve. Always cultivate more than needed to allow for OD_{600} measurements and to monitor the bacterial growth phase effectively.
10. For bacterial cultivation in Erlenmeyer flasks, adhere to the guideline that these should be filled to a maximum of one-fifth of their total volume (e.g., no more than 50 mL in a 250 mL Erlenmeyer). However, for baffled Erlenmeyers, these recommended volumes can be moderately exceeded due to improved aeration for the cultivation of (facultative) anaerobes.

11. When harvesting culture volumes larger than 15 mL, include an additional transfer and centrifugation step (10 min at 6000× *g*, 4 °C) to minimize the risk of detaching and resuspending the bacterial pellet, reducing bacterial loss. After centrifugation in a 50 mL tube, partially remove the supernatant, retaining about 10 mL of the supernatant to resuspend the bacterial pellet. The adjusted volume is then suitable for transfer to a 15 mL centrifuge tube, whose sharper tip typically aids in better bacterial pellet cohesiveness following centrifugation.
12. In the absence of liquid nitrogen, sample freezing can be performed by placing them in a −80 °C freezer or on dry ice. Although these methods are slower than using liquid nitrogen, they are effective.
13. Avoid handling urea-based protein extracts below room temperature to prevent crystal formation. If crystals appear after freezing, vigorously vortex the sample to redissolve the crystals.
14. The dye front from bromophenol blue in the sample buffer may interfere with protein ladder imaging in the 700 nm channel. To avoid this, use protein standards with differently colored fragments, like the red fragments in the Precision Plus Protein™ Dual Xtra Prestained Protein Standards, which can be imaged in the 600 nm channel. Alternatively, use a dye-free or a fluorescent-compatible sample buffer (e.g., Invitrogen™; cat n° LC2570) (Fig. 6) to ensure accurate imaging in alternative wavelength channels.
15. Achieving optimal blotting results often requires adjusting various parameters, including the type and pore size of the membrane, the duration of blotting, and the blotting intensity (e.g., constant voltage or constant amperage). Careful optimization of these factors is essential to ensuring reliable and reproducible results. After optimizing the (semidry) blotting parameters to achieve optimal transfer and membrane retention of microproteins, we selected 10 min of blotting at a constant amperage of 2.5 A and a maximal voltage of 25 V onto a 0.2 μm nitrocellulose membrane as the standard protocol for HiBiT blotting-based analysis of bacterial microproteins. From a quantitative perspective, it is important to note that optimal blotting parameters are MW-dependent, signifying that quantitative conclusions about differences in the abundance of proteins with varying MWs based on blotting detection are not straightforward. While the selected blotting settings (method 1) are beneficial for the maximal retention of microproteins on the membrane (Fig. 3a), they are less effective for transferring high MW proteins (Fig. 3b) compared to 20 min of blotting at a constant amperage of 2.5 A and a

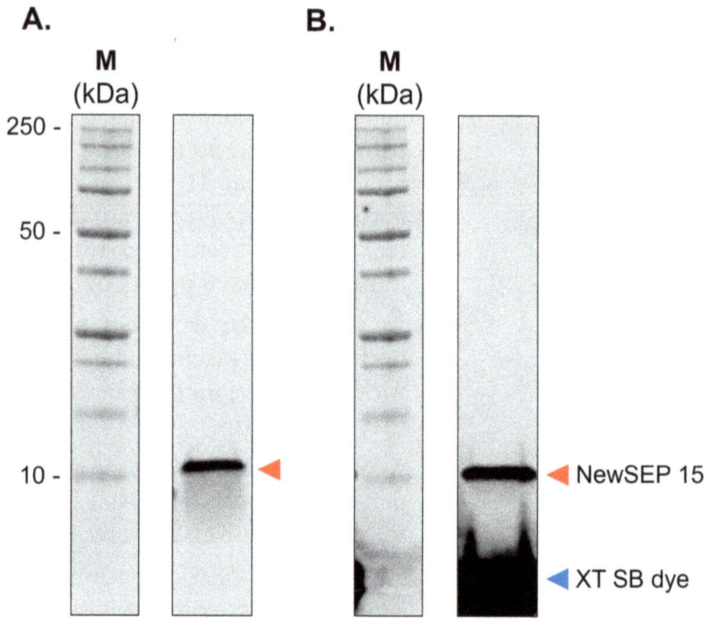

Fig. 6 Use of fluorescent-compatible sample buffer for the preparation of protein samples for SDS–PAGE. The protein fraction was extracted from an endogenously HiBiT-tagged *Salmonella* strain expressing the HiBiT-tagged SEP NewSEP 15 (6.8 kDa), harvested at the stationary growth phase (~ $OD_{600} = 3.6$), and diluted to a concentration of 2.7 mg/ml in urea lysis buffer, 20X XT reducing agent, and either 4X fluorescent-compatible sample buffer (**a**) or 4X XT sample buffer (**b**). The dye front in the XT sample buffer interferes with fluorescence measurement. A total of 50 μg of protein was loaded for fractionation by 1D SDS-PAGE alongside 5 μL of Precision Plus Plus Protein™ All Blue Prestained Protein Standards as molecular weight marker (M). *SB* sample buffer

maximal voltage of 25 V onto a 0.45 μm nitrocellulose membrane (method 2). Using method 1, NewSEP 37 and NewSEP 15 ranked among the top three most abundant proteins in the complete protein target set, whereas with method 2, they appeared among the least abundant (Fig. 3c). Conversely, for the high MW reference protein MrcB (95 kDa), there was a shift from being among the most abundantly detected proteins with method 2 to the least abundantly detected proteins with method 1.

16. According to the Nano-Glo® HiBiT Blotting System technical manual, the most intense luminescence signals are typically observed after overnight incubation of the blots in LgBiT solution at 4 °C with gentle agitation. However, this trend does not universally apply. For instance, for NewSEP 33, higher sensitivity was observed after just 1 h of incubation compared to overnight incubation (Fig. 4). This variability may be attributed to the differing intrinsic properties of the microproteins.

17. Unlike fluorescence-based detection, the intensity of the luminescence signals during HiBiT blotting is not an absolute measure of protein abundance, even at saturation. This limitation is due to the gradual decay of luminescence. To ensure comparability between different blots, it is critical to maintain uniform incubation times with furimazine and consistent timing in the subsequent imaging process. Delaying imaging significantly after the 5-min furimazine incubation can lead to a substantial loss of luminescence signal and, consequently, sensitivity.
18. HiBiT blotting can be combined with antibody-based blotting, allowing for the concurrent analysis of additional proteins that can serve as loading controls or normalization factors, as shown in Figs. 1, 3, 4, and 5. The following antibodies and detection agents were used: Streptavidin, Alexa Fluor™ 680 Conjugate (Thermo Scientific; cat n° S32358) at a dilution of 1:10,000; mouse anti-HiBiT monoclonal antibody (Promega; cat n° CS2006A01) at 1:2000; and IRDye® 800 CW goat anti-mouse secondary antibody (LI-COR; cat n° 926-32210) at 1:5000. Antibodies and detection reagents are diluted in 1/1 (v/v) TBS-T and Intercept® blocking buffer (LI-COR; cat n° 927–70001).

Acknowledgments

This work was made possible through the support of the European Research Council (ERC) under the European Union's Horizon 2020 research and innovation program (grant agreement no. 803972, PROPHECY) and the Fonds Wetenschappelijk Onderzoek (FWO-Vlaanderen, project number G051120N), both awarded to PVD. LS received support from a personal doctoral fellowship granted by FWO-Vlaanderen, with grant number 1119925N. The funders had no role in study design, data collection and analysis, the decision to publish, or the preparation of the manuscript.

References

1. Weidenbach K, Gutt M, Cassidy L et al (2022) Small proteins in archaea, a mainly unexplored world. J Bacteriol 204:e0031321
2. Simoens L, Fijalkowski I, Van Damme P (2023) Exposing the small protein load of bacterial life. FEMS Microbiol Rev 47:1–13
3. Hassel KR, Brito-Estrada O, Makarewich CA (2023) Microproteins: overlooked regulators of physiology and disease. iScience:26
4. Bhati KK, Blaakmeer A, Paredes EB et al (2018) Approaches to identify and characterize microProteins and their potential uses in biotechnology. Cell Mol Life Sci 75:2529–2536
5. Ingolia NT, Ghaemmaghami S, Newman JRS et al (2009) Genome-wide analysis in vivo of translation with nucleotide resolution using ribosome profiling. Science 324:218–223
6. Li Y, Zhou H, Chen X et al (2021) SmProt: a reliable repository with comprehensive

annotation of small proteins identified from ribosome profiling. Genomics Proteomics Bioinformatics 19:602–610

7. Olexiouk V, van Criekinge W, Menschaert G (2018) An update on sORFs.org: a repository of small ORFs identified by ribosome profiling. Nucleic Acids Res 46:497–502
8. Finkel Y, Stern-Ginossar N, Schwartz M (2018) Viral short ORFs and their possible functions. Proteomics 18:e1700255
9. Leong AZX, Lee PY, Mohtar MA et al (2022) Short open reading frames (sORFs) and microproteins: an update on their identification and validation measures. J Biomed Sci 29:19
10. Fijalkowski I, Willems P, Jonckheere V et al (2022) Hidden in plain sight: challenges in proteomics detection of small ORF-encoded polypeptides. microLife 3:1–17
11. Simoens L, Jonckheere V, Catteeuw D, Van Damme P (2025) Endogenous HiBiT-tagging combined with affinity complementation: A new strategy for small open reading frame-encoded polypeptide detection in bacteria. Advances in Sample Preparation 13, 100143
12. Tomisawa S, Abe C, Kamiya M et al (2013) A new approach to detect small peptides clearly and sensitively by Western blotting using a vacuum-assisted detection method. Biophysics 9:79–83
13. Kurien BT, Hal Scofield R (2015) Western blotting: an introduction. In: Kurien BT, Hal Scofield R (eds) Western blotting: methods and protocols. Springer, pp 17–30
14. Schwinn MK, Machleidt T, Zimmerman K et al (2018) CRISPR-mediated tagging of endogenous proteins with a luminescent peptide. ACS Chem Biol 13:467–474
15. Hall MP, Unch J, Binkowski BF et al (2012) Engineered luciferase reporter from a deep sea shrimp utilizing a novel imidazopyrazinone substrate. ACS Chem Biol 7:1848–1857
16. Dixon AS, Schwinn MK, Hall MP et al (2016) NanoLuc complementation reporter optimized for accurate measurement of protein interactions in cells. ACS Chem Biol 11:400–408
17. Promega Corporation. Nano-Glo® HiBiT Blotting System: TM524
18. Hoiseth SK, Stocker BAD (1981) Aromatic-dependent *Salmonella* Typhimurium are non-virulent and effective as live vaccines. Underw Nat 291:238–239
19. Fijalkowski I, Snauwaert V, Van Damme P (2024) Proteins à la carte: riboproteogenomic exploration of bacterial N-terminal proteoform expression. mBio:15
20. Datsenko KA, Wanner BL (2000) One-step inactivation of chromosomal genes in *Escherichia coli* K-12 using PCR products. Proc Natl Acad Sci USA 97:6640–6645
21. Fijalkowski I, Peeters MKR, Damme P Van (2021) Small protein enrichment improves proteomics detection of sORF encoded polypeptides. Front Genet 12
22. Towbin H, Staehelin T, Gordon J (1979) Electrophoretic transfer of proteins from polyacrylamide gels to nitrocellulose sheets: procedure and some applications. Proc Natl Acad Sci USA 76:4350–4354
23. Schwinn MK, Steffen LS, Zimmerman K et al (2020) A simple and scalable strategy for analysis of endogenous protein dynamics. Sci Rep 10:8953
24. Seiler K, Rafiq S, Tschan MP (2023) Isoform-specific, semi-quantitative determination of highly homologous protein levels via CRISPR-Cas9-mediated HiBiT tagging. Bio Protoc 13:e4777
25. Arora D, Abel NB, Liu C et al (2020) Establishment of proximity-dependent biotinylation approaches in different plant model systems. Plant Cell 32:3388–3407
26. Gao Z, Chang C, Yang J et al (2019) AP3: an advanced proteotypic peptide predictor for targeted proteomics by incorporating peptide digestibility. Anal Chem 91:8705–8711

Chapter 17

Gain-of-Functional Genomic Screening for Microproteins Essential for Tumorigenesis

Anni Ge, Yuan Gao, and Xiaolong Yang

Abstract

Mounting studies suggest that the protein-coding potential of the human genome is underestimated, and previously unannotated open reading frames (ORFs) have been revealed with technological advances. Although the coding potential of many non-canonical ORFs (ncORFs) is now recognized, their functional prevalence remains to be characterized, partly due to the technical difficulty and labour intensity in screening numerous ncORFs at once. Here, we describe a gain-of-function genomic screen to identify functional ncORF-encoded proteins responsible for breast cancer tumorigenesis. This method is thought to improve the efficiency of characterizing previously neglected ncORFs and reveal potential targets for cancer treatment.

Key words Functional genomic screen, Gain-of-function screen, Microproteins, Micropeptides, ncORF, smORF, Cancer biology, Tumorigenesis

1 Introduction

The number of canonical proteins increases each year, with more than 17,400 protein-coding genes identified by the Human Proteome Project NeXtProt database [1]. A novel class of proteins translated from previously overlooked non-canonical open reading frames (ncORFs) is identified with improved high-throughput profiling techniques [2–4]. Many ncORFs encode proteins with fewer than 100 amino acids (aa), which are named small ORFs (smORF) or microproteins (miPs). Recently, several miPs have been shown to play important roles in tumorigenesis and metastasis [3, 4]. However, many miPs remain to be discovered, and their biological functions are largely unknown.

Gain- or loss-of-functional genomics screening is commonly performed to understand the relevance of any genes of interest in cancer biology, enabling the discovery of numerous genetic targets responsible for tumour progression [2, 5–8]. While loss-of-

Stephan Wenkel (ed.), *Microproteins: Methods and Protocols*, Methods in Molecular Biology, vol. 2992, https://doi.org/10.1007/978-1-0716-5013-4_17,

functional screenings for miPs involved in tumorigenesis have been reported [9, 10], gain-of-functional screening for miPs important for tumorigenesis has not been reported. We have recently performed a gain-of-function screening for protein kinases involved in tumorigenesis using a soft-agar assay and identified PIK3CA and PIK3CB as important regulators of the Hippo pathway during mammary tumorigenesis [8]. Here, we describe how miPs that may be involved in tumorigenesis can be readily identified by a gain-of-function screen. A cDNA expression library of 553 smORFs encoding mostly miPs is subjected to the screen for detecting functional miPs essential for tumorigenesis [2]. Soft agar assay is chosen to evaluate the tumorigenic potency of each ncORF-encoded protein since it selects cells that exhibit anchorage-independent growth, which is a critical characteristic of malignant cells in disease progression [11, 12]. Unlike normal adherent cells, which undergo anoikis after detaching from the extracellular matrix, transformed cells expressing certain oncoproteins will survive, proliferate, and eventually form a colony in the soft agar [11]. The colonies can then be derived from the agar to extract genomic DNA (gDNA) for recovering gene candidates from the library. Thus, any smORF-encoded miPs that induce anchorage-independent survival in non-tumorigenic cells will be identified, indicating their implications in tumorigenesis.

2 Materials

2.1 Library Preparation

Prepare all reagents with the ultraPure DNase/RNase-free distilled dH_2O. Store all prepared reagents at room temperature (RT) unless noted otherwise.

1. XL10 gold ultracompetent cells (Agilent, Cat. No. 200315).
2. LB medium: Dissolve 25 g of LB Broth, Miller (Fisher Scientific, Cat. No. BP146-2) in 800 mL of dH_2O and bring the volume to 1 liter (L). Sterilize the solution by autoclaving. Keep the media at RT.
3. Ampicillin (Amp) stock (100 mg/mL): Weigh 1 g of ampicillin, Sodium Salt (BioShop, Cat. No. AMP201, stored at 4 °C) in 8 mL of sterile dH_2O in a sterile 15 mL tube. Gently mix until it fully dissolved. Adjust final volume to 10 mL with sterile dH_2O. Filter the stock solution through a 0.22 μm syringe filter. Aliquot the stock solution into smaller volumes (e.g., 1 mL) to avoid repeated freeze-thaw cycles. Store the stock solution at −20 °C.
4. LB medium with ampicillin (LB-Amp, 100 μg/mL): dilute the Amp stock 1:1000 in LB medium.

5. 13 mL round-bottom culture tube (Sarstedt, Cat. No.50-809-216).
6. LB plate with ampicillin (LB plate-Amp): Weigh 20 g LB agar and transfer it into a 1 L flask with 500 mL dH_2O. Sterilize and solubilize the agar by autoclaving. After autoclaving, immediately place the flask in a 42 °C water bath. Add ampicillin to the required concentration (100 μg/mL) in dissolved agar when equilibrated to 42 °C. Immediately dispense 20 mL LB-Amp agar into each of the 100 mm Petri dishes. Leave the dish covers ajar until the agar solidifies. Store the plates at 4 °C for up to 3 months.
7. QIAGEN Plasmid Midi Kit (Cat. No. 12143): Buffer P1, Buffer P2, Buffer P3, Buffer QBT, Buffer QC, Buffer QF, QIAfilter cartridge, QIAGEN-tip.
8. Isopropanol: store at RT in a dark bottle.
9. 70% ethanol.
10. Buffer EB: 10 mM Tris-HCl, pH 8.5.
11. New Brunswick Innova 4230 Incubator (Cat. No. 2926048).
12. Nanodrop OneC Spectrophotometer (Cat. No. ND-ONEC-W).

2.2 Lentivirus Production and Titration

1. Poly-L-Lysine Hydrobromide (PLL): Dissolve 0.1 g of PLL in 10 mL of sterile 1× PBS. Aliquot into 1 mL of 10 mg/mL stock and store at −20 °C.
2. 1× PBS.
3. HEK293T cells.
4. DMEM medium containing 10% Fetal Bovine Serum (FBS) and 1% penicillin/streptomycin (S/P).
5. pMD2.G (Addgene, Cat. No. 12259).
6. psPAX2 (Addgene, Cat. No. 12260).
7. PolyJet In Vitro DNA Transfection Reagent (SignaGen, Cat. No. SL100688): store at 4 °C.
8. 10 mM sodium butyrate: Dissolve 0.154 g of sodium butyrate in 14 mL of dH_2O. Filter with the 0.22 μm syringe unit. Aliquot in 1 mL of stock and store at −20 °C.
9. LentiX™ concentrator (Clontech, Cat. No. 631231).
10. 0.45 μm filter unit.
11. Polybrene (5 mg/mL): store in the dark at −20 °C.
12. NMuMG mammary cells.
13. NMuMG culture medium: DMEM medium containing 10% FBS and 1% S/P supplemented with 10 μg/mL Insulin and 20 mM HEPES.
14. Puromycin dihydrochloride (1 mg/mL).

2.3 Large-Scale Screening

1. Low melting temperature (LMP) agarose (SeaPlaque agarose, from Cedarlane, #50101).
2. NMuMG cells.
3. NMuMG culture medium (*see* Subheading 2.2).
4. Puromycin dihydrochloride (1 mg/mL).
5. 15 mL tubes (Thermo Fisher, Cat. No. 339650).
6. 50 mL tubes (Thermo Fisher, Cat. No. 339652).

2.4 Genomic DNA Extraction, Amplification of Amplicons, and Next Generation Sequencing (NGS)

1. CTAC buffer: 2% w/v cetyltrimethylammonium chloride (CTAC), 2% w/v polyvinylpyrrolidone (PVP-40), 2 M sodium chloride, 100 mM Tris–HCl at pH 8.0, 20 mM ethylenediaminetetraacetic acid (EDTA) in dH_2O. Add 2% v/v-mercaptoethanol immediately before use.
2. High-performance VWR tubes (Cat. No. 21008-178).
3. Chloroform.
4. Isopropanol.
5. 70% ethanol.
6. Buffer EB (*see* Subheading 2.1).
7. PCR reagents (TaKaRa, Cat. No. R050A): 5× PrimeStar GXL buffer, 2.5 mM dNTP, PrimeStar DNA polymerase (1.25 U/μ L). Store all PCR reagents at −20 °C.
8. Oligos (Universal oligos and TruSeq primers): The oligos/primers were synthesized by Life Technology (Thermo Fisher). Reconstitute 1 nmol oligos or primers in 100 μL of 10 mM Tris (pH 8.0) to make 100 μM stock solutions. Dilute 100 μM stock solution at 1:10 to make 10 μM working primers. Store oligos/primers at −20 °C.
9. The following sequences are primers used for PCR:
   (1) *Universal oligos for amplicon amplification:*
   Forward: 5′-TCTTGTGGAAAGGACGA-3′.
   Reverse: 5′-TTAAAGCAGCGTATCCACATAGCGT-3′.
   (2) *"TruSeq" primes for NGS library preparation:*
   Forward: 5′-ACACTCTTTCCCTACACGACGCTCTTC CGATCTTCTTGTGGAAAGGACGA-3′.
   Reverse: 5′GTGACTGGAGTTCAGACGTGTGCTCTT CCGATCTTTAAAGCAGCGTATCCACATAGCGT-3′.
10. 6× DNA Loading Dye: Mix 1.2 mL of 0.5 M EDTA, 6.0 mL of glycerol, 9 mg of xylene cyanol, and 2.8 mL of dH_2O. Aliquot into small amounts and store at 4 °C.
11. 50× TAE buffer (1 L): Dissolve 242 g of Tris-base, 57.1 mL of glacial acetic acid, and 14.6 g of EDTA in 800 mL of

dH$_2$O. Adjust the pH to 7.6 using 6 N HCl and bring the final volume to 1 L. Pass the mixture through a 0.22 μm filter. Dilute it to 1× TAE as the working solution.

12. 1% agarose gel: Weigh 1.0 g of agarose powder and mix into 100 mL of 1 X TAE buffer. Heat the agarose mixture in a microwave until fully dissolved. Pour the dissolved gel into a 15-well tray of gel electrophoresis apparatus.
13. 1 Kb Plus DNA ladder (Thermo Fisher, Cat. No. 10787018): Mix 60 μL of 1 Kb Plus DNA ladder (0.5 μg/μL) with 100 μL of 6× DNA loading dye and 440 μL of buffer EB. Store at 4 °C.
14. QIAquick PCR Purification Kit (Cat. No. 28104): Buffer PB, Buffer PE, Buffer EB, and QIAquick column.
15. Bio-Rad T100 Thermal Cycler (Cat. No. 1861096).
16. Illumina NextSeq Sequencing kit.
17. Nanodrop OneC Spectrophotometer (Cat. No. ND-ONEC-W).
18. Fisher Sonic 500 Dismembrator (Sonicator).

3 Methods

3.1 Library Construction

3.1.1 Preparation of smORF Library

The smORF cDNA library containing 553 lentiviral plasmids encoding mostly miPs was generated and kindly received from the Broad Institute in the format of 96-well microplates [2]. In short, the smORF candidates were selected based on the published predictions of ORF translation, and the cDNA expression library was generated containing 553 ORFs with a V5 tag for protein detection.

1. Combine 0.5 μL of each plasmid from the smORF library containing 553 lentiviral plasmids into a 1.5 mL Eppendorf tube (see **Note 1**).
2. Measure plasmid concentration and dilute the plasmids with Buffer EB to 10 ng/μL.
3. Thaw 100 μL of XL10 Gold ultracompetent cells on ice and transfer the completely thawed competent cells into a 13 mL round-bottom tube (*see* **Note 2**). Mix 50 ng of library plasmids into the competent cells and incubate on ice for 30 min.
4. After incubation, heat the competent cells in a 42 °C water bath for 45 seconds and place the tube on ice for 2 min before adding 900 μL of 37 °C pre-warmed LB media.
5. Shake the tube in the New Brunswick Innova 4230 Incubator at 37 °C for 1 hour at 225 rpm. During shaking, take out two

LB plates-Amp from the 4 °C cold room and pre-warm the plates in the 37 °C incubator.

6. Spread transformed bacteria evenly on the pre-warmed LB-Amp plates at 1:1000 or 1:50000 dilution. Place the plates back in the incubator. Inoculate the remaining transformed bacteria into 100 mL of LB-Amp medium in a 1 L flask. Shake the flask at 37 °C at 250 rpm for 16 hours.
7. The next day, check the transformation efficiency of each plate and calculate the coverage by the equation below (see **Note 3**).

$$\text{Coverage} = \text{number of colonies} \times \text{dilution factor}/\text{number of plasmids from a library}.$$

8. Purify the inoculated bacteria with plasmid QIAGEN Plasmid Midi Kit according to manufacturer's protocol if the plate coverage exceeds 30.
9. Measure the purified plasmid library by using Nanodrop Spectrometer. Dilute the library plasmid DNA to 0.5 μg/mL.

3.1.2 Lentiviral Production, Concentration, and Titration

1. Coat two 150 mm plates with 4 mL of diluted PLL (10 μg/mL) per plate at RT for 10 min. Aspirate the solution and wash each plate once with 4 mL of sterile 1× PBS. Place the coated plates in a 37 °C, 5% CO_2 incubator until use (see **Note 4**).
2. To produce lentivirus expressing smORF library, seed 1.5×10^7 of HEK293T cells in 25 mL of DMEM/10%FBS/1% P.S per 150 mm plate pre-coated with PLL 24 hours before transfection.
3. Change the medium with fresh medium (see **Note 5**). For a 150 mm plate, transfect the cells as follows:

Solution A:	
smORFoeme plasmids (@0.5 μg/μL)	15.00 μL (7.5 μg)
Packaging plasmid (psPAX@0.2 μg/μL)	28.15 μL (5.63 μg)
Envelop plasmid (pMD2G@0.2 μg/μL	9.35 μL (1.87 μg)
Serum-free DMEM medium	572.5 μL
Total volume	625 μL (12.5 μg)
Solution B:	
PolyJet reagent (1:3)	45 μL
DMEM (no antibiotics and FBS)	580 μL
Total volume	625 μL

Immediately add Solution B into Solution A and never in reverse order. Do not keep the transfectant mixture longer than 20 min.

4. Replace the media with complete DMEM containing 10 mM sodium butyrate after 24 h of transfection. Collect the media twice in a 15 mL tube after 48 h and 72 h of transfection. Centrifuge the collected media at 1000× *g* for 5 min. Transfer and pass the supernatant through a 0.45 μm filter unit into a new 50 mL tube and mix with 1/3 volume to the collected supernatant of the Lenti-X™ Concentrator for virus concentration. Incubate the mixture overnight at 4 °C (*see* **Note 6**).
5. Centrifuge the mixture at 1500× *g* at 4 °C for 45 min. Carefully remove the supernatant without disturbing the off-white pellet. Gently resuspend the pellet in 1/50 of the original volume using complete DMEM media. Aliquot the virus into 10 μL per tube and flash freeze each tube in liquid nitrogen. Store the aliquoted virus at −80 °C (*see* **Note 7**).
6. To titer the virus for obtaining MOI (multiplicity of infection) of 0.3 for GOF screening (*see* **Note 8**), seed 5 × 10^5 cell/well of NMuMG cells into each well of a 6-well plate (No. 1–6) and a 35 mm plate (No. 7) so that the cell confluency reaches 40–50% the next day.
7. Infect the cells in the 6-well plate with different amounts of the viruses (Nos. 1–5 for 0, 1, 2, 5.0, 10, 20 μL) except one well (No. 6). Replace the media 24 h later with fresh NMuMG culture media containing 1 μg/μL of puromycin (Nos. 1–6) except for the 35 mm plate (No. 7; *see* **Note 9**).
8. Once all puromycin-treated, uninfected cells died (No. 6), aspirate floating dead cells and count the remaining alive cells in other wells to calculate the percent transduction as follows: % transduction (MOI) = cell number with puromycin treated (No. 1–6)/cell number without puromycin (No. 7) × 100.
9. Plot the graph for % transduction with increasing amounts of virus (0–20 μL). Determine the volume of virus that give ~30% transduction (MOI).

3.2 Functional Genomic Screen

3.2.1 Large-Scale Infection and Functional Scrooning (Fig. 1)

1. Seed NMuMG cells into 100 mm plates to reach 30–40% confluency the next day. Infect the cells with the library virus at an MOI of 0.3 (Fig. 1).
2. 24 hours later, select the infected cells with puromycin (1 μg/μL). The next day, trypsinize and neutralize the trypsin with culture media to count the remaining cells for soft agar assays.
3. Weigh LMP agarose and transfer the agarose powder into a clean 150 mL glass bottle. Sterile the powder by autoclaving at 121 °C for 15 min.

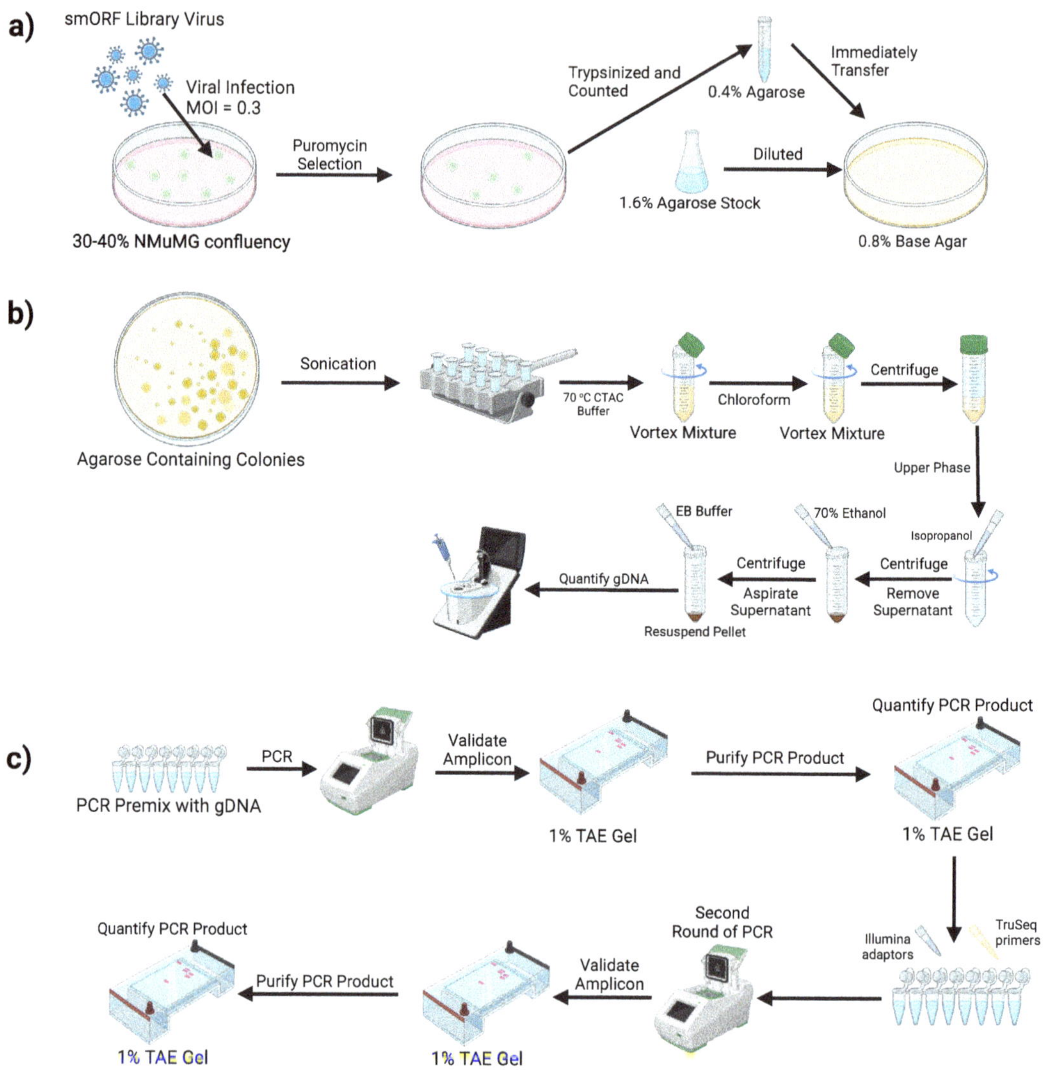

Fig. 1 Schematic overview of the gain-of-function genomic screening workflow to identify tumor-associated microproteins encoded by smORFs. (**a**) Virus infection and soft-agar assays. Initial screening phase involving viral infection of NMuMG cells with a smORF library at MOI 0.3, followed by puromycin selection and soft agar colony formation assay. (**b**) Genomic DNA (gDNA) extraction and purification. Genomic DNA extraction protocol using CTAC buffer-based method with sequential chloroform extraction, isopropanol precipitation, and ethanol washing steps. (**c**) Amplicon amplification by PCR. Two-stage PCR amplification protocol for preparing sequencing libraries, including initial amplicon generation, validation by gel electrophoresis, and subsequent addition of Illumina adaptors and TruSeq primers for NGS

4. Resuspend the agarose powder in NMuMG culture media to reach 1.6% agarose stock. Incubate at 65 °C until the agarose is completely dissolved.
5. Transfer some of the dissolved agarose into another clean 150 mL glass bottle and equilibrate it at 37 °C for at least 30 min. Remain the rest of the dissolved agarose at 65 °C.

6. Mix equilibrated 1.6% agarose with the same volume of culture media to get 0.8% agarose. Dispense 9.6 ml of diluted agarose into each 100 mm plate for a total of six plates. Wait until the agarose becomes solid at RT.
7. Once the base agar is set, transfer the counted and diluted cell suspension into a 15 mL tube and mix well with equilibrated 1.6% agarose to reach 0.4% agarose (*see* **Note 10**). Immediately transfer the mixture to the plate with 0.8% base agar. Prepare each plate individually (see **Note 11**).
8. Once all cells are seeded, temper the plates at 4 °C for 15 min and transfer the plates to the 37 °C, 5% CO_2 incubator. Leave the plates overnight and add 5 mL of culture media with 0.5 μg/μL of puromycin the next day (see **Note 12**). Refresh the media every 3 days and check colony growth every day.

3.2.2 Genomic DNA Extraction (Fig. 1)

The procedures for extracting gDNA from the agar are adapted from Lau et al. (2021) [13].

1. Once visible colonies can be easily observed, remove culture media from each plate with a P1000 pipette (*see* **Note 13**). Collect the agar into 50 mL tubes. Combine and mix all agar thoroughly by sonication (see **Note 14**).
2. Pre-warm CTAC buffer to 70 °C. Take 8 mL of sonicated agar into a 50 mL high-performance VWR tube and mix the agar with 40 mL of pre-warmed CTAC buffer. Vigorously mix and vortex the mixture.
3. Split 24 mL of the mixture into another high-performance tube. Add 24 mL of chloroform to each tube and mix vigorously.
4. Centrifuge at 15000× *g* for 5 min at RT. Collect the resultant clear upper phase into another set of tubes (see **Note 15**).
5. Add an equal volume to the collected upper layer of isopropanol into each tube and mix thoroughly.
6. Centrifuge the mixture at 15000× *g* for 15 min at RT. Remove the supernatant and wash the pellet with 10 mL of 70% ethanol. Centrifuge at 15000× *g* for 5 min.
7. Aspirate the supernatant and air-dry the pellet. Resuspend the pellet in 50 μL of EB with a broad tip (see **Note 16**).
8. Measure the concentration of the purified gDNA by Nanodrop Spectrophotometer.

3.2.3 Amplification of Amplicons (Fig. 1)

1. Set up PCR reaction in a 0.2 mL PCR tube on ice as the following:

gDNA (50 ng/μL)	1 μL
5 × PrimeStar buffer	10 μL
2.5 mM dNTP mix	4 μL
Universal forward primer (10 μM)	1 μL
Reverse primer (10 μM)	1 μL
PrimeSTAR DNA polymerase (1.25 U/μL)	1 μL
dH_2O	32 μL
Total	50 μL

2. Run PCR with Bio-Rad T100 Thermal Cycler using the following thermocycling parameters: 98 °C for 10 s; 60 °C for 15 s; and 68 °C for 30 s, in a total set of 35 cycles (see **Note 17**).
3. Validate the amplicons (233 bp) by running on 1% TAE gel with 5 μL of the PCR products and 1 μL of DNA loading dye (*see* **Note 18**). Once validated, purify the amplicons with the PCR Purification Kit according to the manufacturer's protocol.
4. Quantify the eluted PCR product by running 2 μL of PCR product along with 2, 4, and 8 μL of 1Kb DNA ladder.
5. Attach the amplicons with Illumina adaptors and indexes with TruSeq primers (*see* Subheading 2.4, **step 9**) by a second round of PCR (Fig. 1) using 50 ng of PCR product with reagents and conditions described in **steps 1** and **2** above.
6. Validate and purify the final amplicon products as described in **step 3**.

3.3 Next-Generation Sequencing (NGS) and Bioinformatic Analysis

1. Quantify and prepare the purified amplicons before sending for pair-end Illumina NextSeq sequencing (Genome Quebec in Montreal) following the company's instructions.
2. Once the NGS data is available, download the FASTQ files from the corresponding website.
3. The data is processed in MATLAB with Biopipeline Designer using SeqSplit and SeqFilter functions (*see* **Note 19**). A separate barcode file is prepared and uploaded into MATLAB for identifying smORF from the sequences. The abundance of each identified smORF is then measured and ranked based on the number of reads in the sequencing sample. The top candidate with >5, 000 reads was used as a cutoff for positive hits (Fig. 2).

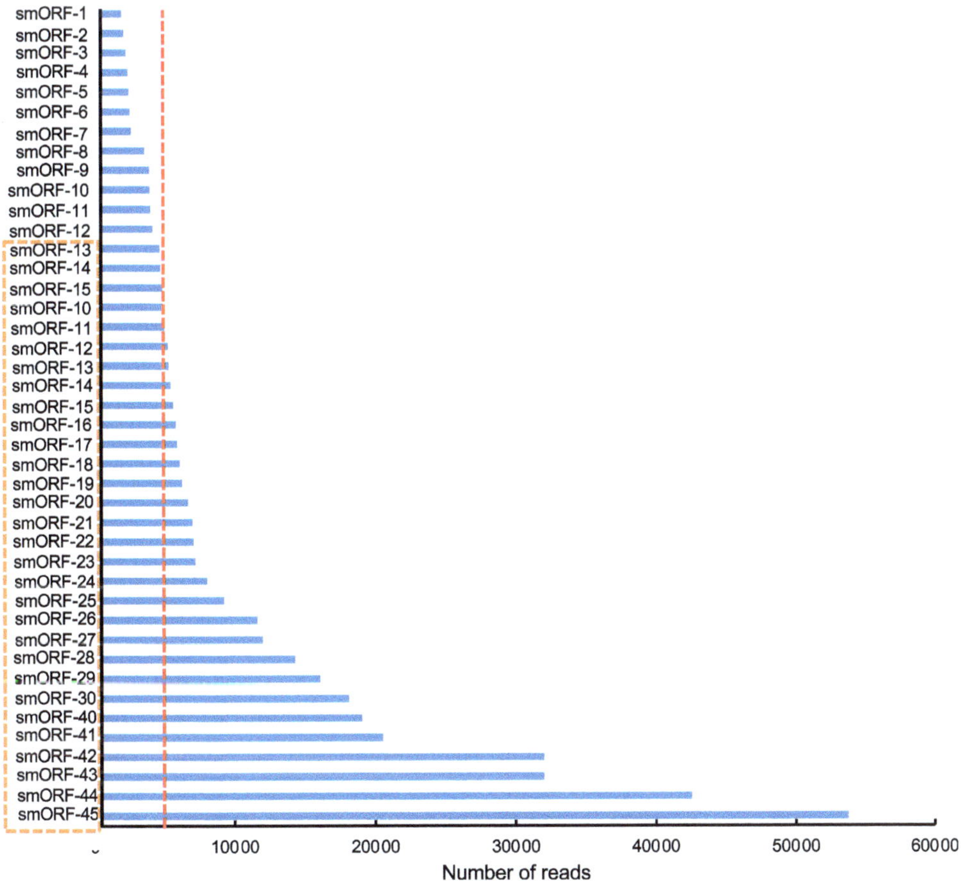

Fig. 2 Ranking of smORF reads from NGS. 5000 reads is regarded as the cutoff value as shown in red dotted line. The positive hits are indicated in orange color dotted lines

4 Notes

1. For long-term storage, the 96-well microplates are placed at −20 °C. Thaw the microplates on ice and centrifuge briefly once thawed to ensure that the content is at the bottom of each well before taking it.
2. The competent cells can easily lose their activities, so it is critical to constantly keep the bacteria on ice.
3. To make sure that a library is successfully transformed, the coverage should be more than 30.
4. HEK293T cells are easily detached and lost from the plate, so it is important to use the pre-coated plate for better attachment.
5. Serum- and antibiotic-free media are required for preparing the transfectant mixture with the PolyJet:DNA ratio of 3:1.

6. The viruses should be always kept on ice while handling with proper PPEs.
7. The flash-frozen virus can be stored for up to a year at −80 °C.
8. The multiplicity of infection (MOI) is calculated based on the plasmids within the library that need to be integrated into the cells. An MOI of 0.3 is chosen to ensure that the infected cell gets transduced with only one viral particle expressing one of the cDNAs from the library. The amount of virus that causes 30% cell survival in selected cells compared to unselected cells will be used for subsequent functional genomic screening.
9. The concentration of puromycin used to select the cells is predetermined with the optimal survival curve. The optimal concentration is then decided with the lowest dose required to kill all cells after one day of antibiotic selection.
10. The cell number for seeding into each soft agar plate is decided based on the entire library to be covered. In our case, approximately 9.2×10^4 cells are seeded into each 100 mm plate for a total of six plates to ensure that each library plasmid gets integrated into 1000 selected cells.
11. The low-melting agar solidifies easily at RT, so we advise mixing the cell suspension per sample individually with equilibrated agar and immediately dispensing the mixture on the plate.
12. To avoid loss of the integrated gene, half the concentration of the antibiotic used for selecting the infected cells is supplemented to the culture media.
13. Remove the culture media from the top agar carefully with a P1000 pipette since it may be easily disturbed.
14. For the optimal recovery of gDNA, the agars need to be mixed thoroughly by sonication. Specific parameters should be adjusted based on the amount of agar used for each experiment.
15. It is important not to disturb the lower phases while transferring the upper supernatant into another tube. The residual upper layer may be lost to avoid accidental mixing with lower layers.
16. The board tip is critical while resuspending the air-dried pellet to avoid tearing the gDNA.
17. The PCR parameters can be adjusted based on the specific length of the amplicons that need to be pooled from the gDNA.
18. Depending on the PCR efficiency and abundance of the extracted genome, the amplicon may not be visible on the gel after the initial run. We advise that another round of PCR may

be performed using the initial PCR mixture to ensure an amplicon band visible on the gel.

19. Preprocessing of the NGS data is essential during data analysis so that the low-quality sequences are filtered from the data. Any base with a quality score less than 20 is removed from the dataset.

References

1. Omenn GS et al (2018) Progress on identifying and characterizing the human proteome: 2018 metrics from the HUPO human proteome project. J Proteome Res 17:4031–4041
2. Prensner JR, Enache OM, Luria V (2021) Noncanonical open reading frames encode functional proteins essential for cancer cell survival. Nat Biotechnol 39:697–704
3. Ge A, Chan C, Yang X (2024) Exploring the dark matter of human proteome: the emerging role of non-canonical open Reading frame (ncORF) in cancer diagnosis, biology, and therapy. Cancers (Basel) 16(15)
4. Posner Z, Yannuzzi I, Prensner JR (2023) Shining a light on the dark proteome: non-canonical open reading frames and their encoded miniproteins as a new frontier in cancer biology. Protein Sci 32(8):e4708
5. Przybyla L, Gilbert LA (2022) A new era in functional genomics screens. Nat Rev Genet 23:89–103
6. Azad T et al (2020) A gain-of-functional screen identifies the hippo pathway as a central mediator of receptor tyrosine kinases during tumorigenesis. Oncogene 39:334–355
7. Zhang DY, Gui X, Yang X (2021) Mapping genetic interactions in human cancer cells using a one-step tRNA-CRISPR system. Methods Mol Biol 2381:175–187
8. Zhao Y et al (2018) PIK3CB positively regulate YAP and TAZ in mammary tumorigenesis through multiple signaling pathways. Mol Cancer Res 16(N/A):1046–1058
9. Prensner JR et al (2021) Noncanonical open reading frames encode functional proteins essential for cancer cell survival. Nat Biotechnol 39(6):697–704
10. Chen J et al (2020) Pervasive functional translation of noncanonical human open reading frames. Science 367:1140–1146
11. Nakamura D (2023) The evaluation of tumorigenicity and characterization of colonies in a soft agar colony formation assay using polymerase chain reaction. Sci Rep 13:5405
12. Deng Z et al (2021) Comprehensive understanding of anchorage-independent survival and its implication in cancer metastasis. Cell Death Dis 12:629
13. Lau HY et al (2021) Evaluating the epithelial-mesenchymal program in human breast epithelial cells cultured in soft agar using a novel macromolecule extraction protocol. Cancer 13:807

Chapter 18

Proteomic Identification of Microproteins Translated from Long Non-coding RNA

Jing Zhang, Zi-Hong Fan, Qing-Yu He, and Yang Wang

Abstract

The identification and exploration of microproteins encoded by long noncoding RNA (lncRNA) genes have the potential to revolutionize our understanding of human biology and expedite medical breakthroughs. Proteomics enables the identification and quantification of thousands of proteins present in cells, tissues or organisms, with LC-MS/MS serving as a pivotal technique in current proteomic research. Here, we describe a comprehensive suite of integrated methods to enhance proteomic analysis for the identification of lncRNA-encoded microproteins. This approach combines tricine gel electrophoresis, in-gel digestion, the establishment of a customized microprotein reference database, and LC–MS/MS analysis.

Key words Microproteins, Proteomics, LncRNA, Tricine gel electrophoresis, In-gel digestion, Customized microprotein reference database

1 Introduction

Microproteins, encoded by long noncoding RNA (lncRNA) genes, constitute a fascinating and novel category of proteins that play crucial roles in various biological processes [1–3]. Despite their potential importance in human biology and medicine, the task of identifying and characterizing these microproteins has proven to be exceptionally challenging [4]. Factors such as their small size, low abundance, and the fact that they were not included in standard protein reference databases have collectively contributed to the difficulty in detecting and studying them.

Liquid chromatography–tandem mass spectrometry (LC–MS/MS) has become a pivotal technique in proteomic research, enabling the identification and quantification of thousands of proteins present in cells, tissues, or organisms. In the context of microprotein identification, effective protein separation and digestion, coupled with the availability of accurate microprotein reference databases, form the essential foundation for proteomics analysis

Stephan Wenkel (ed.), *Microproteins: Methods and Protocols*, Methods in Molecular Biology, vol. 2992,
https://doi.org/10.1007/978-1-0716-5013-4_18,

[5]. Tricine gel electrophoresis is a specialized electrophoresis technique that is particularly suited for the separation of small proteins [6]. Thus, we employed tricine gel electrophoresis to separate proteins according to their size, with a special focus on those within the 2–20 kDa low molecular weight range. Subsequently, in-gel digestion [7, 8] was conducted to enzymatically decompose the proteins into peptides, facilitating their subsequent analysis by LC–MS/MS. In addition, the establishment of a customized microprotein reference database, which encompasses both canonical proteins and theoretical proteins generated from the translating lncRNAs, is crucial for the precise identification of microproteins during the database searching phase of proteomic analysis (NAR).

Collectively, we have developed a suite of methods that integrates tricine gel electrophoresis, in-gel digestion, the creation of a customized microprotein reference database, and LC–MS/MS analysis to facilitate the accurate identification of lncRNA-encoded microproteins.

2 Materials

Prepare all solutions utilizing optima LC/MS grade reagents. Store all reagents at room temperature unless otherwise specified. For stock solutions without explicit formulas, commercial purchase is acceptable.

2.1 Tricine Gel Electrophoresis

1. Deionized water.
2. 10–20% Tricine gels.
3. Protein samples and standards.
4. 10× Tricine SDS running buffer (for 1 L add 121.14 g of Tris base, 179.2 g of Tricine, 10 g of SDS).
5. 2× Tricine SDS sample buffer (for 1 L add 19.7 g of Tris–HCl, 40 g of SDS, 252 g glycerol, and 0.1 g Bromophenol blue).
6. 10× sample reducing agent (for 1 L add 77.1 g DTT).
7. Novex power supply adapters.

2.2 Trypsin Digestion and LC–MS/MS

1. Trypsin.
2. Acetonitrile (ACN).
3. Ammonium bicarbonate (NH_4HCO_3).
4. Reductant: A solution of 25 mM NH_4HCO_3 containing 10 mM dithiothreitol (DTT). For a 4 mL preparation, dissolve 6.2 mg of DTT in 4 mL of 25 mM NH_4HCO_3.
5. Alkylation reagent: A solution of 25 mM NH_4HCO_3 containing 55 mM iodoacetamide (IAA). For a 4 mL preparation, dissolve 40.69 mg of IAA in 4 mL of 25 mM NH_4HCO_3.

6. Covering solution 10% ACN, 25 mM NH_4HCO_3: Mix 500 μL of 100 mM NH_4HCO_3, 200 μL of 100% ACN, and 1.3 mL of deionized water.
7. Enzyme solution: To obtain a stock solution of 1 μg/μL, add 100 μL of 1 mM HCl to the trypsin. Aliquot 2 μL per tube and store at −80 °C. Dilute each 2 μL of enzyme solution with 98 μL of covering solution to achieve a concentration of 0.02 μg/μL.
8. Extraction solution: mix 75 μL of formic acid, 2010 μL ACN, and 915 μL deionized water.
9. Buffer A: 0.1% (v/v) formic acid in deionized water.
10. Buffer B: 80% ACN with 0.1% formic acid.
11. LC–MS/MS system equipped with an EASY-nLC 1200 HPLC system.
12. Orbitrap Fusion Lumos mass spectrometer.

3 Methods

Unless otherwise specified, perform all procedures at room temperature.

3.1 Tricine Gel Electrophoresis

1. Prepare 1X Tricine SDS Running Buffer by diluting 100 mL of 10X buffer with 900 mL of deionized water.
2. Remove the comb and rinse the gel wells with 1X Running Buffer. Remove the white tape near the bottom of the gel cassettes. Place the gels in the XCell SureLock mini-cell gel running tank (*see* **Note 1**). Fill the gel wells and buffer chambers with 1X running buffer.
3. Add Tricine SDS sample buffer (2X) 5 μL, NuPAGE1 reducing agent (10X) 1 μL to 4 μL protein samples (10 μg/μL, dilute samples with 1X lysis buffer, if necessary) and up to 10 μL volume. Heat the mixed protein samples at 85 °C for 2 min. Centrifuge the heated samples at 3000 × *g* for 30 s to bring down the condensate. Load protein samples 10 μL/well and standard 5 μL/well on the gel.
4. Run at 125 V constant till the dye front (from the SDS sample buffer) reaches 1 cm from the bottom of the gel. After electrophoresis, delicately separate the gel plates with a spatula, ensuring that the gel adheres to one of the glass plates. Excise the lane corresponding to the 2–20 kDa range, then carefully cut the resulting strips into small pieces approximately the size of rice grains. Finally, place 3–5 of these grains into each Eppendorf tube.

3.2 In-Gel Digestion

1. Wash the gel pieces thoroughly with deionized water twice to remove any impurities. Then, introduce 50% ACN and gently blow and rinse them two additional times.
2. Dehydrate the gel pieces using 100% ACN until they turn completely white. If the gel block does not turn white after blowing, remove the ACN, add fresh 100% ACN, and dehydrate once more.
3. Based on the size of the gel particles in the sample, add 50–100 μL of reductant, seal the container, and incubate in a 57 °C water bath (*see* **Note 2**) for 1 h.
4. After removing the sample from the water bath, allow it to cool to room temperature. Remove any excess liquid and promptly add an equal volume of alkylation reagent. Place the sample at room temperature in the dark for 30 min. Terminate the alkylation reaction and wash away the IAA by rinsing the gel block twice with 25 mM NH_4HCO_3.
5. Dehydrate the gel pieces again with 100% ACN until they turn white. Add 25 mM NH_4HCO_3, pipette to ensure even coverage, and let it stand until the gel particles become transparent. Dehydrate once more with 100% ACN.
6. Add an appropriate volume of enzyme solution to each tube and place them on ice for 30 min. Check for adequate swelling of the gel pieces in the enzyme solution. Aspirate any excess enzyme solution. If the gel block remains white, indicating insufficient enzyme solution, add more as needed.
7. Add 20 μL of covering solution, seal the tubes, and incubate them in a 37 °C incubator for 16–18 h.
8. Remove the samples from the 37 °C incubator and centrifuge them at 5000× *g* for 30 s. Transfer the supernatant to clean 200 μL Eppendorf tubes.
9. Add extraction solution to the remaining gel pieces and incubate at 37 °C for 30 min. Sonicate for 10 min, then repeat with a 5-min interval for another 10 min, ensuring the water temperature remains below 40 °C. Centrifuge at 5000× *g* for 30 min, combine the supernatants, and subject the combined solution to freeze-drying.

3.3 Mass Spectrometry Analysis

1. The tryptic peptide is reconstituted with mobile phase A and subsequently analyzed using a data-dependent acquisition (DDA) mode on an LC-MS/MS system equipped with an EASY-nLC 1200 HPLC system and Orbitrap Fusion Lumos mass spectrometer (*see* **Note 3**).
2. For liquid chromatography (LC) separation, the tryptic peptides are separated on a custom-made analytical column with a 150 μm inner diameter fused silica capillary tubing featuring a

hand-pulled spray tip and packed with 25 cm of 1.9 μm/120 Å C18-AQ beads. The flow rate of the mobile phases is maintained at 600 nL/min, with the following gradient profile: 15 min of equilibration with 100% Buffer A; a gradient of Buffer B from 5% to 12% over 0 to 14 min; from 12% to 44% over 14 to 44 min; and from 44% to 98% over 44–65 min, followed by a 10-min wash with 98% Buffer B.

3. MS/MS analysis is performed using the Orbitrap mass analyzer, configured with a nanospray ionization source operating at 2.3 kV in positive mode and an ion transfer tube temperature of 320 °C. The default charge state is set to 2, and the Orbitrap detector resolution is 120,000 at m/z 200, with a scan range of 350–1800 m/z. Fragmentation is achieved using a 1.6 Da isolation window and higher-energy collisional dissociation (HCD) with a normalized collision energy (NCE) of 30. Additional MS parameters include an S-lens RF level of 40%, an AGC target of 4.0e5, a maximum injection time of 50 ms, and charge states ranging from 2 to 7. Data-dependent acquisition is conducted in top speed mode, with a dynamic exclusion time of 60 s and a 10 ppm tolerance around the selected precursor. Parent ions are selected for fragmentation based on a signal intensity threshold of 2.4e4.

3.4 Sample-Specific Microprotein Reference Database

1. Obtain full-length translating mRNA-seq data (*see* **Note 4**) and use it to identify translating lncRNAs.
2. Conduct a comprehensive 3-frame translation analysis of these lncRNAs, encompassing all potential reading frames.
3. Process the resulting amino acid sequences to exclude any sequences located downstream of stop codons.
4. Select and retain only those sequences that are between 20 and 200 amino acids in length, assigning a distinct identifier to each to create a theoretical proteome derived solely from translating lncRNAs.
5. Download the canonical proteome database from UniProt (www.uniprot.org), including proteins that are less than 200 amino acids in length.
6. Merge this customized lncRNA-derived proteome database with the canonical proteome in FASTA file format to facilitate the identification of microproteins.

3.5 Database Search

1. The raw MS data were searched against the combined database using MaxQuant (*see* **Note 5**).
2. The search parameters were configured as follows: trypsin was selected as the enzyme; carbamidomethyl (C) was set as the fixed modification; variable modifications included oxidation (M), Gln → pyro-Glu (N-terminus), and acetyl (N-terminus);

the tolerance was set to 10 ppm for MS and 0.02 Da for MS/MS, with up to two missed cleavage sites allowed.

3. Protein identifications were validated according to criteria such as a protein-level false discovery rate (FDR) of $\leq 1\%$, a minimum of two identified peptides, at least one unique peptide, and a peptide length of ≥ 7 amino acids.
4. The identified microprotein-related protein dataset, peptide dataset, and PSM (peptide-spectrum match) dataset were obtained for further analysis.

4 Notes

1. The electrophoresis equipment and stock solutions used in our method were purchased from Thermo Scientific; however, alternative products and equipment are also viable options.
2. Adjust the water bath to 57 °C beforehand to eliminate any waiting time during the reduction step.
3. We employed the EASY-nLC 1200 (Thermo Scientific) HPLC system and Orbitrap Fusion Lumos (Thermo Scientific) mass spectrometer. Additionally, other advanced instruments such as the Thermo Scientific Orbitrap Astral, Bruker timsTOF Pro, Bruker timsTOF Ultra 2, etc., are also highly recommended.
4. Prepare full-length translating mRNA-seq according to the method described by Wang et.al [9].
5. MaxQuant is a user-friendly and free search engine software for protein databases. Other search engines, including Proteome Discoverer, PFind, Mascot, X!Tandem, etc., are also acceptable, with attention to parameter settings.

References

1. Lu S, Wang T, Zhang G et al (2020) Understanding the proteome encoded by "non-coding RNAs": new insights into human genome. Sci China Life Sci 63:986–995
2. Pennisi E (2024) 'Dark proteome' survey reveals thousands of new human genes. Science 386: 951–952
3. Hofman DA, Prensner JR, van Heesch S (2024) Microproteins in cancer: identification, biological functions, and clinical implications. Trends Genet. https://doi.org/10.1016/j.tig.2024.09.002
4. Chothani SP, Adami E, Widjaja AA et al (2022) A high-resolution map of human RNA translation. Mol Cell 82:2885–2899 e2888
5. Lu S, Zhang J, Lian X et al (2019) A hidden human proteome encoded by 'non-coding' genes. Nucleic Acids Res 47:8111–8125
6. Schagger H (2006) Tricine-SDS-PAGE. Nat Protoc 1:16–22
7. Shevchenko A, Tomas H, Havlis J et al (2006) In-gel digestion for mass spectrometric characterization of proteins and proteomes. Nat Protoc 1:2856–2860
8. Chen Y, Li Y, Zhong J et al (2015) Identification of missing proteins defined by chromosome-centric proteome project in the cytoplasmic detergent-insoluble proteins. J Proteome Res 14:3693–3709
9. Wang T, Cui Y, Jin J et al (2013) Translating mRNAs strongly correlate to proteins in a multivariate manner and their translation ratios are phenotype specific. Nucleic Acids Res 41:4743–4754

Chapter 19

Urinary Micropeptides: Integrative In Silico Profiling and LC–MS/MS Analysis for Prostate Cancer Biomarker Discovery

Rita Ferreira and Rui Vitorino

Abstract

Micropeptides encoded by small open reading frames (sORFs) represent a newly recognized class of biomolecules involved in important cellular functions such as metabolism, signal transduction and homeostasis. In the past, these peptides have been overlooked due to annotation and technical challenges. Recent advances in high-throughput techniques such as ribosome profiling, liquid chromatography-tandem mass spectrometry (LC–MS/MS) and bioinformatics tools have enabled the systematic identification and characterization of these peptides. In prostate cancer (PCa), urinary micropeptides have been shown to be promising noninvasive biomarkers, as the prostate is located close to the urinary tract and urine can be easily collected. In this chapter, we discuss the methods used to detect micropeptides, incorporating in silico predictions, LC–MS/MS-based proteomic analyzes, and metabolic pathway enrichment approaches. We highlight the recent discoveries of micropeptides with differential expression in PCa and their potential as diagnostic and prognostic biomarkers. These findings highlight the importance of micropeptide research and pave the way for new strategies in PCa diagnosis and personalized medicine.

Key words Micropeptides, sORFs, Noncoding RNA, Proteomics, Therapeutic targets, Systems biology

1 Introduction

Prostate cancer (PCa) is the most commonly diagnosed noncutaneous cancer in men and is the fifth leading cause of cancer-related mortality worldwide. In 2020, approximately 1.4 million new cases were diagnosed, and the disease was responsible for 375,000 deaths worldwide [1]. Despite advances in diagnostic and therapeutic strategies, primary screening methods such as the prostate specific antigen (PSA) test and digital rectal examination (DRE) remain limited by their low specificity and high false-positive rates. These shortcomings often lead to unnecessary prostate biopsies, which are invasive and carry the risk of bleeding, infection and sampling

Stephan Wenkel (ed.), *Microproteins: Methods and Protocols*, Methods in Molecular Biology, vol. 2992,
https://doi.org/10.1007/978-1-0716-5013-4_19,

errors [1, 2]. Therefore, there is an urgent need for non-invasive, sensitive biomarkers that can improve the early detection, diagnosis and monitoring of PCa [3].

Urine has shown promise as a biofluid for biomarker discovery due to its proximity to the prostate and its high content of prostate secretions [1]. Urine diagnostics offer several advantages over blood-based tests, including the ease of collection, the stability of biomolecules in urine and the possibility of repeated sampling [2]. Urine proteomics has proven to be particularly valuable for the identification of proteins and peptides that are differentially expressed in PCa patients compared to healthy controls [1, 3]. By using advanced analytical techniques such as liquid chromatography-tandem mass spectrometry (LC–MS/MS), researchers can build a comprehensive profile of the urinary proteome, facilitating the discovery of potential diagnostic and prognostic biomarkers [2].

Among the emerging fields of proteomics, the study of micropeptides—small proteins encoded by short open reading frames (sORFs) of less than 100 amino acids—has gained increasing attention [4, 5]. Micropeptides, previously overlooked due to technical limitations in detection and annotation, have now been identified as important regulators of various cellular processes, including metabolism, signal transduction and stress response [4]. Micropeptides can be derived from various genomic regions, including long non-coding RNAs (lncRNAs), mitochondrial RNAs and untranslated regions (UTRs) of messenger RNAs [4, 6]. In cancer biology, micropeptides can function as oncogenes, tumor suppressors or modulators of the tumor microenvironment, making them potential candidates for PCa biomarkers [5, 6].

Ribosome profiling (Ribo-seq) has been used to identify actively translated sORFs and the peptides they encode. This high-resolution technique captures ribosome-protected RNA fragments (RPFs) that provide direct evidence of translation [7]. Ribo-seq allows the detection of low-abundance peptides that may be missed by conventional proteomic techniques. By integrating Ribo-seq data with validation by MS, confidence in the identification of functional micropeptides is significantly increased [5, 6]. Bioinformatics tools such as MiPepid and SmProt facilitate the identification of coding sORFs by predicting their translational potential from genomic sequences [5, 8]. Ribosome profiling provides experimental evidence for active sORF translation by capturing ribosome-bound RNA fragments [5]. This technique is complemented by LC–MS/MS, which confirms the presence of micropeptides in biological samples and provides insights into their abundance and post-translational modifications [2, 5].

Recent studies combining these techniques have identified micropeptides that are differentially expressed in PCa tissue and urine [1]. For example, mitochondria-derived peptides (MDPs) such as humanin and MOTS-c regulate mitochondrial function and cellular metabolism, processes that are often dysregulated in cancer [4]. In addition, micropeptides derived from lncRNAs have been shown to affect tumor growth, immune evasion and drug resistance [4, 6]. These findings challenge the traditional classification of non-coding regions and emphasize the need to re-evaluate the human genome for hidden coding potential [9].

Urinary micropeptides hold promise as non-invasive biomarkers for PCa, and their small size and stability make them suitable for detection in urine, where they can reflect molecular changes associated with tumor development and progression [1, 10]. The integration of proteogenomics, which combines proteomic data with genomic and transcriptomic information, enhances the ability to identify novel biomarkers and understand the molecular heterogeneity of PCa [1]. Exploratory studies applying proteogenomics to urine samples have uncovered mutant protein isoforms that may play a protective or deleterious role in PCa, providing additional insights into tumor biology [1].

Despite these advances, there are still some challenges in the detection and characterization of micropeptides. Their low abundance and small size make it difficult to identify using conventional proteomic techniques [5]. Improved sample preparation, enrichment strategies and high-resolution MS are required to overcome these limitations [2]. In addition, functional validation of the identified micropeptides is crucial to confirm their relevance to PCa pathology. This can be achieved by a combination of in vitro and in vivo studies. For in vitro validation, experiments such as overexpression or knockdown of the micropeptides in PCa cell lines can be performed to observe their effects on cell proliferation, migration, apoptosis and signaling pathways [4, 11–13]. For in vivo validation, mouse xenograft models or genetically engineered mouse models can be used to study the effects of micropeptide expression in a more complex biological environment [14, 15].

The aim of this chapter is to present a peptide-centric approach for the analysis of urinary proteomic data to identify micropeptides with potential diagnostic and prognostic value for PCa. The final goal is to translate these discoveries into clinical practice and develop reliable tests for the detection of micropeptides in urine and other biofluids. This approach integrates proteomics, micropeptide analysis and bioinformatics to gain a comprehensive understanding of the molecular landscape of PCa.

2 Materials

2.1 Biological Material and Equipment

- Fresh urine samples—collected under sterile conditions from human subjects.

- Centrifuge—Capable of 4000× *g*, 4 °C (e.g., Eppendorf 5810 R or equivalent).
- Ultrafiltration devices—Vivaspin 500, 10 kDa MWCO (Sartorius Biotech, Germany).
- −80 °C freezer—Ultra-low temperature freezer (e.g., Thermo Scientific).
- SDS-PAGE system—Vertical electrophoresis system (e.g., Bio-Rad Mini-PROTEAN Tetra System).
- High-resolution mass spectrometer—e.g., Orbitrap Q Exactive HF-X, Thermo Fisher Scientific, USA.
- C18 reverse-phase LC column—e.g., Acclaim PepMap RSLC, 75 μm × 50 cm, Thermo Fisher Scientific.
- RC-DC™ Protein Assay Kit—Bio-Rad Laboratories, USA.
- Tris buffer (0.5 M, pH 6.8) with 4% SDS—Prepared in-house or sourced from Sigma–Aldrich.
- Dithiothreitol (DTT, 10 mM)—Sigma-Aldrich/Merck KGaA, Germany.
- Iodoacetamide (IAA, 55 mM)—Sigma-Aldrich/Merck KGaA, Germany.
- Sequencing-grade trypsin—Promega Corporation, USA.
- Solvents for LC–MS/MS:
  - Acetonitrile (HPLC grade)—Fisher Chemical.
  - Formic acid (LC–MS grade)—Sigma–Aldrich.
  - Ammonium bicarbonate—Sigma–Aldrich.

2.2 Software and Databases

- MaxQuant v2.1.0.0—Developed by the Cox Lab, Max Planck Institute of Biochemistry.
- PEAKS Studio—Bioinformatics Solutions Inc., Canada.
- HMPA (Human MicroPeptide Atlas)—Online database (academic source).
- FuncPEP & SmProt—Academic databases for small open reading frame (sORF) peptides.
- MiPepid—Web-based prediction tool for micropeptides.
- SmProt Analysis Suite—Associated with the SmProt database.
- STRING Database—Functional protein association networks, developed by EMBL (https://stitch.embl.de/).
- KEGG (Kyoto Encyclopedia of Genes and Genomes)—Pathway mapping database by Kyoto University Bioinformatics Center (https://www.genome.jp/kegg/).

3 Methods

3.1 Urine Collection and Processing

A stepwise proteomic pipeline was implemented to identify and characterize urinary micropeptides in PCa (Fig. 1).

1. Collect fresh urine samples from subjects under sterile conditions (*see* **Note 1**).
2. Centrifuge samples at 4000× *g*, 4 °C, for 20 min to remove cellular debris (*see* **Note 2**).
3. Carefully transfer the clarified supernatant into clean tubes.
4. Freeze immediately at −80 °C to prevent protein degradation (*see* **Note 3**).
5. Concentrate samples using Vivaspin 500 ultrafiltration units (10 kDa MWCO, Sartorius Biotech) (*see* **Note 4**).
6. Determine protein concentration using the RC-DC™ Protein Assay Kit (Bio-Rad), following manufacturer's instructions (*see* **Note 5**).

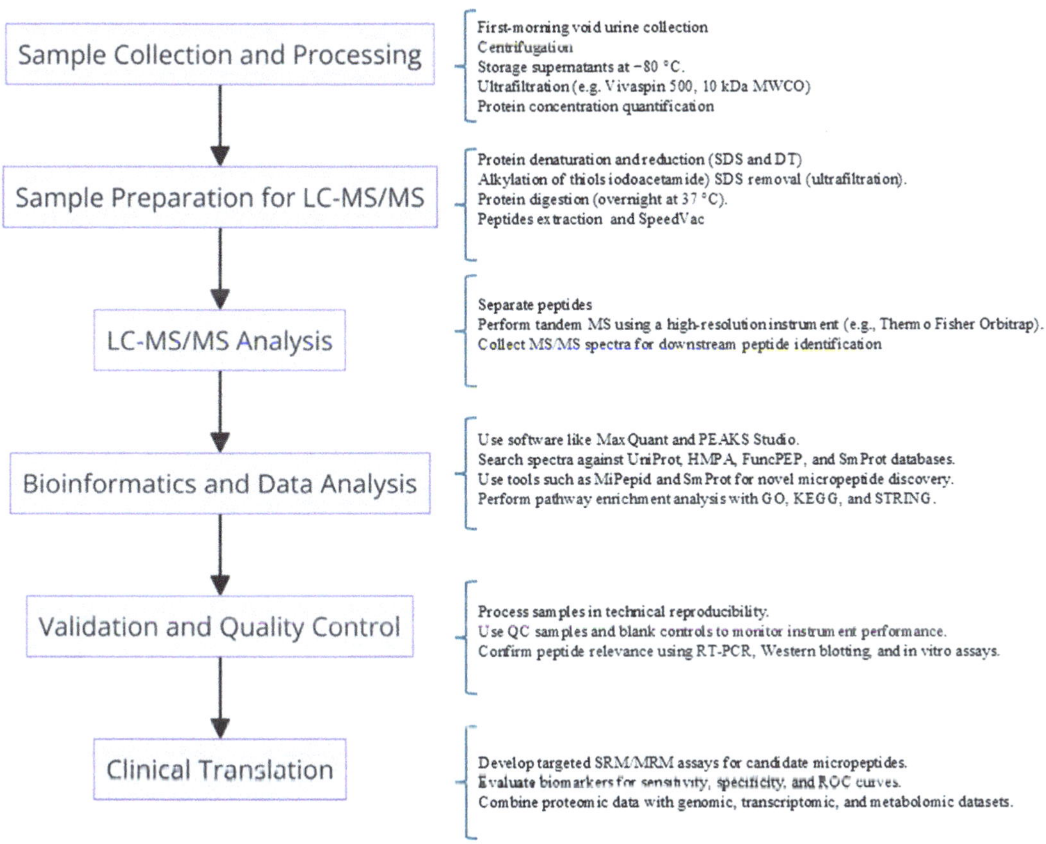

Fig. 1 Proteomic pipeline for the detection of urinary micropeptides in prostate cancer. This schematic illustrates the workflow for the identification and characterization of urinary micropeptides as potential biomarkers for prostate cancer (PCa). It includes urine collection and processing, LC–MS/MS proteomic analysis, computational prediction of micropeptides using curated databases and machine learning tools, and functional annotation by pathway enrichment analysis

3.2 Sample Preparation for Proteomics

1. Denature proteins using 0.5 M Tris buffer (pH 6.8) with 4% SDS.
2. Reduce disulfide bonds with 10 mM dithiothreitol (DTT, Sigma–Aldrich).
3. Alkylate thiol groups with 55 mM iodoacetamide (IAA, Sigma–Aldrich) to prevent reformation of disulfide bonds.
4. Digest proteins overnight at 37 °C with sequencing-grade trypsin (Promega) (*see* **Note 6**).
5. Extract peptides, dry using vacuum centrifugation, and resuspend in 0.1% formic acid for MS analysis (*see* **Note 7**).

3.3 LC–MS/MS Proteome Profiling

1. Inject resuspended peptides onto a C18 reverse-phase analytical column (Thermo Fisher PepMap RSLC) (*see* **Note 7**).
2. Separate peptides using a linear gradient of acetonitrile in 0.1% formic acid.
3. Analyze peptides using Orbitrap Q Exactive HF-X (Thermo Fisher Scientific) in data-dependent acquisition (DDA) mode (*see* **Note 7**).

   Peptide Identification Settings

- Software: MaxQuant (v2.1.0.0), PEAKS Studio (Bioinformatics Solutions Inc.).
- Variable Modifications:
  - Methionine oxidation (+16 Da).
  - Protein N-terminal acetylation (+42 Da).
  - Phosphorylation (+80 Da).
- Fixed Modifications:
  - Cysteine carbamidomethylation (+57 Da).
- Mass Tolerances:
  - Precursor ion: ±20 ppm.
  - Fragment ion: ±0.15 Da (*see* **Note 9**).
- *Example Spectra.*

  Spectra data for representative peptides, such as **pep7-alt-ALB** (*PADLPSLAADFVESKDVCK*) and **pep2-alt-A1BG** (*LETPDFQLFK*), demonstrated clear fragmentation patterns and high-intensity signals (Fig. 2).

3.4 Computational Prediction of Micropeptides

To identify and characterize micropeptides encoded by small open reading frames (sORFs), a multistep computational strategy was employed:

3.4.1 Database Integration and Prediction Workflow

1. Database Integration

   Data from the following curated databases were integrated to establish a comprehensive sORF reference set:

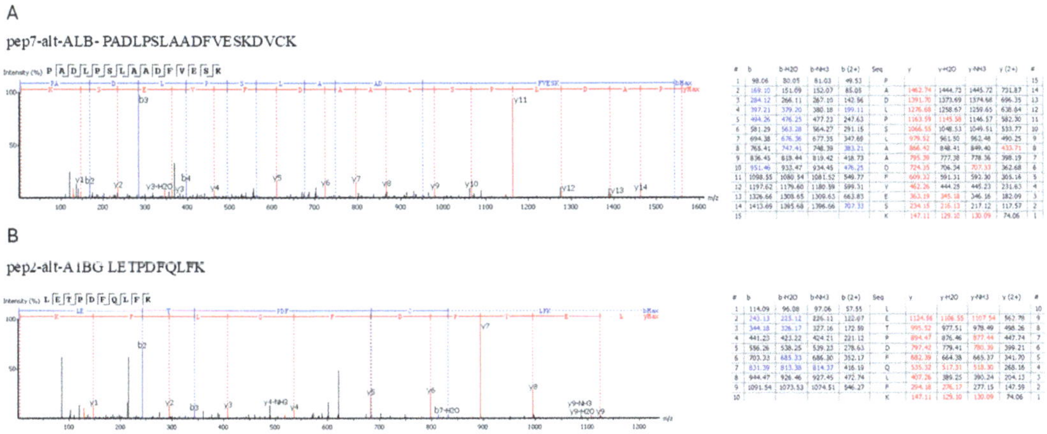

Fig. 2 Spectra for representative peptides

- HMPA—Human MicroPeptide Atlas.
- FuncPEP—Functional peptides database containing MS-validated micropeptides.
- SmProt—Small protein database with entries derived from lncRNAs and other noncanonical coding regions.

2. Prediction Tools

   Micropeptides were predicted using:

- MiPepid—A machine learning-based tool employing logistic regression and k-mer sequence features [6].
- SmProt Analysis Suite—For annotation based on ribosome profiling, MS data, and conservation [8].

3. Experimental Validation via Translation Evidence

   Ribosome profiling (Ribo-Seq) data were used to detect ribosome-protected mRNA fragments, thereby confirming active translation of candidate sORFs (*see* **Note 10**).

4. Filtering Criteria for High-Confidence Micropeptides

   Only micropeptides meeting the following criteria were retained:

- Sequence length ≤ 100 amino acids.
- Conservation across multiple species.
- Evidence of translation from Ribo-Seq data [16].
- Detection in biological samples via mass spectrometry.

The following table summarizes representative micropeptides identified through this workflow (Table 1). The identified micropeptides were further characterized for key properties such as the following: Start Codon: Commonly used start codons (ATG, CTG, and TTG); Hydrophobicity: Calculated using the grand average of

Table 1
Identified micropeptides

Transcript	Gene symbol	chromosome	Start	End	Localization	Start codon	Proteins	Sequence	Length	Missed cleavages	Mass (Da)
ENST00000427391	FIRRE	chrX	131783961	131796174	Cytoplasm	CTG	pep5-nc-FIRRE	SRQPAPPTLQPQR	13	1	1474.8005
ENST00000653715	PROX1-AS1	chr1	213894685	213965359	Cytoplasm	ATG	pep7-nc-PROX1-AS1	LSWRQRMPR	9	2	1228.6611
ENST00000525057	HPX	chr11	6437652	6441519	Mitochondrion	CTG	pep1-alt-HPX	TSAHGNVAEGETK PDPDVTER	21	1	2209.0247
ENST00000510166	ALB	chr4	73405148	73408806	Mitochondrion	TTG	pep10-alt-ALB	LGEENFK	7	0	835.40758
ENST00000676612	CTSC	chr11	88294009	88300596	Extracellular	TTG	pep2-d-CTSC	LPTSWDWR	8	0	1059.5138
ENST00000412181	AL445931.1	chr9	134058400	134058468	Cytoplasm	TTG	pep1-nc-AL445931.1	LDIQNMSAQNK	11	0	1260.6132
ENST00000567259	AL096828.2	chr20	63127716	63128090	Mitochondrion	TTG	pep5-nc-AL096828.2	SFIKLTGTQNQNMK	14	1	1608.8294
ENST00000531372	IGLL5	chr22	22895587	22895691	Extracellular	CTG	pep-d-IGLL5	LSLTPEQWK	9	0	1100.5866
ENST00000505649	ALB	chr4	73413552	73418100	Mitochondrion	TTG	pep2-d-ALB	VEEPQNLIK	9	0	1068.5815
ENST00000424664	FAF1-AS1	chr1	50462218	50462262	Mitochondrion	ATG	pep1-nc-FAF1-AS1	LSPTNLPR	8	0	896.50797
ENST00000659672	AC013486.1	chr15	74040211	74040390	Extracellular	TTG	pep-nc-AC013486.1	KMGNRALSLR	10	2	1144.6499
ENST00000476441	ALB	chr4	73418202	73420295	Cytoplasm	CTG	pep4-d-ALB	LEVDETYVPK	10	0	1191.6023
ENST00000555115	AL358333.1	chr14	51982750	51982845	Cytoplasm	CTG	pep-nc-AL358333.1	LNSRMDHFSIIEK	13	1	1588.8032
ENST00000604178	AC134879.2	chrY	11473740	11474039	Cytoplasm	ATG	pep-alt-AC134879.2	MTQSPSSLSASVGDR	15	0	1521.7093
ENST00000453955	TTTY19	chrY	8704482	8704538	Cytoplasm	TTG	pep-nc-TTTY19	LSNTSMNAPLQIEK	14	0	1544.7868
ENST00000652970	FTX	chrX	74203331	74203435	Mitochondrion	TTG	pep15-nc-FTX	LGKQGLDPSCLWTK	14	1	1601.8236
ENST00000435524	AC068533.3	chr7	66120563	66120730	Cytoplasm	ATG	pep-nc-AC068533.3	MWNLQIWR	8	0	1145.5804
ENST00000633227	HBB	chr11	5225601	5226747	Cytoplasm	CTG	pep4-alt-HBB	LSTPDAVMGNPK	12	0	1228.6122
ENST00000595014	A1BG	chr19	58352146	58353111	Cytoplasm	CTG	pep2-alt-A1BG	LETPDFQLFK	10	0	1236.639
ENST00000505649	ALB	chr4	73413633	73418100	Mitochondrion	ATG	pep9-alt-ALB	FLYEYAR	7	0	960.47052
ENST00000669680	AL353732.2	chr9	21416855	21418058	Cytoplasm	ATG	pep1-nc-AL353732.2	KIFCMSKMEK	10	2	1300.6342
ENST00000476441	ALB	chr4	73419568	73420295	Extracellular	ATG	pep5-d-ALB	DDFAAFVEK	9	0	1040.4815
ENST00000493011	TF	chr3	133753649	133755496	Extracellular	CTG	pep4-alt-TF	LAPNNLKPVVAEF YGSK	17	1	1845.9989
ENST00000581510	SUPT6H	chr17	28697918	28700662	Nucleus	CTG	pep-alt-SUPT6H	TPASINATPANIN LADLTR	19	0	1952.0327
ENST00000520057	IGHV3-62	chr14	106643291	106643416	Mitochondrion	CTG	pep-alt-IGHV3-62	LVESGEGLVQPGGSLR	16	0	1596.8471
ENST00000553189	TMTC1	chr12	29755788	29755865	Extracellular	TTG	pep7-alt-ALB	PADLPSLAADFVESKDVCK	19	1	2061.0089
ENST00000608412	AC058791.1	chr7	130938864	130939046	Mitochondrion	TTG	pep2-nc-AC058791.1	LVTNLEK	7	0	815.47527
ENST00000505649	ALB	chr4	73413540	73418100	Mitochondrion	ATG	pep7-alt-ALB	PADLPSLAADFVESK	15	0	1558.7879
ENST00000289779	AL591806.2	chr1	160998874	161001083	Cell membrane	GTG	pep-d-AL591806.2	ITASYEDRVTFLPTGITFK	19	1	2158.131

Start position	End position	PEP	Score	Intensity AC381	Intensity AC384	Intensity AC386	Intensity AC392	Intensity AC399	Intensity UT805	Intensity UT812	Intensity UT815	Intensity UT821	Intensity UT822	GRAVY
49	61	4.12E-02	6.53E+01	0.00E+00	0.00E+00	0.00E+00	0.00E+00	2.25E+05	0.00E+00	0.00E+00	0.00E+00	1.58E+07	0.00E+00	- 1.68
18	26	2.53E-02	4.57E+01	3.59E+07	0.00E+00	0.00E+00	0.00E+00	0.00E+00	0.00E+00	0.00E+00	1.25E+08	0.00E+00	0.00E+00	- 1.62
26	46	2.85E-03	4.04E+01	1.80E+06	0.00E+00	0.00E+00	0.00E+00	0.00E+00	0.00E+00	3.77E+06	0.00E+00	2.85E+06	0.00E+00	- 1.31
1	7	1.29E-02	8.97E+01	0.00E+00	0.00E+00	0.00E+00	0.00E+00	6.68E+06	0.00E+00	0.00E+00	2.54E+07	0.00E+00	0.00E+00	- 1.17
1	8	5.68E-03	8.70E+01	0.00E+00	5.17E+06	0.00E+00	0.00E+00	0.00E+00	0.00E+00	3.99E+06	0.00E+00	4.13E+06	0.00E+00	- 1.14
1	11	3.12E-02	4.26E+01	3.09E+07	0.00E+00	0.00E+00	0.00E+00	0.00E+00	0.00E+00	4.35E+07	0.00E+00	0.00E+00	0.00E+00	- 0.93
52	65	3.11E-03	5.36E+01	1.74E+06	0.00E+00	0.00E+00	0.00E+00	0.00E+00	2.59E+07	0.00E+00	0.00E+00	0.00E+00	0.00E+00	- 0.81
1	9	2.16E-02	5.76E+01	0.00E+00	0.00E+00	0.00E+00	0.00E+00	1.62E+06	1.98E+05	0.00E+00	0.00E+00	0.00E+00	0.00E+00	- 0.81
1	9	7.52E-04	9.65E+01	0.00E+00	0.00E+00	0.00E+00	1.07E+07	0.00E+00	1.38E+07	0.00E+00	5.98E+07	6.16E+06	7.80E+06	- 0.78
2	9	2.54E-02	5.36E+01	0.00E+00	0.00E+00	2.35E+07	0.00E+00	0.00E+00	0.00E+00	0.00E+00	0.00E+00	5.36E+06	0.00E+00	- 0.64
12	21	3.94E-02	4.10E+01	2.42E+06	0.00E+00	0.00E+00	0.00E+00	0.00E+00	0.00E+00	2.48E+06	0.00E+00	0.00E+00	0.00E+00	- 0.63
1	10	1.44E-02	5.77E+01	0.00E+00	0.00E+00	2.18E+07	0.00E+00	0.00E+00	0.00E+00	0.00E+00	1.19E+07	0.00E+00	0.00E+00	- 0.58
1	13	1.42E-02	4.14E+01	0.00E+00	0.00E+00	0.00E+00	0.00E+00	3.35E+06	0.00E+00	0.00E+00	0.00E+00	3.42E+06	0.00E+00	- 0.48
1	15	1.16E-70	1.95E+02	9.29E+06	1.00E+07	1.35E+07	6.18E+06	7.61E+06	0.00E+00	4.70E+06	3.66E+06	2.53E+07	1.39E+07	- 0.43
1	14	6.03E-03	5.39E+01	0.00E+00	1.09E+07	0.00E+00	0.00E+00	0.00E+00	0.00E+00	0.00E+00	5.61E+07	0.00E+00	0.00E+00	- 0.43
17	30	4.69E-02	5.28E+01	0.00E+00	1.27E+08	0.00E+00	0.00E+00	0.00E+00	0.00E+00	0.00E+00	7.32E+06	0.00E+00	0.00E+00	- 0.41
1	8	2.56E-02	4.31E+01	2.06E+06	0.00E+00	0.00E+00	0.00E+00	0.00E+00	0.00E+00	0.00E+00	0.00E+00	2.46E+04	0.00E+00	- 0.39
1	12	7.70E-03	5.76E+01	0.00E+00	0.00E+00	0.00E+00	0.00E+00	0.00E+00	2.15E+06	0.00E+00	0.00E+00	0.00E+00	1.88E+06	- 0.36
1	10	1.12E-13	1.67E+02	6.08E+07	4.78E+07	2.75E+08	8.53E+07	8.26E+07	1.40E+07	2.22E+07	1.15E+08	6.72E+07	3.94E+08	- 0.35
2	8	1.12E-02	9.81E+01	0.00E+00	0.00E+00	6.79E+06	0.00E+00	2.59E+07	3.41E+06	0.00E+00	1.12E+08	0.00E+00	1.74E+07	- 0.31
59	68	5.05E-03	6.64E+01	0.00E+00	0.00E+00	4.65E+06	0.00E+00	0.00E+00	2.35E+07	0.00E+00	0.00E+00	0.00E+00	0.00E+00	- 0.24
2	10	9.74E-03	6.84E+01	0.00E+00	0.00E+00	0.00E+00	0.00E+00	5.35E+06	0.00E+00	0.00E+00	2.50E+07	0.00E+00	0.00E+00	- 0.11
1	17	1.30E-02	3.95E+01	0.00E+00	0.00E+00	0.00E+00	0.00E+00	0.00E+00	0.00E+00	0.00E+00	0.00E+00	3.35E+06	2.63E+06	- 0.09
87	105	3.46E-02	4.15E+01	2.54E+06	0.00E+00	0.00E+00	0.00E+00	0.00E+00	4.47E+06	0.00E+00	0.00E+00	0.00E+00	8.41E+06	- 0.04
1	16	2.37E-03	6.24E+01	0.00E+00	0.00E+00	0.00E+00	0.00E+00	0.00E+00	3.45E+06	1.33E+06	0.00E+00	0.00E+00	0.00E+00	0.00
2	20	2.48E-08	1.23E+02	5.64E+06	0.00E+00	0.00E+00	8.53E+06	1.42E+07	6.40E+06	4.33E+06	1.04E+08	9.02E+06	0.00E+00	0.01
1	7	5.90E-03	9.63E+01	4.62E+06	0.00E+00	0.00E+00	0.00E+00	0.00E+00	0.00E+00	0.00E+00	0.00E+00	1.03E+07	0.00E+00	0.03
2	16	3.27E-12	1.34E+02	6.44E+07	2.26E+07	1.56E+07	0.00E+00	2.16E+07	3.13E+07	3.84E+07	5.99E+08	7.42E+07	4.01E+07	0.05
20	38	3.15E-03	4.67E+01	3.53E+07	1.47E+07	0.00E+00	4.93E+06	2.90E+06	0.00E+00	0.00E+00	0.00E+00	0.00E+00	0.00E+00	0.11

(continued)

Table 1
(continued)

Transcript	Gene symbol	chromosome	Start	End	Localization	Start codon	Proteins	Sequence	Length	Missed cleavages	Mass (Da)
ENST00000553189	TMTC1	chr12	29755788	29755865	Extracellular	TTG	pep-alt-TMTC1	LGEVSLPR	8	0	869.49707
ENST00000570589	SCPEP1	chr17	56987719	56997060	Cytoplasm	GTG	pep1-alt-SCPEP1; pep2-d-SCPEP1	VAEQVLNAVNK	11	0	1183.6561
ENST00000520721	IGHV3-22	chr14	106257881	106258054	Mitochondrion	CTG	pep-alt-IGHV3-22	LVESGGGLVQPGGSLR	16	0	1524.826
ENST00000505649	ALB	chr4	73413540	73418100	Mitochondrion	ATG	pep8-alt-ALB	LPSLAADFVESKDVCK	16	1	1777.892
ENST00000505649	ALB	chr4	73413552	73418100	Mitochondrion	TTG	pep8-alt-ALB	LPSLAADFVESK	12	0	1275.6711
ENST00000476441	ALB	chr4	73417565	73420295	Mitochondrion	GTG	pep3-d-ALB	VSTPTLVEVSR	11	0	1186.6558
ENST00000517927	MIR3142HG	chr5	160486529	160486699	Cytoplasm	ATG	pep1-nc-MIR3142HG	ISVNLTK	7	0	773.4647
ENST00000604178	AC134879.2	chrY	11473740	11474039	Cytoplasm	ATG	pep-alt-AC134879.2	LLIYAASSLQSGIPTR	16	0	1688.9461
ENST00000520721	IGHV3-22	chr14	106257881	106258054	Mitochondrion	CTG	pep-alt-IGHV3-22	GLEWVGFIR	9	0	1075.5815
ENST00000425591	AC006019.1	chr7	154058438	154058689	Nucleus	TTG	pep3-nc-AC006019.1	ILGSTMWVLSWGGK	14	0	1533.8014
ENST00000659729	AL137139.2	chr13	99853850	99853933	Extracellular	ATG	pep5-nc-AL137139.2	ILLKMSR	7	1	859.53134
ENST00000570589	SCPEP1	chr17	56987719	56997060	Cytoplasm	GTG	pep1-alt-SCPEP1	MAAGIGLELYK	11	0	1164.6213
ENST00000420195	SH3BP5-AS1	chr3	15254638	15254706	Cytoplasm	CTG	pep-nc-SH3BP5-AS1	LCLLIAHSFSNSK	13	0	1488.7759
ENST00000546453	KRT79	chr12	52821875	52824287	Cytoplasm	CTG	pep1-alt-KRT79	LDSIIAEVK	9	0	986.56481
ENST00000566159	UBE2I	chr16	1314579	1314974	Extracellular	GTG	pep-alt-UBE2I	VTAVSLPR	8	0	841.50215
ENST00000289779	AL591806.2	chr1	160998874	161001083	Cell membrane	GTG	pep-d-AL591806.2	VTFLPTGITFK	11	0	1222.6962
ENST00000606379	AL137127.1	chr1	19073012	19073110	Mitochondrion	TTG	pep-nc-AL137127.1	LLLLCVCGCPLSAGLR	16	0	1800.9413

Start position	End position	PEP	Score	Intensity AC381	Intensity AC384	Intensity AC386	Intensity AC392	Intensity AC399	Intensity UT805	Intensity UT812	Intensity UT815	Intensity UT821	Intensity UT822	GRAVY
1	8	2.11E-04	1.31E+02	3.03E+08	4.45E+08	7.57E+08	1.14E+08	2.40E+08	1.27E+08	6.28E+08	1.24E+08	5.93E+08	5.57E+08	0.13
120	130	7.94E-06	1.00E+02	0.00E+00	0.00E+00	0.00E+00	4.93E+06	0.00E+00	0.00E+00	0.00E+00	0.00E+00	1.24E+07	0.00E+00	0.19
1	16	4.25E-12	1.35E+02	1.74E+07	1.29E+07	2.08E+07	1.65E+07	1.62E+07	0.00E+00	2.87E+07	1.92E+07	9.11E+07	2.00E+07	0.19
1	16	1.06E-04	1.03E+02	0.00E+00	0.00E+00	0.00E+00	0.00E+00	4.54E+06	0.00E+00	0.00E+00	4.62E+07	0.00E+00	0.00E+00	0.21
1	12	2.64E-16	1.53E+02	2.19E+07	2.27E+07	2.41E+07	3.75E+07	9.31E+07	1.60E+07	6.39E+07	4.23E+08	1.21E+08	7.15E+07	0.34
1	11	1.08E-02	6.18E+01	0.00E+00	0.00E+00	0.00E+00	0.00E+00	7.35E+06	0.00E+00	0.00E+00	2.46E+07	0.00E+00	5.19E+06	0.35
2	8	2.28E-03	8.78E+01	0.00E+00	3.55E+07	0.00E+00	9.21E+06	1.88E+07	0.00E+00	1.10E+08	1.16E+08	1.02E+08	1.10E+08	0.51
43	58	1.58E-02	4.95E+01	0.00E+00	0.00E+00	1.02E+07	0.00E+00	0.00E+00	0.00E+00	4.57E+07	0.00E+00	0.00E+00	0.00E+00	0.60
41	49	7.38E-06	1.43E+02	8.04E+06	4.09E+06	8.06E+06	0.00E+00	0.00E+00	0.00E+00	6.05E+06	0.00E+00	8.37E+06	3.30E+06	0.62
15	28	1.76E-02	4.60E+01	4.18E+06	0.00E+00	2.45E+05	0.00E+00	0.00E+00	0.00E+00	0.00E+00	0.00E+00	0.00E+00	0.00E+00	0.64
18	24	5.21E-02	4.16E+01	3.45E+09	0.00E+00	0.00E+00	0.00E+00	4.25E+05	0.00E+00	0.00E+00	3.23E+08	0.00E+00	0.00E+00	0.69
59	69	4.93E-02	4.94E+01	0.00E+00	0.00E+00	0.00E+00	0.00E+00	0.00E+00	1.15E+06	0.00E+00	5.68E+08	0.00E+00	0.00E+00	0.74
1	13	0.00E+00	0.00E+00	0.00E+00	2.36E-02	4.43E+01	3.86E+06	0.00E+00	0.00E+00	0.00E+00	0.00E+00	4.11E+07	0.00E+00	0.77
1	9	0.00E+00	0.00E+00	0.00E+00	3.83E-02	5.26E+01	0.00E+00	2.58E+07	7.68E+05	0.00E+00	0.00E+00	0.00E+00	0.00E+00	0.79
1	8	0.00E+00	0.00E+00	0.00E+00	5.07E-03	9.08E+01	2.50E+09	0.00E+00	0.00E+00	0.00E+00	0.00E+00	4.89E+09	0.00E+00	0.80
28	38	3.43E-03	0.00E+00	0.00E+00	0.00E+00	0.00E+00	0.00E+00	0.00E+00	7.57E+01	3.99E+06	2.31E+06	5.68E+06	3.11E+06	0.92
1	16	0.00E+00	0.00E+00	0.00E+00	0.00E+00	0.00E+00	0.00E+00	0.00E+00	5.03E-03	5.02E+01	0.00E+00	3.68E+07	6.12E+06	1.79

hydropathy (GRAVY) score; Missed Cleavages: Evaluated to assess the efficiency of tryptic digestion; and Intensity Values: Quantified across different conditions to identify differential expression patterns.

3.5 Pathway Enrichment Analysis

3.5.1 Functional Annotation

Identified micropeptides were annotated based on their involvement in:

- Cellular components (e.g., mitochondria, cytoplasm, and extracellular space).
- Molecular functions (e.g., binding activity and catalytic function).
- Biological processes (e.g., immune response, apoptosis, and metabolic regulation).

3.5.2 Pathway Mapping

Pathway analysis was conducted using:

- KEGG (Kyoto Encyclopedia of Genes and Genomes) for pathway classification and disease relevance.
- STRING database for protein–protein interaction prediction and functional enrichment.

3.6 Spectral Data Analysis and Validation

3.6.1 LC–MS/MS Setup and Data Acquisition

Peptide samples were analyzed using:

- C18 reverse-phase chromatography for peptide separation.
- High-resolution Orbitrap mass spectrometry for data-dependent acquisition (DDA) of peptide spectra.

3.6.2 Spectral Matching and Software Tools

- Fragment ion spectra were matched to predicted peptides using MaxQuant (v2.1.0.0) and PEAKS Studio for both database-dependent and de novo sequencing.

3.6.3 Validation Criteria

To ensure high-confidence identifications, the following criteria were applied:

- False Discovery Rate (FDR) $\leq 1\%$ at both peptide and protein levels.
- Peptide score > 50.
- Posterior Error Probability (PEP) < 0.05.

3.7 Peptide Quantification

Normalized intensity values were calculated to quantify peptide abundance. Differential expression analysis was performed to identify micropeptides with significant expression changes across prostate cancer (PCa) and control samples.

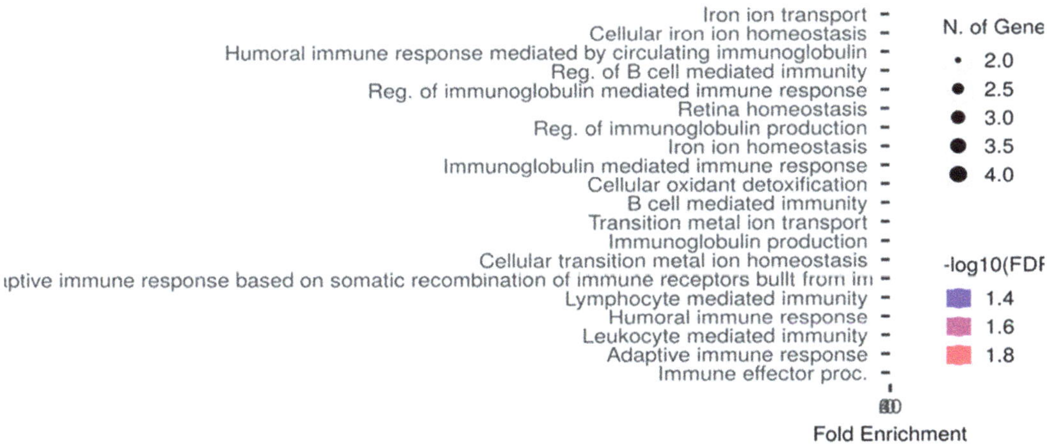

Fig. 3 Biological meaning through pathway enrichment analysis

3.8 Functional Analysis of Immune and Iron-Related Pathways

Targeted enrichment analysis was performed to investigate the involvement of the identified micropeptides in immune and iron-related biological processes. The results revealed significant enrichment in immune response pathways, including those associated with humoral immunity, B-cell activation, and adaptive immune mechanisms. In parallel, several micropeptides were mapped to pathways involved in iron metabolism, such as iron ion transport, iron homeostasis, and cellular metal ion regulation. These findings align with the known biological demand of tumor cells for iron to support their rapid proliferation. Furthermore, the data suggest a complex interplay between immune modulation and iron metabolism, which may contribute to prostate cancer (PCa) progression. Additionally, enrichment of oxidative stress-related pathways, particularly those involved in the detoxification of reactive oxygen species (ROS), was observed, highlighting potential mechanisms of redox balance and cellular adaptation in the tumor microenvironment (Fig. 3).

4 Notes

1. Ensure all urine samples are collected at the same time of day (e.g., first-morning void) to minimize variability due to circadian rhythms and hydration levels.
2. Consistently maintain the centrifugation speed and temperature (4000× *g* for 20 min at 4 °C) to ensure reproducibility. Deviations may lead to incomplete removal of particulates or loss of low-abundance proteins.
3. Immediately process and freeze urine samples after collection to minimize protein degradation. Delays can lead to the breakdown of peptides and proteins, affecting the integrity of the

results. Avoid repeated freeze-thaw cycles, as this can degrade proteins and peptides, impacting data quality.

4. Monitor the concentration process carefully; overconcentration may result in protein precipitation.
5. Dilute samples appropriately to ensure measurements fall within the linear range of the DC™ Protein Assay Kit (Bio-Rad). High concentrations may saturate the assay, leading to inaccurate quantification.
6. Thoroughly denature and reduce proteins to ensure efficient tryptic digestion. Incomplete reduction and alkylation can lead to missed cleavages, affecting peptide identification. When performing tryptic digestion, ensure the enzyme-to-substrate ratio (1:50 w/w) is maintained for optimal digestion efficiency.
7. Ensure complete extraction of peptides from gel pieces by performing multiple extraction steps with 50% acetonitrile/5% formic acid. Avoid over-drying peptides during vacuum concentration, as this may lead to difficulties in resuspension.
8. Regularly clean and calibrate the liquid chromatography and mass spectrometry systems to avoid signal loss and ensure high-resolution data acquisition.
9. Use multiple software platforms (e.g., MaxQuant and PEAKS Studio) for comprehensive peptide identification and validation. Apply a stringent 1% false discovery rate (FDR) to minimize false positives and enhance data reliability.
10. Use the most up-to-date databases (e.g., UniProt, HMPA, FuncPEP, and SmProt) to improve peptide identification and minimize false negatives.

Acknowledgments

This work was supported by the Portuguese Foundation for Science and Technology (FCT), European Union, QREN, FEDER, and COMPETE for funding the Institute of Biomedicine (iBiMED; UIDB/04501/2020, POCI-01-0145-FEDER-007628), LAQV-REQUIMTE (UIDB/50006/2020 and UIDP/50006/2020), Instituto Português de Oncologia do Porto Francisco Gentil, EPE (CI-IPOP-134-2020).

Declaration of Competing Interest The authors declare that there are no known competing financial interests or personal relationships that could have appeared to influence the work reported in this paper.

Ethics Statement The study was accepted by the IPO-Porto Ethics Committee (Comissão de Ética para a Saúde, Reference 282R/2017) and performed according to Declaration of Helsinki.

Data Availability Statement The MS proteomics data have been deposited in the ProteomeXchange Consortium via the PRIDE [30] partner repository with the data set identifier PXD017902.

References

1. Lima T et al (2022) Application of proteogenomics to urine analysis towards the identification of novel biomarkers of prostate cancer: an exploratory study. Cancer 14(8):2001
2. Lima T et al (2023) A peptide-centric approach to analyse quantitative proteomics data- an application to prostate cancer biomarker discovery. J Proteome 272:104774. https://doi.org/10.1016/j.jprot.2022.104774
3. Elguoshy A et al (2024) Investigating and annotating the human peptidome profile from urine under normal physiological conditions. Proteomes 12(3):18
4. Yuanyuan J, Xinqiang Y (2022) Micropeptides identified from human genomes. J Proteome Res 21(4):865–873. https://doi.org/10.1021/acs.jproteome.1c00889
5. Fijalkowski I et al (2022) Hidden in plain sight: challenges in proteomics detection of small ORF-encoded polypeptides. microLife:3. https://doi.org/10.1093/femsml/uqac005
6. Zhu M, Gribskov M (2019) MiPepid: micro-Peptide identification tool using machine learning. BMC Bioinform 20(1):559. https://doi.org/10.1186/s12859-019-3033-9
7. Su D et al (2024) Ribosome profiling: a powerful tool in oncological research. Biomark Res 12(1):11. https://doi.org/10.1186/s40364-024-00562-4
8. Hao Y et al (2018) SmProt: a database of small proteins encoded by annotated coding and non-coding RNA loci. Brief Bioinform 19(4):636–643. https://doi.org/10.1093/bib/bbx005
9. Choteau SA et al (2021) MetamORF: a repository of unique short open reading frames identified by both experimental and computational approaches for gene and metagene analyses. Database 2021. https://doi.org/10.1093/database/baab032
10. Murgan SS et al (2020) Ovarian cancer: tumor-specific urinary micro-peptides profiling as potential biomarkers for early diagnosis. Proteomes 8(4):32
11. Namekawa T, Ikeda K, Horie-Inoue K, Inoue S (2019) Application of prostate cancer models for preclinical study: advantages and limitations of cell lines, patient-derived xenografts, and three-dimensional culture of patient-derived cells. Cells 8(1). https://doi.org/10.3390/cells8010074
12. Zhou H et al (2024) Micropeptides: potential treatment strategies for cancer. Cancer Cell Int 24(1):134. https://doi.org/10.1186/s12935-024-03281-w
13. Fernando A et al (2023) Identification of a micropeptide linked to cancer stem cell regulation and chemoresistance. bioRxiv:2023.2003.2014.532696. https://doi.org/10.1101/2023.03.14.532696
14. Richmond A, Su Y (2008) Mouse xenograft models vs GEM models for human cancer therapeutics. Dis Model Mech 1(2–3):78–82. https://doi.org/10.1242/dmm.000976
15. Vrbnjak K, Sewduth RN (2024) Multi-omic approaches in cancer-related micropeptide identification. Proteomes 12(3):26
16. Ingolia NT (2016) Ribosome footprint profiling of translation throughout the genome. Cell 165(1):22–33. https://doi.org/10.1016/j.cell.2016.02.066

Chapter 20

Using Untargeted Mass Spectrometry-Based Proteomics to Identify Putative Microprotein Biomarkers from Patient Tampons

Gordon T. Luu, Conrad Fihn, and Laura M. Sanchez

Abstract

Mass spectrometry-based proteomics is a versatile technique that facilitates the study of microproteins as biomarkers and potential translational applications by utilizing liquid chromatography–trapped ion mobility–tandem mass spectrometry (LC–TIMS–MS/MS) and matrix assisted laser desorption/ionization-time-of-flight mass spectrometry (MALDI–TOF MS). Importantly, MS has been shown to be compatible with complex biological samples (i.e., mammalian biofluids and patient samples). Here we describe a workflow combining: (1) top–down proteomics for analysis of intact microproteins via MALDI–TOF MS and (2) bottom–up proteomics for analysis of enzymatically digested microproteins via LC–TIMS–MS/MS to annotate putative microproteins of interest from a complex biological sample: patient-derived tampons.

Key words Microproteins, MALDI, Mss spectrometry, LC–MS/MS, Bioinformatics, Cheminformatics

1 Introduction

Microproteins (also known as peptides or micropeptides) are small proteins (typically less than or equal to 100–150 amino acids in length) that are translated from mRNA originating from small open reading frames (sORFs) [1]. They remain understudied due to historic rules in which ORFs under 300 base pairs were considered noncoding RNAs to prevent false positive ORF assignment [2–4]. As a result, the biological significance of microproteins has yet to be fully elucidated; recent work has shown that they are often involved in regulatory functions across a variety of cellular domains [5]. Additionally, microproteins are often less abundant than larger proteins [6, 7], and in some cases, poorly conserved across evolutionarily related species due to being evolutionarily young [8, 9], presenting difficulty for those studying their mechanisms of action. However, advances in different omics technologies (i.e., ribosome

Stephan Wenkel (ed.), *Microproteins: Methods and Protocols*, Methods in Molecular Biology, vol. 2992,
https://doi.org/10.1007/978-1-0716-5013-4_20,

profiling, proteogenomics, and mass spectrometry) in conjunction with software developments have allowed for further study of sORFs and subsequently microproteins [10]. In this chapter, we focus on the use of mass spectrometry as a technique for microprotein biomarker discovery and potential translational applications.

Mass spectrometry (MS) is a versatile technique that facilitates the interrogation of various "-omes" of an organism. Furthermore, due to the regulatory nature of many microproteins, MS provides information on how changes to these biological processes (i.e., in different disease states) at different stages of the central dogma can affect an organism. Through the lens of MS, new insights into microproteins have resulted in discoveries in the basic functionality of various biological systems as well as revealing diagnostic and therapeutic applications. MS-based proteomics utilizes liquid chromatography-tandem mass spectrometry (LC–MS/MS) to analyze proteins (top–down proteomics) or peptides (bottom–up proteomics). Pan et al. have demonstrated the use of bottom–up proteomics to annotate tissue-specific mouse microproteins from the brain, heart, liver, kidney, and spleen and posited the functions of select microproteins [11]. Similarly, Cassidy et al. have demonstrated the utility of top–down proteomics in microprotein studies by annotating 36 proteoforms derived from 12 microproteins in the archaea *Methanosarcina mazei* in addition to highlighting optimizations to improve the top–down workflow [12]. Lastly, Li et al. demonstrated the complementarity of bottom–up and top–down proteomics to annotate 87 microproteins and detect 570 nonannotated microproteins associated with various stages of the mitotic cell cycle in human Hep3B liver cells [13]. Furthermore, advances in instrumentation such as the inclusion of ion mobility–mass spectrometry (IM–MS) have allowed for further development of separation and analysis methodologies in proteomics [14–16]. However, despite the utility of LC–MS/MS based proteomics, two significant challenges were posed as noted by Vitorino et al. and Yeasmin et al. [6, 7]: (1) proteins that are not abundant, such as microproteins, can evade detection during analysis and (2) this is then further compounded by the need for (sometimes extensive) sample preparation for compatibility with MS hardware, resulting in the loss of biomaterial and further reducing the amount of microproteins available for MS analysis.

Matrix-assisted laser desorption/ionization time-of-flight (MALDI-TOF) MS-based proteomics can alleviate the above challenges due to the relatively miniscule amount of sample preparation required prior to analysis. MALDI-TOF MS has had a long history of use in both bottom–up and top–down proteomics [17], and more recently, it has been increasingly adopted as a screening technique. For example, clinical chemists have adopted the use of biotyping to be able to quickly, efficiently, and effectively diagnose patients presenting with infectious agents to determine the genera, and sometimes species, of the infectious microorganism in a more specific and time-sensitive manner. Biotyping works through profiling a subset of intact microbial microproteins in the 2–20 kDa range, which includes highly conserved ribosomal proteins that prove useful for microbial strain identification. Intact (micro)protein profiles in the form of mass spectra can then be compared to databases containing validated spectra derived from microbial isolates to identify the infectious agents grown from patient samples. Faron et al. and Wilson et al. have previously demonstrated these workflows with the intention of gaining FDA clearance for clinical use [18, 19]. MALDI profiling has also been applied to plasma cell dysplasia through the 'Mass-Fix' workflow [20–22]. In Mass-Fix, a modified microprotein mass range (9–32 kDa) is interrogated to screen for plasma cell dyscrasias. Following acquisition and analysis of patient samples, the Mass-Fix workflow has been able to provide further insight into M proteins involved in plasma cell dyscrasias through retrospective analysis [23, 24]. As shown in these workflows, MALDI–TOF MS is a powerful technique to analyze microproteins, which fall into the "Goldilocks" zone of mass-to-charge ratio (*m/z*) values in which MALDI is well suited to analyze. Unlike ESI, intact proteins are not prone to generating multiply charged ions, instead appearing as singly charged ions in MALDI spectra [25]. As a result, smaller mass ranges that are prone to convolution due to background (i.e., MALDI matrix) signals can be omitted during data acquisition. Furthermore, microproteins are small enough that the inherent disadvantages of the microchannel plate (MCP) detector used by MALDI–TOF MS instruments for high masses are generally a nonissue [26–30].

Recently, we have sought to apply a similar MALDI–TOF MS-based workflow to ovarian cancer screening [31, 32]. Vaginal lavages from a murine model of ovarian cancer were analyzed using MALDI–TOF MS to acquire intact microprotein profiles. Combined with orthogonal LC–MS/MS analysis, six microproteins were putatively annotated, and cystatin A was shown to be a potential biomarker for ovarian cancer screening. The workflow was then applied to deidentified patient derived tampons as a proof of concept, which allowed for the putative annotation of cystatin A among other as yet to be annotated microprotein signals. In this chapter, the workflow is described in detail as it was applied to the patient derived tampons. Data acquisition will be covered with a focus on MALDI–TOF MS and LC–TIMS–MS/MS mass spectrometers, which utilizes trapped ion mobility spectrometry (TIMS) to allow for separation of isomers and isobars. Data analysis will cover open source software where available. Importantly, while the sample preparation will primarily focus on patient derived tampons, it should be noted that data acquisition and analysis methodologies are not limited to such samples and could apply to a number of complex biofluids.

2 Materials

2.1 Extraction of Microproteins from Tampons

1. Patient tampons.
2. Sterile work area (i.e., biosafety cabinet).
3. 50 mL conical tubes.
4. Dry ice.
5. 20 mL plastic syringe.
6. Razor blade or box cutter or knife.
7. Forceps.
8. LC–MS grade or 18.2 MΩ·cm Milli-Q deionized H_2O.
9. −80 °C freezer.
10. 15 mL conical tubes.

2.2 Removal of Unlysed Cells and Larger Proteins from Protein Extracts

1. P1000 micropipette and pipette tips.
2. LC–MS grade or 18.2 MΩ·cm Milli-Q deionized H_2O.
3. Amicon ultra 30 kDa centrifugal filter devices (MilliporeSigma).
4. Centrifuge.
5. 20 mL scintillation vials.
6. −20 °C freezer.
7. Lyophilizer.
8. −80 °C freezer.

2.3 Bicinchoninic Acid (BCA) Assay

1. P10, P200, and P1000 micropipette and pipette tips.
2. Pierce BCA Protein Assay Kit (Thermo Fisher Scientific, Catalog Number 23225).
3. LC–MS grade or 18.2 MΩ·cm Milli-Q deionized H_2O.
4. 96 well plate.
5. Microplate reader.
6. Scintillation vials.

2.4 MALDI–TOF MS

1. P2 or P10 micropipette (single or multichannel) and pipette tips.
2. Sinapinic acid.
3. LC–MS grade acetonitrile (ACN).

4. LC–MS grade or 18.2 MΩ·cm Milli-Q deionized H_2O.
5. LC–MS grade trifluoroacetic acid (TFA).
6. 20 mg/mL sinapinic acid in 70:30 ACN:Milli-Q H_2O w/ 0.1% TFA.
7. Ice or 4 °C refrigerator.
8. 384 spot MALDI target plate.
9. MALDI target plate adapter.
10. MALDI calibrant for microprotein range (i.e., red phosphorus and protein calibration standard I).
11. MALDI matrix for intact proteins (20 mg/mL sinapinic acid in 70:30 ACN:Milli-Q H_2O with 0.1% TFA).
12. flexControl method optimized for microprotein range.
13. MALDI–TOF mass spectrometer.
14. Acquisition computer with flexControl installed.

2.5 Data Analysis

1. Data processing computer.
2. R.
3. RStudio or integrated development environment (IDE) of choice.
4. MALDIquant.
5. Web browser (for MetaboAnalyst).
6. ProteoWizard MSConvert.

2.6 From Protein Extracts to Protein Annotation

1. P10, P200, and P1000 micropipette and pipette tips.
2. S-Trap Micro Column (ProtiFi).
3. 1.5 mL microcentrifuge tubes.
4. Lyophilizer.
5. 10% sodium dodecyl sulfate (SDS) solution with protease inhibitor cocktail (Roche, 2.6.6 Manufacturer Number 11697498001).
6. Dithiothreitol (DTT).
7. Heating block.
8. Iodoacetamide.
9. Microcentrifuge.
10. 12% aqueous phosphoric acid.
11. triethylammonium bicarbonate in 90:10 methanol:Milli-Q H_2O (S-Trap buffer).
12. Digestion enzyme (i.e., trypsin, LysC, etc.).
13. Water bath.

14. 50 mM triethylammonium bicarbonate in LC–MS grade or 18.2 MΩ·cm Milli-Q deionized H_2O.
15. 0.2% formic acid in LC–MS grade or 18.2 MΩ·cm Milli-Q deionized H_2O.
16. 60:40 ACN:Milli-Q H_2O.
17. Speedvac with microcentrifuge adapter.

2.7 Desalting and Cleanup of Enzymatic Digests for MS Analysis

1. P10 and P200 micropipette and pipette tips.
2. 0.1% formic acid in LC–MS grade or 18.2 MΩ·cm Milli-Q deionized H_2O.
3. C18 Resin ZipTip pipette tips (Millipore Sigma).
4. 1.5 mL microcentrifuge tubes.
5. 80:20 ACN:Milli-Q H_2O w/ 0.1% TFA.
6. 40:60 ACN:Milli-Q H_2O w/ 0.1% TFA.
7. 5:95 ACN:Milli-Q H_2O w/ 0.1% TFA.
8. Micropipette.
9. Solvent waste receptacle.
10. Speedvac with microcentrifuge adapter.

2.8 Acquisition of Bottom–up Proteomics Datasets Using LC–TIMS–MS/MS

1. P10, P200, and P1000 micropipette and pipette tips.
2. LC–MS grade ACN.
3. LC–MS grade or 18.2 MΩ·cm Milli-Q deionized H_2O.
4. LC–MS grade formic acid.
5. 1.5 mL microcentrifuge tubes.
6. Microcentrifuge.
7. 0.1% formic acid in LC–MS grade or 18.2 MΩ·cm Milli-Q deionized H_2O.
8. Glass LC vials and caps.
9. LC glass inserts (optional, used if the available sample volume is low).
10. Pierce HeLa Protein Digest Standard (Thermo Fisher Scientific, Catalog Number 88328) or equivalent quality control sample.
11. ESI-L LCMS Tuning Solution (Agilent Technologies Inc., Part Number G1969–85000).
12. C18 UPLC column, i.e., Phenomenex Luna Omega C18 column (particle size: 1.6 μm, pore size: 100 A, inner diameter: 2.1 mm, and length: 150 mm).
13. LC–MS.
14. Acquisition computer with Compass HyStar and timsControl installed.

2.9 Validation and Annotation of Bottom-up Proteomics LC–TIMS–MS/MS Datasets

1. Data processing computer.
2. FragPipe.
3. Python or R.
4. RStudio or PyCharm or integrated development environment (IDE) of choice.

3 Methods

3.1 Extraction of Microproteins from Tampons

The following protocol has been optimized to balance ease of sample preparation with untargeted protein extraction efficiency. Due to the potential presence of human blood, samples are treated as biosafety level 2 (BSL-2) and all work should be performed in a biosafety cabinet.

1. Patient tampons should be stored in 50 mL conical tubes at −80 °C prior to preparation for mass spectrometric analysis, detailed below.
2. Remove the tampons from the −80 °C freezer and transport them to the biosafety cabinet or sterile benchtop for dissection. Transportation should occur with dry ice to prevent excessive thawing of the tampon. Depending on the amount of frozen fluid (i.e., blood, mucus, and vaginal fluid) on the tampon, a minor amount of thawing is required due to the difficulty of cutting through ice, but too much thawing can result in an excessive amount of biofluid loss and fraying of the tampon resulting in more difficulty when cutting the tampon (*see* **Notes 1** and **2**).
3. Prepare a 20 mL plastic syringe by removing the plunger and placing the syringe tube on an appropriately sized plastic rack.
4. Cut the top ¼ of the tampon using a razor blade, box cutter, or knife and place the tampon section inside the 20 mL plastic syringe using forceps (*see* **Note 3** and **4**).
5. Replace the plunger into the syringe but do not compress the tampon section yet.
6. Replace the blade of the cutting utensil or sterilize the blade by dipping it in 70% ethanol and carefully wiping it. Sterilize the forceps by dipping it in 70% ethanol and carefully wiping it. Alternatively, ensure that a sterile pair of forceps is available per tampon.
7. Repeat **steps 2–6** for each tampon sample to be prepared. To prevent excessive thawing, as noted above, it can be helpful to remove the tampons from the −80 °C freezer in batches if a large number of samples are to be processed. However, care should still be taken to prevent excessive opening and closing of the freezer in a short amount of time to prevent excessive frost buildup.

8. Use the syringe to draw up 2–5 mL of LC–MS grade or 18.2 MΩ·cm Milli-Q deionized H_2O to extract cells and proteins from the tampon. Water is used to induce osmosisc-based cell lysis. While water may result in less efficient protein extraction compared to other commonly used protein extraction buffers (i.e., phosphate buffered saline; PBS, tris (hydroxymethyl)aminomethane; Tris, 4-(2-Hydroxyethyl)-1-piperazineethanesulfonic acid; HEPES, ammonium bicarbonate, triethanolamine bicarbonate), avoiding the use of buffers prevents the need for additional sample cleanup prior to MALDI-TOF MS analysis [33] (*see* **Note 5**).
9. Following a 10 min incubation, compress the tampon to extract a crude extract containing cells and proteins into a 15 mL conical tube.
10. Repeat **steps 8** and **9** for each sample. This second round of biomaterial extraction can be added to the conical tube used in the first round of extraction. For tampons with large amounts of biomaterial, more than 2 rounds of extraction can be performed, but sample filtration and concentration may take longer due to the increased sample volume.
11. Once complete, ensure that the biosafety cabinet surface, plastic racks, forceps, and blades are properly sterilized and/or disposed of.

3.2 Removal of Unlysed Cells and Larger Proteins from Protein Extracts

Following protein extraction, non-protein biomass and proteins that are larger than the desired mass range of interest should be removed to improve sensitivity of microprotein detection. While whole cells and larger proteins can be profiled via MALDI-TOF MS [34], their inclusion in the extract can result in lower concentration, and therefore sensitivity, of microproteins.

1. Dilute each crude extract with LC–MS grade or 18.2 MΩ·cm Milli-Q deionized H_2O to 4 mL (or an appropriate volume; *see* **Note 6**).
2. Pre-wet the filters in the Amicon Ultra 30 kDa centrifugal filter devices (MilliporeSigma) by adding 4 mL (or an appropriate volume) of LC–MS grade or 18.2 MΩ·cm Milli-Q deionized H_2O to the filter and centrifuging at 14,000× *g* for 10 min per the manufacturer documentation. Be sure to note the orientation of the filters. Amicon Ultra centrifugal filter devices are manufactured as an insert with a V shape containing 2 filter membranes. Each filter membrane should be facing the inside of the rotor in the same orientation as shown in Fig. 1, regardless of filter device capacity.

Fig. 1 Optimal orientation of the Amicon Ultra 0.5 mL centrifugal filter. Amicon Ultra centrifugal filters should be oriented as shown, where the filter membrane is facing the center of the rotor

3. Transfer diluted crude extracts from the 15 mL conical tubes to Amicon Ultra 30 kDa centrifugal filter devices (Millipore-Sigma). As noted in Step 2, be sure to note the orientation of the filters.
4. Centrifuge at 14,000× *g* for 10 min per the manufacturer documentation. The filtrate, which is collected into the conical tube provided by the Amicon kit, will contain proteins less than or approximately equal to 30 kDa and will be the primary solution of interest. Please note that Amicon Ultra centrifugal filters are not hard cutoffs and some proteins just slightly larger than the stated filter size can pass through the filter depending on the centrifugation parameters and duration, resulting in their presence in the filtrate.
5. Collect the retentate by inverting the filters in a new, clean conical tube provided by the Amicon kit and centrifuging at 1000× *g* for 2 min per the manufacturer documentation. It will contain proteins larger than 30 kDa and unlysed cells, which can be stored for future analyses.
6. Transfer all filtrate and retentate samples to appropriately sized scintillation vials (i.e., 20 mL) in which the filtrate and retentate samples do not fill more than two-thirds of the vial. Label all vials appropriately.
7. Freeze all filtrate and retentate samples at −20 °C.
8. Once frozen, concentrate all filtrate and retentate samples via lyophilization and store the concentrated filtrate and retentate samples long term at −80 °C.

3.3 Determining Protein Concentration for Protein Extract Normalization Using the Bicinchoninic Acid (BCA) Assay

While not strictly necessary, determining protein concentration prior to intact microprotein analysis via MS is recommended since extracts may contain biological material in addition to proteins (i.e., metabolites, lipids, cells, etc.) Performing normalization to the protein concentration as opposed to only the biomass ensures accurate relative quantitation.

1. Remove crude protein extracts containing the microproteins (filtrate from previous section) from the −80 °C freezer.
2. Dilute each crude protein extract with LC–MS grade or 18.2 MΩ·cm Milli-Q H_2O to a concentration of 10 mg/mL.
3. Use the Pierce BCA Protein Assay Kit (Thermo Fisher Scientific, Catalog Number 23225) to determine the protein concentration in each 10 mg/mL extract per the manufacturer's instructions.
4. Dilute crude protein extracts with LC–MS grade or 18.2 MΩ·cm Milli-Q H_2O as needed to normalize the protein concentration in each extract (i.e., 10 μg/μL). Normalizing the protein concentration of all extracts ensures that relative quantitation from the following MS experiments can be performed with confidence (See Note 7).

3.4 Acquisition of Intact Protein Profiles from Tampon Protein Extracts via MALDI-TOF MS

MALDI-TOF MS in linear mode facilitates profiling intact proteins due to its ability to capture signals in the microprotein range or larger while ignoring confounding signals (i.e., MALDI matrix peaks). Here, some general guidelines are outlined for acquiring high quality spectra using Bruker axial TOF instruments for microproteins.

1. Mix BCA-normalized protein extracts from the previous section with equal volume of MALDI matrix for intact proteins. Chill the mixture on ice or at 4 °C for 10–30 min to facilitate co-crystallization; the optimal amount of time to chill the mixture will be sample dependent.
2. At least one calibrant spot co-crystallized with MALDI matrix should be present on the MALDI target plate. Ideally, calibrant spots should also be spotted throughout the plate to allow the acquisition software to periodically calibrate the mass range during data acquisition (see **Notes 8** and **9**).
3. Spot at least three technical replicates of each protein extract/ MALDI matrix mixture using a single or multichannel micropipette onto the 384 spot MALDI target plate. Each spot should be 1–1.5 μL. Allow all spots to air dry in a fume hood or under a low flow inert pure gas (i.e., high purity nitrogen or argon; *see* **Note 10–12**).
4. In flexControl, load an optimized linear mode method tuned for the microprotein range (anywhere between 1.5–30 kDa).

Table 1
Select suggested parameters for flexControl v3.4 method to be used for microprotein data acquisition. The method is built off of the default linear mode method supplied by the user's field service engineers

Parameter	Value	Notes
General MALDI settings		
Laser power	Should be optimized to ensure spectral intensity is at least 10^4–10^5 detector counts	The ideal value will be instrument specific and dependent on the age of the hardware
Number of laser shots	4000 shots	
Laser frequency	Maximum value	Set to the maximum value to minimize acquisition time; the maximum value will depend on the instrument used
Sample carrier tab		
Random walk	Complete sample	Ensures unbiased sampling during data acquisition
Shots at raster spot	500 shots	
Detection tab		
Detector gain	Should be optimized to ensure spectral intensity is at least 10^4–10^5 detector counts	The ideal value will be instrument specific and dependent on the age of the hardware
Laser size	Should be optimized to ensure spectral intensity is at least 10^4–10^5 detector counts	The ideal laser size depends on the quality of the co-crystallization between the sample and the MALDI matrix, though larger laser sizes tend to work better for non-imaging applications
Sample rate	Maximum value	The maximum value ensures high mass resolution
Realtime smoothing	Off or low	While not strictly necessary if baseline processing will be performed post-acquisition, realtime baseline smoothing can sometimes result in better data quality
Baseline offset adjustment	0%	
Spectrometer tab		
Matrix suppression	Enabled	
Suppress up to	Low mass range—20%	For example, if the low mass range is set to 2000 Da, the matrix suppression cutoff should be set to 1600 Da

See Table 1 for an example of select flexControl method parameters. Most other settings in the method will be instrument specific and dependent on the age/state of the current hardware. For a full description of flexControl method parameters, please refer to the Help menu in flexControl.

5. Perform calibration using a calibrant of choice. The calibrant used can be a purchased commercial mixture or validated in house mixture that covers the entire mass range of interest for the loaded flexControl method.
6. Set up AutoXecute methods for calibration (if applicable) and sample spectra acquisition (*see* **Note 13** for descriptions of select AutoXecute method parameters).
7. Set up an AutoXecute run by clicking on the "New..." button in the Run section of the AutoXecute tab in flexControl and follow the sequence setup wizard.
8. Select the MALDI target plate geometry. Click Next.
9. On the Laser Power Tuning Page, no action is necessary. Laser power tuning will not be used. Click next.
10. Select the calibration template based on the calibration spots that were prepared in **step 2** (*see* **Notes 8** and **9**). Select the spots (both calibrant and sample spots) that will be analyzed during this run by clicking and dragging over the plate map. Hold the "Ctrl" key and click and drag to select multiple groups of disconnected spots. When all desired spots have been selected for analysis, click next.
11. Select the Calibration and Sample AutoXecute methods to be used for data acquisition and select the output Data Directory. No MS/MS method is needed. Under sample name, a generic name can be selected (i.e., "sample"), and comments can be added to add any desired user defined metadata. Ensure that "Use Container Format" remains unchecked.
12. When finished, the AutoXecute Run Editor will appear. Here, the Sample Name column for each spot can be modified as needed. When finished, save the AutoXecute sequence.
13. Start the acquisition by clicking the "Start automatic Run" button in the AutoXecute tab in flexControl.

3.5 Identifying Putative Biomarkers from MALDI-TOF MS Protein Profiles Using MALDIquant and MetaboAnalyst

Analysis of profiling data, regardless of mass range of interest, can be difficult due to the lack of available software. MALDIquant is an open-source R package that is commonly used for MALDI-TOF MS data analysis [35]. While the R programming environment provides much freedom for users to create their own data analysis pipeline, including one suited for microproteins, it requires a basic level of programming and statistical knowledge. In this section, an example of how MALDIquant can be used for preprocessing,

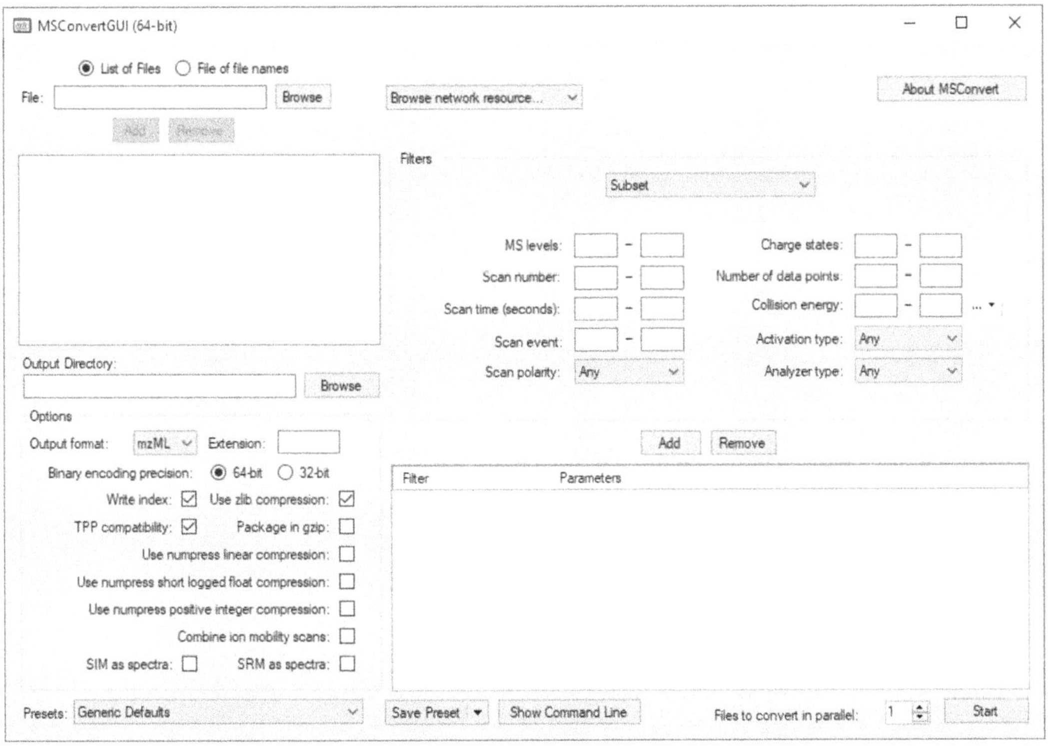

Fig. 2 Parameters (default) for ProteoWizard MSConvert to convert Bruker MALDI–TOF MS spot data to mzML files. The same parameters can be used for mzXML files by selecting "mzXML" under the "Output format" dropdown menu

followed by statistical analysis to identify putative biomarkers using the MetaboAnalyst platform is detailed.

1. Convert the raw intact protein data to mzML or mzXML format using ProteoWizard MSConvert in profile mode using the following settings in Fig. 2 [36]. Please note that the entire sample name directory will be treated as a file and can be loaded into MSConvert for conversion (*see* **Note 14**).
2. Load the data into MALDIquant and perform preprocessing and peak picking using instrument appropriate settings to obtain a feature table for each spectrum. Refer to the MALDIquant documentation for available processing functions and examples of usage. An example script for axial TOF data acquired on the Bruker autoflex (Billerica, MA) can be found at https://github.com/gtluu/mimb_microproteins_maldi_analysis named "MSV000091443.R". The example MALDI dataset can be found at https://massive.ucsd.edu/ProteoSAFe/dataset.jsp?task=130ae77d0205489eb2f61477490656c9 (*see* **Notes 15** and **16**).

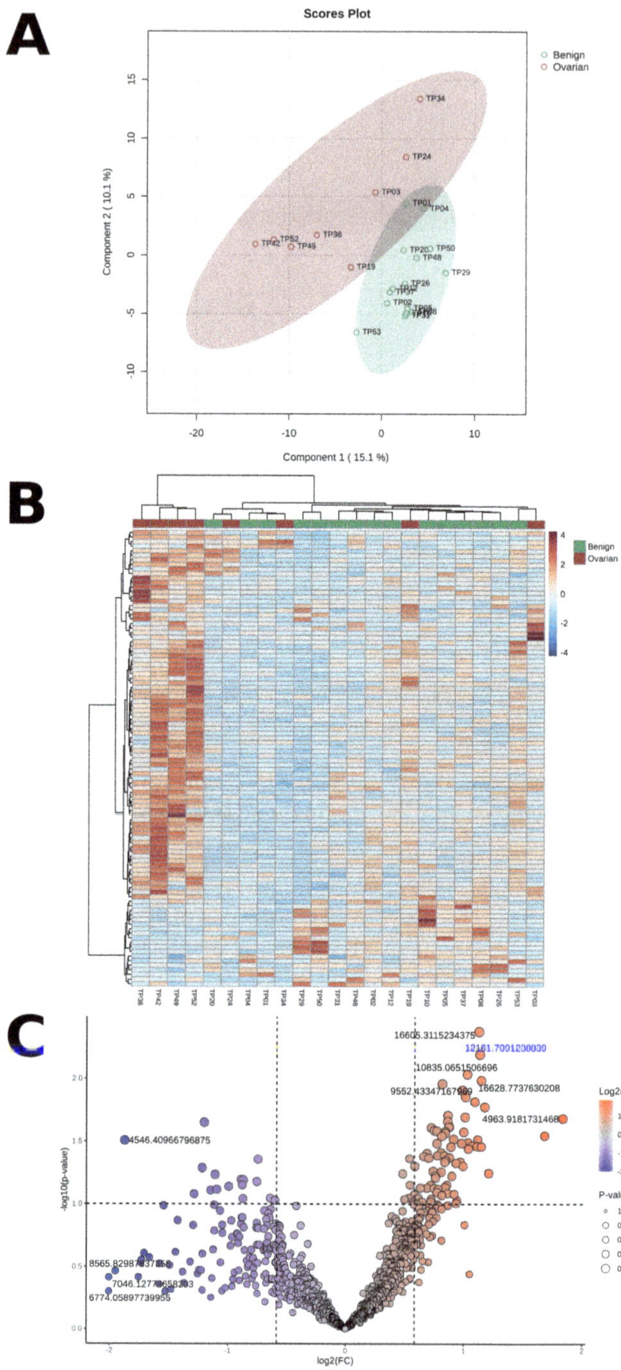

Fig. 3 (**a**) sPLSDA clustering comparing benign vs ovarian cancer samples using 100 variables per component and 5 components. Component 1 vs Component 2 is shown which shows distinct clustering of benign samples and clustering of ovarian cancer samples with more variability. (**b**) Heatmap clustering analysis via the Ward algorithm using euclidean distance as the distance metric based on the top 100 most relevant features determined via t-test. Certain ovarian cancer samples remain unclustered, possibly due to variations in the type and stage of ovarian cancer diagnosed. (**c**) Volcano plot indicating the presence of

3. For the example Bruker autoflex dataset shown, preprocessing steps include (1) mass range trimming, (2) intensity transformation, (3) baseline smoothing, (4) baseline removal, and (5) intensity normalization.
4. Technical replicates from each protein extract are then averaged, and peak picking followed by peak binning are performed on each mean spectrum.
5. Finally, a feature matrix in the form of a dataframe is created in which each row represents the mean spectrum of an intact microprotein extract and each column represents a spectral feature (i.e., *m/z* value). A column containing the disease state of each patient derived extract is also added to the feature matrix, which is then exported as a *.csv file with no index column.
6. Each spectrum can be classified based on the origin and disease state of the patient tampon (i.e., benign vs malignant ovarian cancer). Statistical tests can then be used to compare classes and determine which features are more or less abundant in each class, where each feature is a putative biomarker.
7. MetaboAnalyst is an online platform that provides an easy to use interface to perform various statistical analysis workflows and visualize their results [37].
8. For feature matrices such as the one derived in **steps 2–5** from MALDI-TOF MS profiles, the "Statistical Analysis [one factor]" workflow can be used.
9. Upload the feature matrix *.csv file to the section titled "A plain text file (.txt or .csv)". The "Data Type" should be set to "Peak Intensities" and the "Format" should be set to "Samples in rows (unpaired)". Once the upload completes, click "Submit".
10. MetaboAnalyst will allow for editing group labels, performing data filtering, missing value imputation, and normalization if needed. In this analysis, filtering was performed using the interquartile range (IQR) method set to filter out 25% of samples. No other missing value imputation or normalization was performed.
11. In the left panel, select the statistical analysis workflow of interest. In this analysis, sparse partial least squares

Fig. 3 (continued) significantly up and downregulated features when comparing ovarian cancer vs benign microprotein profiles. Using a $\log_2$ fold change cutoff of 1.5 and a $-\log_{10}$ p-value threshold of 0.1, 71 features were found to be upregulated, while 16 features were found to be downregulated in ovarian cancer samples

discriminant analysis (sPLSDA) and Heatmap (clustering) analysis are used to determine how similar the microprotein profiles are for each diagnostic class. Fold change analysis via the volcano plot analysis is then used to determine which features are highly upregulated or downregulated between ovarian cancer vs benign microprotein profiles (Fig. 3).

12. Putative biomarkers that are shown to be highly upregulated or downregulated in the class of interest can be noted for further investigation using orthogonal mass spectrometry experiments.

3.6 Enzymatic Digestion of Crude Protein Extracts to Allow for Protein Annotation

While MALDI-TOF MS is a powerful technique for global profiling, it is limited in its ability to annotate microproteins due to the lack of structural information available from MS^1 level data. Therefore, orthogonal experiments using liquid chromatography-trapped ion mobility-tandem mass spectrometry (LC–TIMS–MS/MS) are crucial for providing more high quality microprotein annotations from complex biological fluids such as our protein extracts. The following section details the use of S-Traps (ProtiFi) for enzymatic protein digestion to allow for downstream data acquisition and annotation. It should be noted that some sample preparation steps (i.e., resuspension in 10% SDS solution) are incompatible with in solution enzymatic digest protocols that do not utilize S-Traps.

1. A modified filter-aided sample preparation protocol using the S-Trap (ProtiFi) can be used for enzymatic digests of protein extracts.
2. Aliquot less than or equal to 100 μg of the microprotein extract into a microcentrifuge tube and concentrate it via lyophilization.
3. Resuspend the protein extract in 25 μL 10% SDS solution containing a protease inhibitor cocktail (Roche, Manufacturer Number 11697498001) prepared according to the manufacturer's instructions.
4. Reduce disulfide bonds by adding DTT to the protein extract to a final concentration of 20 mM and heat at 95 °C for 10 min.
5. After cooling to room temperature, alkylate cysteines by adding iodoacetamide to a final concentration of 40 mM and incubate in the dark at room temperature for 30 min. This incubation can be done in a cabinet drawer as an example.
6. After incubation, centrifuge each protein extract and remove any undissolved matter by transferring the supernatant to a new microcentrifuge tube.
7. Add 12% aqueous phosphoric acid at a 1:10 ratio to a final concentration of 1.2% phosphoric acid.

8. Add 165 μL of triethylammonium bicarbonate in 90:10 methanol:Milli-Q H_2O (S-Trap buffer) and mix, then add the mixture to the S-Trap and centrifuge at 2000× *g* for 30 sec to elute the S-Trap buffer and retain proteins in the S-Trap column.
9. Add 150 μL of the S-Trap buffer to the S-Trap, rotate the S-Trap column 180, and centrifuge again at 2000× *g* for 30 s. Repeat this process five times, rotating the S-Trap column 180 between each repeated centrifugation step. It can be helpful to mark the S-Trap column to ensure adequate rotation as long as the mark does not come into contact with any solvent, buffer, or sample.
10. Place the S-Trap column in a clean microcentrifuge tube and add the digestion enzyme of choice at a 1:50 weight:weight ratio to the S-Trap column. Commonly used digestion enzymes include trypsin, LysC, or a combination of the two (*see* **Note 17**).
11. Incubate the S-Trap overnight at 37 °C in a water bath to perform enzymatic digestion of proteins (*see* **Note 18**).
12. After overnight digestion, add 40 μL 50 mM triethylammonium bicarbonate in Milli-Q H_2O to the S-Trap column and centrifuge at 4000× *g* for 1 min. This first elution will capture ~85% of peptides on the S-Trap column.
13. Add 40 μL 0.2% formic acid in Milli-Q H_2O and centrifuge at 4000× *g* for 1 min. This second elution will capture ~10% of peptides on the S-Trap column.
14. Add 35 μL of 60:40 ACN:Milli-Q H_2O and centrifuge at 4000× *g* for 1 min. This last elution will capture the remainder of peptides on the S-Trap column, primarily hydrophobic peptides.
15. Concentrate the eluted peptide solution in vacuo at 60 °C for 1 h or until dry.

3.7 Desalting and Cleanup of Enzymatic Digests for MS Analysis

Following enzymatic digestion of proteins, desalting is required to prevent incompatible and insoluble material from entering the mass spectrometer, which can adversely affect the instrument's performance and longevity.

1. After enzymatic digestion, resuspend digested protein extracts in 100 μL 0.1% formic acid in Milli-Q H_2O.
2. Prior to MS analysis, samples will need to be desalted using C18 Resin ZipTip pipette tips (Millipore Sigma) per the manufacturer's protocol to prevent contamination of the LC column and mass spectrometer (*see* **Note 19**).
3. Concentrate the desalted digested protein extracts in vacuo at 45 °C for 45 min or until completely dry.

3.8 Acquisition of Bottom-up Proteomics Datasets Using LC–TIMS–MS/MS

Bottom-up proteomics allows for annotation of microproteins based on peptide sequencing via database searching. The following section has been optimized for the timsTOF fleX mass spectrometer coupled to the Bruker Elute UPLC, which is an analytical flow LC utilizing Compass Hystar for LC control and timsControl for MS control (*see* **Note 7**).

1. Resuspend the desalted digested protein extracts from the previous section in 0.1% formic acid in Milli-Q H_2O to a concentration of 0.1 μg/μL.
2. Centrifuge in a microcentrifuge at 10,000× *g* for 2 min to remove any non-soluble material.
3. Transfer 50–100 μL to a μL pickup insert placed inside an LC vial.
4. Bottom-up proteomics experiments can typically be performed on a C18 column. Here, a Phenomenex Luna Omega C18 column (Particle Size: 1.6 μm, Pore Size: 100 A, Inner Diameter: 2.1 mm, Length: 150 mm) is used for analytical flow ultra-high performance LC (UPLC).
5. In Compass Hystar, an LC method should be established for bottom-up proteomics applications. An example is provided here for analytical flow instruments. However, the exact method may need to be optimized depending on the exact LC system, column, and amount of sample being used and analyzed.
6. Example LC method for Bruker Elute UPLC and Phenomenex Luna Omega C18 column:
   - (a) Flow Rate: 0.3 mL/min.
   - (b) Column Oven Temperature: 60 C.
   - (c) Injection Mode: Partial Loop.
   - (d) Injection Volume: 10–15 μL.
   - (e) Flush Volume: 50 μL.
   - (f) Needle Height: 4.0 mm.
   - (g) The LC solvent gradient can be found in Table 2.
7. Similarly, in timsControl, an ESI-MS method should be established for bottom-up proteomics applications. See Table 3 for example ddaPASEF parameters and Fig. 4/Table 4 for example diaPASEF parameters. These values should be used as a starting point and further optimized to suit the specific instrument being used as the hardware configuration may be slightly different.
8. In Compass HyStar under the “Acquisition” tab, create a new sample table using the “New” button.

Table 2
Solvent A = 0.1% formic acid in Milli-Q H_2O. Solvent B = ACN w/0.1% formic acid

A (%)	B (%)	Time (min)
95	5	0.0
95	5	2.0
76	24	35.0
67	33	45.0
5	95	45.1
5	95	50.0
95	5	50.1
95	5	55.0

9. Enter all sample names, LC vial positions in the autosampler, desired output data path, and file paths to separation and MS methods. The Injection Method should be set to "Standard", and no Processing Method is necessary. The first samples run should be at least one solvent blank (negative control) followed by a QC sample (positive control, i.e., Pierce HeLa Protein Digest Standard; Thermo Fisher Scientific Catalog Number 88328). At least two biological or technical replicates of each sample should be acquired in the dataset, and samples should be randomized to avoid batch effects/variability. Additionally, blanks and/or QC samples should be run every 5–10 injections depending on desired sample throughput. Lastly, blank and QC samples should be run at the end of the run.
10. Save the sample table when done. Figure 5 contains an example of a sample table with three distinct biological samples.
11. Equilibrate the LC system by right clicking on the first line sample table and selecting "Upload Sample Conditions" (*see* **Note 20**).
12. In timsControl, check the TIMS In pressure to ensure that the TIMS gas flow pressure has not drifted over time; if drift has been observed, nitrogen gas flow to the TIMS can be adjusted with the black knob on the front of the instrument. The optimal TIMS In pressure value is dependent on the instrument's hardware configuration (132 ± 0.5 V for TIMS cartridge #1875139 or 164 ± 0.5 V for TIMS cartridge #894787).
13. Move the calibrant syringe line in the front of the timsTOF from the divert valve to the ESI source nebulizer and calibrate the *m/z* and TIMS dimensions under the Calibration Tab in

Table 3
Example of select ddaPASEF parameters for timsControl method

Parameter	Value
TIMS	On
MS settings	
Scan begin	100
Scan end	1700
Ion polarity	Positive
Scan mode	PASEF
TIMS settings	
Mode	Custom
1/K0 start	0.60
1/K0 end	1.60
Ramp time	100.0 ms
Accumulation time	100.0 ms
Duty cycle	100%
MS averaging	1
Source tab	
End plate offset	500 V
Capillary	4500 V
Nebulizer	0.4 bar
Dry gas	3.0 L/min
Dry temp	200 °C
Tune tab—General	
Deflection 1 delta	70.0 V
Funnel 1 RF	300.0 Vpp
isCID energy	0.0 eV
Funnel 2 RF	300.0 Vpp
Multipole RF	500.0 Vpp
Collision energy	10.0 eV
Collision RF	1500.0 Vpp
Ion energy	5.0 eV
Low mass	100.00 m/z
Transfer time	80 μs

(continued)

Table 3
(continued)

Parameter	Value
Pre pulse storage	5 μs
High sensitivity detection	Disabled
Tune tab—TIMS	
Δt1	−20.0 V
Δt2	−160.0 V
Δt3	110.0 V
Δt4	110.0 V
Δt5	0.0 V
Δt6	55.0 V
Collision cell in	300.0 V
Lock accumulation to mobility range	Enabled
MS/MS tab	
Number of PASEF ramps	10
Charge minimum	0
Charge maximum	5
Precursor repetitions	Simple
Target intensity	10,000
Intensity threshold	1000
Active exclusion	Enabled
Release after	0.40 min
TIMS stepping	Disabled
Collision energy settings	
1/K0 [V * s / cm^2]	Collision energy [eV]
0.60	20.00
1.60	59.00
Isolation width settings	
Mass [m/z]	Width [m/z]
700	2.00
800	3.00

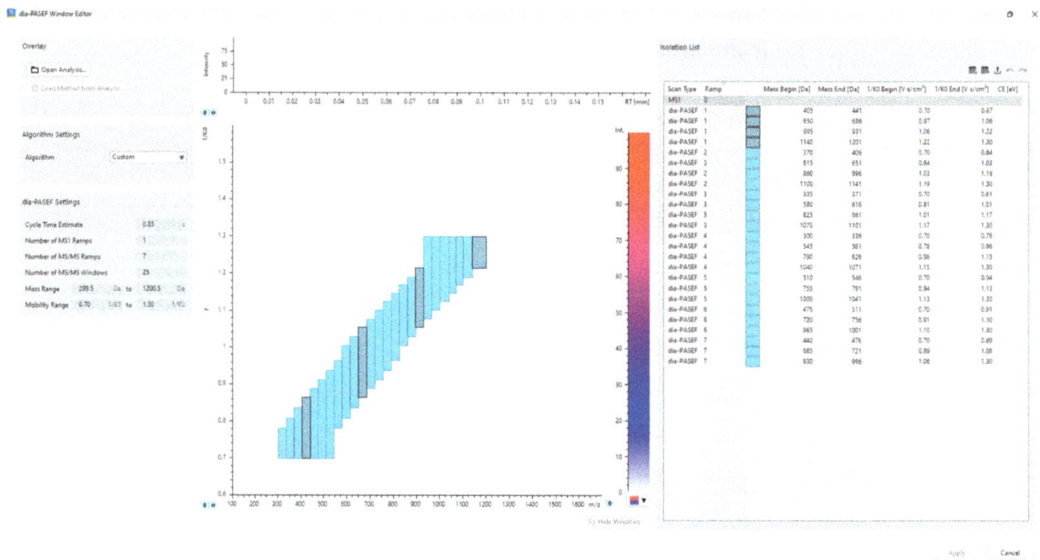

Fig. 4 Example of dia-PASEF windows for LC–TIMS–MS/MS proteomics data acquisition

timsControl using ESI-L LCMS Tuning Solution (Agilent Technologies Inc., Part Number G1969-85000) to ensure mass and mobility accuracy during data acquisition (*see* **Note 21**).

14. Once complete, replace all LC and syringe lines to their original positions.

15. Once the LC system is equilibrated and the MS system is calibrated, in Compass HyStar, ensure all checkboxes in the sample table are enabled. Click the "Start" button and select "Start Sequence" to begin data acquisition.

3.9 Validation and Annotation of Putative Biomarkers from Intact Protein Profiles Using LC–TIMS–MS/MS

Over the years, a variety of software have been developed and iterated upon to perform annotation, label-free quantitation, and label dependent quantitation, including MaxQuant [38], FragPipe [39], DIA-NN [40], Spectronaut [41], Proteome Discoverer [42], and more. In this section, we focus on the use of FragPipe to annotate LC–TIMS–MS/MS datasets from the Bruker timsTOF fleX MS. The following section is meant to be a brief guide on the basic operation of FragPipe and how to integrate its results with MALDI data processed using MALDIquant. For full details on each workflow and its parameters, please see FragPipe's documentation.

1. Download and install FragPipe and all desired dependencies. Here, we use FragPipe v22.0.

2. Under the Workflow tab, load the dataset. FragPipe should automatically detect whether the dataset was acquired using data-dependent acquisition (DDA) or data-independent

Table 4
Example of select diaPASEF parameters for timsControl method

Parameter	Value
TIMS	On
MS settings	
Scan begin	100
Scan end	1700
Ion polarity	Positive
Scan mode	dia-PASEF
TIMS settings	
Mode	Custom
1/K0 start	0.60
1/K0 end	1.60
Ramp time	100.0 ms
Accumulation time	100.0 ms
Duty cycle	100%
MS averaging	1
Source tab	
End plate offset	500 V
Capillary	4500 V
Nebulizer	0.4 bar
Dry gas	3.0 L/min
Dry temp	200 °C
Tune tab—General	
Deflection 1 delta	70.0 V
Funnel 1 RF	300.0 Vpp
isCID energy	0.0 eV
Funnel 2 RF	300.0 Vpp
Multipole RF	500.0 Vpp
Collision energy	10.0 eV
Collision RF	1500.0 Vpp
Ion energy	5.0 eV
Low mass	100.00 m/z
Transfer time	80 μs

(continued)

Table 4
(continued)

Parameter	Value
Pre pulse storage	5 μs
High sensitivity detection	Disabled
Tune tab—TIMS	
Δt1	−20.0 V
Δt2	−160.0 V
Δt3	110.0 V
Δt4	110.0 V
Δt5	0.0 V
Δt6	55.0 V
Collision cell in	300.0 V
Lock accumulation to mobility range	Enabled
MS/MS tab	
Dia-PASEF windows	See Fig. 3
Collision energy settings	
1/K0 [V × s/cm^2]	Collision energy [eV]
0.60	20.00
1.60	59.00

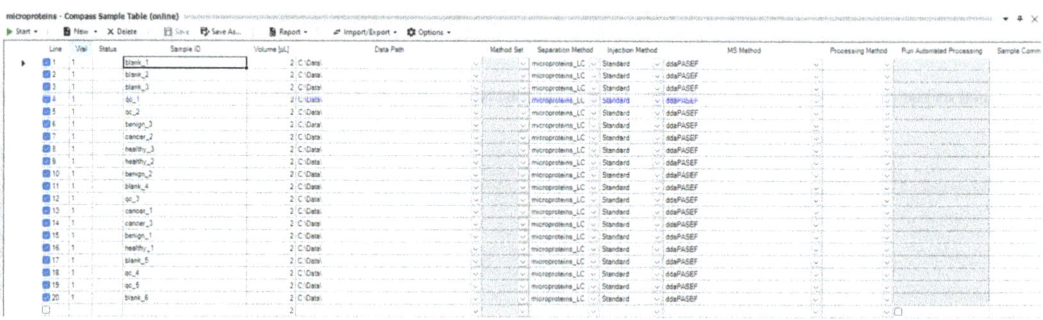

Fig. 5 Example sample table containing triplicate acquisitions for healthy, benign, and cancer samples

acquisition (DIA) and whether TIMS was enabled or disabled. Different experiments can be used to group samples for quantification (*see* **Note 22** for manual parameter selection).

3. Under the DIA Pseudo MS2 tab, diaTracer can be used for diaPASEF Spectrum Deconvolution if processing a TIMS-

Table 5
Example of output after correlating MALDIquant features to protein annotations from FragPipe. In this example, a mass tolerance window of ±0.5 Da is used to correlate autoflex and timsTOF fleX data

MALDIquant feature	Protein. ID	Protein.Description	FragPipe annotated protein mass
7933.54	P56385	ATP synthase subunit e, mitochondrial	7934.0078
8218.75	P63173	Large ribosomal subunit protein eL38	8219.0078
10192.80	P14854	Cytochrome c oxidase subunit 6B1	10193.0078
11531.32	P20962	Parathymosin	11531.0078
12784.72	P62888	Large ribosomal subunit protein eL30	12785.0078
19398.80	Q9GZZ1	N-alpha-acetyltransferase 50	19399.0078

enabled dataset; otherwise, DIA-Umpire SE can be used for DIA Spectrum Deconvolution.

4. Under the Database tab, select a FASTA file containing protein sequences to be used for annotation. FASTA format databases can be downloaded directly from FragPipe using the Download button and specifying the Uniprot organism/proteome ID. The download tool also has options to include only reviewed protein sequences and add decoy sequences, common contaminants, isoform sequences, and iRT sequences (*see* **Note 23**).
5. Under the MSFragger tab, settings for the generation of a spectral library from the in silico digestion of the protein sequences in the protein sequence database can be modified. Users should be aware of the typical properties of the enzyme used during protein digestion for their bottom-up proteomics experiments such as which amino acids are cut by the enzyme, the typical number of allowable missed cleavages, and common variable and fixed modifications resulting from the digestion process.
6. Under the Validation tab, select any desired validation workflows including deep learning based rescoring, PSM validation, PTM site localization (if performing PTM analysis), protein inference, and FDR analysis.
7. Under the PTMs tab, PTM-Shephard can be enabled and configured if performing post-translational modification (PTM) analysis on microproteins.

8. Under the Glyco tab, PTM-Shephard or O-Pair can be enabled and configured for performing glycan or O-glycan analysis, respectively, on microproteins.
9. Under the Quant (MS1) tab, quantification between groups of DDA samples defined under the Workflow tab can be enabled and configured using label free quantification (LFQ) or labeling (i.e., stable isotope labeling; SILAC). For the purposes of microprotein analysis, quantification is not strictly necessary but can be used as an orthogonal resource to further support differential expression revealed in MALDI-TOF MS experiments on intact microproteins.
10. Alternatively, under the Quant (Isobaric) tab, quantification between groups of DDA samples defined under the Workflow tab can be enabled and configured when using isobaric tags as labels (i.e., tandem mass tags; TMT). Similarly, this data can further support previously observed differential expression.
11. Under the SpecLib tab, spectral libraries generated during FragPipe processing can be saved as pepXML or TSV format among other FragPipe result files for further use in other software.
12. Under the Quant (DIA) tab, quantification between groups of DIA samples defined under the Workflow tab can be enabled and configured using DIA-NN.
13. Under the Skyline tab, a Skyline document can be generated for further analysis in Skyline MS. [43]
14. Under the Run tab, select the output directory to export FrapPipe results to and configure any other desired processing settings. Once configured, click Run.
15. FragPipe will generate a directory containing results including but not limited to a table of peptide and protein hits in TSV format (peptide.tsv and protein.tsv, respectively).
16. FragPipe protein annotations can be correlated to MALDIquant features using the scripting language of choice. In short, both tables are loaded as dataframes and table merging functions (i.e., pandas.DataFrame.merge_asof() in Python or fuzzyjoin::difference_inner_join() in R) are used to match MALDI-TOF MS features to LC–TIMS–MS/MS annotations based on similarity of the mass-to-charge ratio (*m/z*). An example script is included at https://github.com/gtluu/mimb_microproteins_maldi_analysis named MALDIquant_FragPipe_Validation.R. Table 5 contains an example of the output using select features and protein groups detected by MALDIquant and FragPipe, respectively.
17. This method allows for discovery of putative biomarkers in a completely untargeted manner. However, please note that

although orthogonal MS methods are utilized for annotation, this yields at best Level 2a identification according to Schymanski et al. [44] For confirmation of a given protein/ biomarker in each sample, methods such as immunohistochemistry, MS/MS and RT and CCS matching to pure proteins/standards, or NMR structure elucidation should be performed.

4 Notes

1. All patient samples are treated at biosafety level 2 (BSL2). Two pairs of gloves can be worn over each other to prevent infection in the case of ripped gloves.
2. Study design when using tampons should take menstruation cycle and status into consideration during sample collection. The presence of menstrual biomaterial (i.e., blood) in only a subset of samples can act as a confounding factor when comparing disease states during downstream analysis.
3. Utilizing easily replaceable, disposable, or sterilizable blades to prevent sample cross contamination can simplify this process.
4. While this section of the tampon will generally have the most biological material, the rest of the tampon can be saved for subsequent (re-)analysis using MS or other orthogonal techniques as necessary. Importantly, the total amount of biomaterial on the tampon will be dependent on the length of use by the patient.
5. The volume here can be variable due to the differences in the physical sizes of different tampons depending on the brand that was used. Ideally, a study should take care to use a consistent brand during sample collection.
6. The exact size of the Amicon Ultra centrifugal filter devices used will depend on the volume of the crude protein extract. For example, the Amicon Ultra-4 centrifugal filter devices contain filters with a 4 mL sample loading volume in a 15 mL conical tube, while the Amicon Ultra-15 centrifugal filter devices contain filters with a 15 mL sample loading volume in a 50 mL conical tube.
7. Please note that the total volume and/or amount of protein material that is necessary for MS analysis ultimately depends on the instrumentation available and the limit of detection for those instrument(s). For MALDI-TOF MS intact microprotein analysis, 10–20 mg/mL should be sufficient. Nano flow LC–TIMS–MS/MS analysis following requires ~300 ng of digested peptides, while standard flow requires at least 1–1.5 μg.

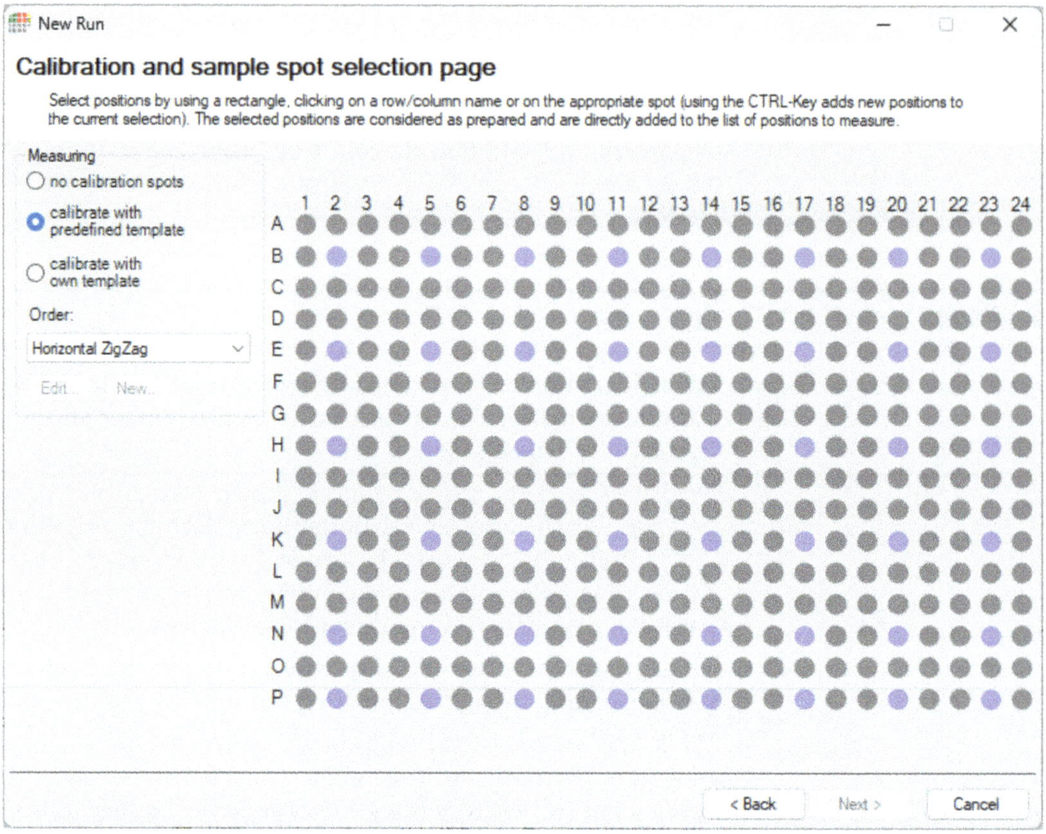

Fig. 6 Default calibration spot plate map in flexControl 3.4. Here, gray spots represent spot coordinates where samples should be deposited; purple spots represent spot coordinates where mass calibrant solutions should be deposited

8. The default sample map including calibration spots splits the plate into 3 × 3 grids where the center spot is used as the calibrant spot and the surrounding 8 spots are samples acquired after mass calibration using the center spot (Fig. 6). Alternatively, a user defined template can be defined in flexControl.
9. At the time of writing, experiments run on the Bruker neofleX MALDI-TOF MS do not utilize calibration spots.
10. While analysis of intact protein profile datasets only requires three technical replicates in the absence of biological replicates, at least 10 technical replicates should be prepared if the spectra are to be used for database generation.
11. Multichannel pipettes deposit samples onto every other spot on standard Bruker 384 spot MALDI target plates.
12. Ground steel, polished steel, or Bruker AnchorChip target plates can be used with minimal changes to data quality. At higher volumes (> 1 μL), depositing extract/matrix mixtures

with higher concentrations of organic solvent may result in samples bleeding into neighboring spots due to lack of surface tension, resulting in cross contamination.

13. Enabling fuzzy laser control in the Laser tab allows for the laser power to fluctuate between laser shots to ensure that the ideal laser power is used for data acquisition. Laser power changes are based on criteria specified in the Evaluation tab. The spectra evaluation criteria can be defined in the Evaluation tab can be defined based on minimum mass resolution or peak width in a targeted or untargeted manner. In the Accumulation and Movement tabs, fuzzy control can be enabled to allow for dynamic acquisition of only high quality spectra. Here, spectra will be evaluated every X (i.e., 500) shots, and if the spectra pass the evaluation criteria, it will be added to a sum spectrum until a sum of Y (i.e., 4000) shots have been accumulated. Importantly, under the Movement tab, the "Quit sample after . . ." option should be set to less than or equal to 10 to ensure that too much time is not spent on a single sample during acquisition. Additionally, the raster mode should be set to random walk under the Movement tab. No Processing or MS/MS configuration is necessary.

14. Bruker axial TOF data in the XMASS format are stored in a nested directory containing the following hierarchy: sample_-name > chip_spot > replicate number > detector mode > raw data. The sample name is defined in the AutoXecute sequence. The chip and spot refer to the sample coordinates on a given MALDI target plate. Multiple directories for each replicate are present if multiple spectra from the same spot (i.e., technical replicates) are given the same sample name. Detector mode refers to whether a linear or reflectron mode method was used during acquisition. Finally, the raw data includes an fid file containing the raw data in addition to various plain text meta-data files. A pdata directory may also be present which contains processed data from flexControl/flexAnalysis. When data processing is performed on the raw data, the raw data is not modified; instead, additional processed files are written to the pdata directory.

15. Not all preprocessing steps may be necessary depending on the quality of the dataset and the capabilities of the instrument used to acquire data.

16. If not using MALDIquant, the processing software of choice should be able to output a table in which each row corresponds to a spectrum, each column corresponds to a feature (i.e., *m/z* value), and each value in the table corresponds to an intensity value.

Table 6
Number of data points required for mass and mobility calibration. Note that only Linear mode is available for mobility calibration

Mode	Minimum number of calibrant points	Number of calibrant points required for good calibration
Linear correction	1	1
Linear	2	3
Quadratic	3	4
Enhanced quadratic	4	5
HPC	7	8

17. Depending on the sample, the ratio of digestion enzyme to protein in sample can be modified to promote more digestion (i.e., 1:20 or even 1:10 ratio of enzyme:sample protein) or less digestion (i.e., 1:100 ratio of enzyme:sample protein). Note that increasing the ratio of digestion enzymes can also lead to overdigestion of sample proteins, while decreasing the ratio can lead to little or no digestion.
18. Typically, an overnight digest will be ~14–18 h.
19. Take care to follow the manufacturer's protocols strictly, as peptides can be lost from the sample during the desalting process, resulting in (sometimes greatly) reduced coverage from bottom-up proteomic MS analysis.
20. LC column equilibration should be done slowly by slowly ramping the flow rate up to the desired flow rate that will be used during data acquisition. In Compass HyStar using the Bruker Elute UPLC system, this can be done by right clicking on the Elute Pump module and selecting "Direct Control". Ensure the "Soft Ramp" option is checked to allow for slow ramping of the LC flow rate.
21. In older versions of timsControl, during mass and mobility calibration, the calibration StdDev and Score may appear as 0.00 ppm and 100%, respectively. Calibration is performed by fitting the detected calibrant points in a spectrum to a calibration curve via a polynomial equation that requires a minimum of N data points (Table 6). However, if only the minimum number of data points is detected, the calibrant points appear to fit perfectly to the calibration curve, which results in a false positive "good calibration" state; in reality, no calibration is performed due to insufficient data points. Therefore, StdDev

values of 0.00 ppm should be considered invalid during calibration; if present, at least one additional calibrant point should be obtained before accepting the calibration.

22. The "Default" workflow can be selected for datasets acquired using data dependent acquisition (DDA) in Auto MS/MS mode or ddaPASEF mode in timsControl. For data independent acquisition datasets, select the appropriate diaPASEF method (i.e., DIA_SpecLib_Quant_diaPASEF). Be sure to select "IM-MS" under "MS data type" if TIMS enabled datasets are being processed.
23. Alternatively, custom FASTA databases can be loaded from a local hard drive. For analyses focused on microproteins, a custom proteome database containing only sequences less than or equal to 30 kDa can be loaded to prevent unwanted hits and decrease processing time.

Acknowledgments

This work was supported by the National Institute of General Medical Sciences K12GM139185 (CF) and National Cancer Institute R01CA240423 (to LMS) of the National Institutes of Health, and by the Laura Crandall Brown Foundation Ovarian Cancer Early Detection Research Grant from the Foundation for Women's Cancer supported by the Laura Crandall Brown Foundation (to LMS)

Conflict of Interest GTL is an employee of Bruker Scientific LLC (Billerica, MA).

References

1. Hassel KR, Brito-Estrada O, Makarewich CA (2023) Microproteins: overlooked regulators of physiology and disease. iScience 26:106781
2. Basrai MA, Hieter P, Boeke JD (1997) Small open reading frames: beautiful needles in the haystack. Genome Res 7:768–771
3. Das S, Yu L, Gaitatzes C et al (1997) Biology's new Rosetta stone. Nature 385:29–30
4. Fickett JW (1995) ORFs and genes: how strong a connection? J Comput Biol 2:117–123
5. Sergiev PV, Rubtsova MP (2021) Little but loud. The diversity of functions of small proteins and peptides—translational products of short reading frames. Biochemistry (Mosc) 86:1139–1150
6. Yeasmin F, Yada T, Akimitsu N (2018) Micropeptides encoded in transcripts previously identified as long noncoding RNAs: a new chapter in transcriptomics and proteomics. Front Genet 9:144
7. Vitorino R, Guedes S, Amado F et al (2021) The role of micropeptides in biology. Cell Mol Life Sci 78:3285–3298
8. Sandmann C-L, Schulz JF, Ruiz-Orera J et al (2023) Evolutionary origins and interactomes of human, young microproteins and small peptides translated from short open reading frames. Mol Cell 83:994–1011.e18
9. Siepel A, Bejerano G, Pedersen JS et al (2005) Evolutionarily conserved elements in vertebrate, insect, worm, and yeast genomes. Genome Res 15:1034–1050

10. Valdivia-Francia F, Sendoel A (2024) No country for old methods: new tools for studying microproteins. iScience 27:108972
11. Pan N, Wang Z, Wang B, et al (2021) Mapping microproteins and ncRNA-encoded polypeptides in different mouse tissues. Front Cell Dev Biol 9:687748
12. Cassidy L, Helbig AO, Kaulich PT, et al (2021) Multidimensional separation schemes enhance the identification and molecular characterization of low molecular weight proteomes and short open reading frame-encoded peptides in top-down proteomics. J Proteomics 230: 103988
13. Li B, Zhang Z, Wan C (2022) Identification of microproteins in Hep3B cells at different cell cycle stages. J Proteome Res 21:1052–1060
14. Meier F, Beck S, Grassl N, et al (2015) Parallel accumulation-serial fragmentation (PASEF): Multiplying sequencing speed and sensitivity by synchronized scans in a trapped ion mobility device. J Proteome Res 14:5378–5387
15. Meier F, Brunner A-D, Koch S, et al (2018) Online parallel accumulation-serial fragmentation (PASEF) with a novel trapped ion mobility mass spectrometer. Mol Cell Proteomics 17: 2534–2545
16. Mendes ML, Borrmann KF, Dittmar G (2024) Eleven shades of PASEF. Expert Rev Proteomics 21:367–376
17. Luu GT and Sanchez LM (2021) Toward improvement of screening through mass spectrometry-based proteomics: ovarian cancer as a case study. Int J Mass Spectrom 469
18. Wilson DA, Young S, Timm K et al (2017) Multicenter evaluation of the Bruker MALDI Biotyper CA system for the identification of clinically important bacteria and yeasts. Am J Clin Pathol 147:623–631
19. Faron ML, Buchan BW, Hyke J, et al (2015) Multicenter evaluation of the bruker MALDI Biotyper CA System for the identification of clinical aerobic gram-negative bacterial isolates. PLoS One 10:e0141350
20. Milani P, Murray DL, Barnidge DR, et al (2017) The utility of MASS-FIX to detect and monitor monoclonal proteins in the clinic. Am J Hematol 92:772–779
21. Dasari S, Kohlhagen MC, Dispenzieri A et al (2022) Detection of plasma cell disorders by mass spectrometry: a comprehensive review of 19,523 cases. Mayo Clin Proc 97:294–307
22. Kohlhagen M, Dasari S, Willrich M, et al (2020) Automation and validation of a MALDI-TOF MS (Mass-Fix) replacement of immunofixation electrophoresis in the clinical lab. Clin Chem Lab Med 59:155–163
23. Mellors PW, Dasari S, Kohlhagen MC, et al (2021) MASS-FIX for the detection of monoclonal proteins and light chain N-glycosylation in routine clinical practice: a cross-sectional study of 6315 patients. Blood Cancer J 11:50
24. Baloda V, Wheeler SE, Murray DL et al (2022) Mu heavy chain disease with MYD88 L265P mutation: an unusual manifestation of lymphoplasmacytic lymphoma. Diagn Pathol 17:63
25. Karas M, Glückmann M, and Schäfer J (2000) Ionization in matrix-assisted laser desorption/ionization: singly charged molecular ions are the lucky survivors. J Mass Spectrom 35:1–12
26. Westman A, Brinkmalm G, and Barofsky DF (1997) Induced Satura- tion Effects in Chevron Microchannel Plate Detectors 169:79–87
27. Coeck S, Beck M, Delauré B, et al (2006) Microchannel plate response to high-intensity ion bunches. Nucl Instrum Methods Phys Res A 557:516–522
28. Gilmore IS and Seah MP (2000) Ion detection Efficiency in SIMS: Dependencies on Energy, Mass, and Composition for Microchannel Plates Used in Mass Spectrometry 202:217–229
29. Chen X, Westphall MS, and Smith LM (2003) Mass Spectrometric Analysis of DNA Mixtures: Instrumental Effects Responsible for Decreased Sensi- tivity with Increasing Mass 75:5944–5952
30. van Remoortere A, van Zeijl RJM, van den Oever N et al (2010) MALDI imaging and profiling MS of higher mass proteins from tissue. J Am Soc Mass Spectrom 21:1922–1929
31. Luu GT, Ge C, Tang Y et al (2023) An integrated approach to protein discovery and detection from complex biofluids. Mol Cell Proteomics 22:100590
32. Galey MM, Young AN, Petukhova VZ et al (2020) Detection of ovarian cancer using samples sourced from the vaginal microenvironment. J Proteome Res 19:503–510
33. Jiang Y, Rex DAB, Schuster D, et al (2024) Comprehensive overview of bottom-up proteomics using mass spectrometry. ACS Meas Sci Au 4:338–417
34. Petukhova VZ, Young AN, Wang J, et al (2019) Whole Cell MALDI Fingerprinting Is a Robust Tool for Differential Profiling of Two-Component Mammalian Cell Mixtures. J Am Soc Mass Spectrom 30:344–354
35. Gibb S, Strimmer K (2012) MALDIquant: a versatile R package for the analysis of mass

spectrometry data. Bioinformatics 28:2270–2271
36. Chambers MC, Maclean B, Burke R, et al (2012) A cross-platform toolkit for mass spectrometry and proteomics. Nat Biotechnol 30: 918–920
37. Xia J, Psychogios N, Young N et al (2009) MetaboAnalyst: a web server for metabolomic data analysis and interpretation. Nucleic Acids Res 37:W652–W660
38. Cox J, Mann M (2008) MaxQuant enables high peptide identification rates, individualized p.p.b.-range mass accuracies and proteome-wide protein quantification. Nat Biotechnol 26:1367–1372
39. Kong AT, Leprevost FV, Avtonomov DM, et al (2017) MSFragger: ultrafast and comprehensive peptide identification in mass spectrometry-based proteomics. Nat Methods 14:513–520
40. Demichev V, Messner CB, Vernardis SI, et al (2020) DIA-NN: neural networks and interference correction enable deep proteome coverage in high throughput. Nat Methods 17: 41–44
41. Bruderer R, Bernhardt OM, Gandhi T, et al (2015) Extending the limits of quantitative proteome profiling with data-independent acquisition and application to acetaminophen-treated three-dimensional liver microtissues. Mol Cell Proteomics 14:1400–1410
42. Orsburn BC (2021) Proteome discoverer—a community enhanced data processing suite for protein informatics. Proteomes 9:15
43. MacLean B, Tomazela DM, Shulman N et al (2010) Skyline: an open source document editor for creating and analyzing targeted proteomics experiments. Bioinformatics 26:966–968
44. Schymanski EL, Jeon J, Gulde R et al (2014) Identifying small molecules via high resolution mass spectrometry: communicating confidence. Environ Sci Technol 48:2097–2098

Index

Stephan Wenkel (ed.), *Microproteins: Methods and Protocols*, Methods in Molecular Biology, vol. 2992, https://doi.org/10.1007/978-1-0716-5013-4,

www.ingramcontent.com/pod-product-compliance
Ingram Content Group UK Ltd.
Pitfield, Milton Keynes, MK11 3LW, UK
UKHW051051310326
469509UK00003B/9
* 9 7 8 1 0 7 1 6 5 0 1 2 7 *